THE LAND IS A MAP

Placenames of Indigenous Origin in Australia

THE LAND IS A MAP

Placenames of Indigenous Origin in Australia

Edited by

LUISE HERCUS, FLAVIA HODGES, JANE SIMPSON

ANU

THE AUSTRALIAN NATIONAL UNIVERSITY

E PRESS

ANU

E PRESS

© ANU E Press

Published by ANU E Press
The Australian National University
Canberra ACT 0200, Australia
Email: anuepress@anu.edu.au
Web: http://epress.anu.edu.au

Previously published by Pandanus Books in association with Pacific Linguistics,
Research School of Pacific and Asian Studies, The Australian National University,
2002.

National Library Catalogue-in-Publication Entry

Title: The land is a map : placenames of indigenous origin in
              Australia / edited by Luise Hercus, Flavia Hodges, Jane Simpson.

ISBN: 9781921536564 (pbk.) 9781921536571 (pdf)

Subjects: Names, Aboriginal Australian.
             Names, Geographical--Australia.

Other Authors/Contributors:
             Hercus, L. A. (Luise Anna), 1926-
             Hodges, Flavia.
             Simpson, Jane Helen.

Dewey Number: 919.4003

Cover design by Emily Brissenden, Photography by Bob Cooper.

CONTENTS OVERVIEW

ASSIGNING AND REINSTATING PLACENAMES

APPENDIX

CONTENTS IN DETAIL

LIST OF MAPS

List of figures and tables

PREFACE

The nucleus of this volume is a group of papers presented at two conferences. The first was a one-day conference on Australian placenames of Indigenous origin held at The Australian National University on 31 October 1999 and sponsored jointly by AUSTRALEX (the Australasian Association for Lexicography), the pilot National Place Names Project, and the Australian Language Research Centre. The second conference was held in Adelaide on 8 April 2000 under the auspices of the fully-fledged Australian National Placenames Survey (ANPS) and the Committee for Geographical Names in Australasia.

The title comes from a remark made to Peter Sutton by Denny Bowenda, a Wik man of Aurukun, Queensland. We chose it as the title because it alludes to important differences between the ways the Australian colonists have named places in Australia and the ways Indigenous Australians name places.

WARNINGS:

1)   The maps in this volume represent the opinions of the authors based on flawed data, and the authors, editors and publishers make no claims as to the historical veracity of the maps in this volume.

2)   This book contains the names of Aboriginal people who may have passed away. Care should be taken in saying these names in the presence of Aboriginal people who may know them.

Our thanks are due in particular to Professor Margaret Clunies Ross and the University of Sydney's Australian Language Research Centre, which generously provided a publication subsidy; to Susan Poetsch, ANPS Research Associate, who compiled the indexes and handled a mass of administrative detail; to Malcolm Ross, John Bowden, Margaret Forster and Julie Manley at Pacific Linguistics; to the team at Pandanus Books; to Kay Dancey at the ANU Cartographic Unit; to Mrs W.E.H. Stanner, and to the participants at the two conferences, whose comments and discussion have enriched these papers.

Luise Hercus
Australian National University

Flavia Hodges
Macquarie University

Jane Simpson
University of Sydney

ILLUSTRATION CREDITS:
Photograph of King Anthony (Chapter 11) courtesy of the State Library of Victoria.
Cartoon of Light's Vision (Chapter 18) courtesy of Michael Atchison, *Adelaide Advertiser*.

NOTES ON CONTRIBUTORS

Barry Alpher has published a dictionary of Yir-Yoront and is currently working on problems of comparative Pama-Nyungan and writing up field materials on Yir-Yoront, Yirrk-Mel, Ogunyjan, Pakanh and Olgol in Cape York Peninsula. He has taught anthropology and linguistics at Arizona State University and at the University of Sydney, taught at and headed the School of Australian Linguistics in Batchelor, Northern Territory, and worked in applied linguistics and anthropology with Native Americans, with Spanish-speaking migrant workers in the United States of America, and with Aboriginal Australians.

Rob Amery, Senior Lecturer, Unaipon School, University of South Australia. Rob completed a PhD in 1998 (published in August 2000) on Kaurna language reclamation. He serves as consultant linguist to the Kaurna language programs in schools and various community projects that incorporate the Kaurna language and has been consulted about numerous placenaming initiatives and other Kaurna naming activity. He works closely with members of the Kaurna community to reclaim the language from historical materials and to develop the language for use in a range of contemporary contexts.

Anna Ash has been working in the area of Aboriginal language maintenance for several years. She worked with *Ngakulmungan Kangka Leman* (the Language Committee of Mornington Island) on the production of the *Lardil Dictionary* and *A Learner's Guide to Lardil*. Anna is currently a Research Associate with the School of Languages, Cultures and Linguistics at the University of New England, Armidale; she is developing a Yuwaalaraay–Yuwaaliyaay–Gamilaraay dictionary database and co-ordinating the production of a poster 'Aboriginal Languages of Northern New South Wales'.

Brett J. Baker is a Lecturer at the Department of Linguistics at New England. Brett completed his PhD thesis 'Word Structure in Ngalakan' in 1999. The thesis was based on ongoing fieldwork in the Roper River region of the Northern Territory since 1994. Brett's current project — in partnership with Diwurruwurru-jaru Aboriginal Corporation (the Katherine Language Centre) — focuses on spoken interaction in traditional Aboriginal languages of the area, particularly the prosodic and referential aspects thereof. The project also has a practical application: to make natural interactive language the basis of teaching methodology and materials production for school language programs, through an electronic database archive.

The Committee for Geographical Names in Australia (CGNA) was formed in 1984 to provide a coordinating role in Australian placenaming activities. CGNA was established within the Intergovernmental Committee on Surveying and Mapping (ICSM) in 1993 with the support of the Australian Surveying and Land Information Group (AUSLIG). Representation on the CGNA comes from the Australian State and Territory naming authorities and other bodies with an interest in nomenclature. In 1998 New Zealand was formally welcomed as a full participating member and the group became the Committee for Geographical Names in

Australasia. Macquarie University also joined CGNA in 1999 to help foster a strong working relationship during the development of the Australian National Placenames Survey (ANPS).

Tamsin Donaldson is currently a Visiting Research Fellow at The Australian National University. Since publishing a Ngiyampaa (Wangaaypuwan) grammar (1980) she has focused mainly on Australian languages and language-related projects, working with Indigenous people all over Australia through appointments at the former School of Australian Linguistics, Batchelor, Northern Territory, and at the Australian Institute of Aboriginal and Torres Strait Islander Studies, and also as an independent consultant. She has taught in the Anthropology and Music departments at the University of Sydney, and as a fellow in 'Race and the Humanities' at Cornell, USA. Her publications concentrate on New South Wales languages, and on social, historical and other topics where linguistics helps, especially songs.

Luise Hercus was a Fellow of St Anne's College, Oxford, having studied both Modern Languages and Oriental Studies. In 1962 she began working independently on salvage work in Aboriginal Languages, studying languages that were on the brink of extinction. She has continued this work ever since. She was Reader in Sanskrit at The Australian National University from 1969 to 1991. Since then she has been Visiting Fellow in the Department of Linguistics, Faculty of Arts, Australian National University, writing up grammars, dictionaries and traditional texts, and continuing fieldwork mainly in the north of South Australia and adjacent areas of New South Wales and Queensland.

Philip Jones has worked as a curator in the Anthropology Department at the South Australian Museum since 1982. His doctoral thesis concerns the history of collecting Australian ethnographic material, against the background of anthropological history. He has produced a number of museum exhibitions, concentrating upon the classical past of Aboriginal societies and their frontier interactions with Europeans. He has undertaken historical research in the eastern Lake Eyre region since 1984, and has a particular interest in the ethnographic collections and records gathered by the Lutherans at Killalpaninna Mission on the Cooper Creek.

Patrick McConvell has a BA(Hons) and PhD in African Studies and linguistics from the School of Oriental and African Studies in London and first worked with the Gurindji people when employed as a Research Fellow by the Australian Institute of Aboriginal Studies in 1974–77. As well as working on several languages he also wrote several land claim books. Later he worked in the Manjiljarra program at Strelley School in the Pilbara before lecturing at the School of Australian Linguistics at Batchelor. Together with Joyce Hudson, he helped establish the Kimberley Language Resource Centre in 1984, and after a couple of years at Ngalangangpum School, Turkey Creek, as a linguist, he joined the Anthropology Department at the Northern Territory University. Most recently he also taught Anthropology at Griffith University, worked on Native Title in North Queensland, and is currently Research Fellow, Language and Society, at the Australian Institute of Aboriginal and Torres Strait Islander Studies, Canberra. He is editor (with Nick Evans) of *Archaeology and Linguistics: Aboriginal Australia in global perspective*.

Paul Monaghan is researching a PhD in linguistics at the University of Adelaide focusing on land–language issues in the north-west of South Australia. Other research interests include early forms of Aboriginal English in South Australia. He has worked as a researcher for the De Rose Hill native title claim in the north-west of South Australia.

Nicholas Reid is a Senior Lecturer at the University of New England, where he teaches a range of units including Aboriginal Linguistics, Aboriginal Languages Today, and a new unit on Race and Racism. He has developed extensive multimedia resources, including award-winning online units and CD-ROMs. Nick's main research interests involve Aboriginal languages and linguistic typology. He also enjoys involvement in applied community linguistics projects, such as dictionary and curriculum development with the Nauiyu Nambiyu School in the Northern Territory, and placenames work with the Armidale City Council. Nick welcomes feedback on his paper and can be contacted on <nreid@metz.une.edu.au>.

Edward Ryan is a postgraduate student in the History Department of Latrobe University and is currently researching the history of the Wergaia and neighbouring peoples of north-west Victoria and south-west New South Wales. Besides placenames, his other research interests include landscape and environment change in the greater Mallee region. In a broader context he pursues similar research interests in Irish and Scottish Gaelic history, language and literature.

Bernhard Schebeck studied ancient Semitic languages and anthropology at Vienna University, general linguistics at the Sorbonne, and Kurdish and north-west Caucasian languages in Paris. On his first visit to Australia, 1964–67, he did a linguistic survey of the northern part of South Australia, began a study of the Yolngu dialects of north-east Arnhem Land, and a study of the Adnyamathanha language of the northern Flinders Ranges, South Australia. After defending his doctoral thesis on the phonological systems of Australian languages in Paris in 1972, he returned to Australia in 1973 to continue his study of Yolngu, and has since worked as a computer programmer as well as a consultant on various language and sociolinguistic projects.

Jane Simpson has worked for many years with Warumungu and Warlpiri people, and has written on both languages. She has a long-standing interest in the languages and societies of southern South Australia, and has collaborated with Luise Hercus on several projects relating to this. She teaches linguistics in the Department of Linguistics, University of Sydney.

Peter Sutton is an independent scholar and consultant in anthropology and linguistics. He has worked with Aboriginal people in a number of different regions since 1969. He has worked on over 50 land claim cases since 1979. His books include *Dreamings: the art of Aboriginal Australia* (ed., 1988), *Wik-Ngathan Dictionary* (1985), *Country: Aboriginal boundaries and land ownership in Australia* (1995), and *Native Title and the Descent of Rights* (1998).

Franca Tamisari trained in Social Anthropology at the London School of Economics and Political Sciences where she obtained a PhD in 1995. Based in Milingimbi, a community in the Crocodile Islands, she has been conducting intensive fieldwork research with Yolngu people in north-east Arnhem Land, Northern Territory, since 1990. She now lectures in Anthropology at the School of Social Science, University of Queensland. Her research focuses on Australian Indigenous cosmology and epistemology, language, ceremonial dance and song performances.

Michael Walsh has carried out research on Australian Aboriginal languages since the early 1970s, mostly in northern Australia in the coastal strip from Darwin to the Western Australian border. One part of this research has focused on Indigenous placenames, especially in connection with land claim and native title work. He is based in the Department of Linguistics at the University of Sydney.

David Wilkins began his fieldwork in Aboriginal Australia in 1982, and has spent approximately 60 months in the field, working mainly in Central Australia on Eastern and Central (Mparntwe) Arrernte and other languages of the Arandic family. He has worked with a number of Aboriginal educational organisations helping to develop and evaluate culturally appropriate language and curriculum materials. These organisations include the Yipirinya School, Intelyape-lyape Akaltye, and the Institute of Aboriginal Development. Wilkins has published over 40 articles and chapters dealing with a wide range of issues in semantics, pragmatics, anthropological linguistics, cognitive linguistics, child language acquisition, gesture and neurolinguistics. He has been a faculty member in Linguistics at UCDavis and SUNY Buffalo. He also worked for seven years as a senior scientific research in the Language and Cognition group (formerly the Cognitive Anthropology Research Group) of the Max Planck Institute for Psycholinguistics, Nijmegen, The Netherlands. While there his focus of research was the relation between language, culture and spatial cognition. He is currently a Scientific Researcher at the Center for Aphasia and Related Disorders, VA Northern California Health Care System, Martinez, California.

Georgina Yambo Williams, Kaurna senior woman, was the first to call for the revival of the Kaurna language as a spoken language in the mid-1980s. In the early 1980s she worked with the South Australian Museum to document and promote the Tjirbruki Dreaming track running from Warriparinga (Sturt River) in Adelaide, down to Cape Jervis in the south. Since then she has worked tirelessly to protect Kaurna heritage and the Tjirbruki story along the coast in connection with the Onkaparinga, Holdfast Bay and Yankalilla councils, and especially at Warriparinga with the Marion Council.

OVERVIEW

1 INDIGENOUS PLACENAMES: AN INTRODUCTION

Luise Hercus and Jane Simpson

1 INTRODUCTION[1]

In Australia we have two sets of placenames, one superimposed over the other. These are the set of networks of placenames that Indigenous Australians developed to refer to places (the Indigenous placename networks), and the set of placenames that Europeans developed to refer to places (the introduced placename system). The placenames bestowed by Europeans have been systematically recorded and they form the bulk of the official placenames of Australia, ranging from names for houses, streets and dams, to names for States and the country itself. Indigenous placenames, on the other hand, have been sporadically recorded by Europeans, varying in accuracy of reproduction of sound, meaning and even referent; there are many examples of Europeans recording words of the ordinary language, thinking that they were placenames (but see Hercus's paper in this volume for apparently 'silly' names that are likely to have been genuine placenames).

The ways of forming Indigenous placenames, their meanings, the features that they refer to, the networks that they form, and the rights to bestowal of names differ greatly from European toponymic practices. These differences in toponymy caused much confusion to European settlers, and, as Paul Carter (1988:61–68) argues, reinforced beliefs that Aborigines had different views of places, and thus of ownership of places.

---

[1] This paper owes a great deal to conversations with David Nash and Harold Koch, neither of whom, however, is responsible for our conclusions. We are also grateful for valuable comments from the audiences at the Second Workshop on Australian Languages, University of Melbourne, 20 December 1997; the AUSTRALEX conference, University of Queensland, 17 July 1998; the Australian Placenames of Indigenous Origin: Interdisciplinary Colloquium, the South Australian Geographical Names Committee Premises, Adelaide, 9 April 2000, and The Australian National University, Linguistics Department Seminar, 17 April 2000.

We use the following conventions for typestyle of placenames of words of Indigenous languages. When a placename in a spelling that is no longer current is discussed as a name, it is given in italics. When a placename in the modern spelling of an Indigenous language is discussed as a name, it is given in italic bold. When a placename is discussed as a place, it is given in plain typestyle. Words of Indigenous languages in modern spelling are given in bold. Words of Indigenous languages in a spelling that is no longer current are given in italics. Glosses for meanings of placenames are given in single inverted commas.

L. Hercus, F. Hodges and J. Simpson, eds, *The Land is a Map: placenames of Indigenous origin in Australia*, 1–23. Canberra: Pandanus Books for Pacific Linguistics, 2002.

Within Indigenous toponymic practices there is much variation, depending both on society and on environment. The papers in this volume attempt to address the variation within Indigenous toponymic practices, as well as other issues of toponymy, including documenting, assigning and reinstating Indigenous placenames. In this introductory paper we outline some of these issues, and we contrast the Indigenous practices with the introduced practices.

The earliest recorded Indigenous placenames are over 40 names recorded as placenames in the Sydney area (Troy 1994). One of those that Troy lists, *Parramatta*, is unequivocally an Indigenous placename that has been taken into the introduced system.[2] Such incorporation of Indigenous placenames into the official introduced placenames set has been sporadic, as Monaghan (this volume) points out. The surveyor Thomas Mitchell's preference for Indigenous placenames is well known, and contrasts with some other explorers (but see Carter (1988:67) for discussion of his motives for using them). Later writers sometimes also expressed the desire to use Indigenous placenames. For example, Edward Stephens (1889:498) wrote in an essay on Aborigines:

> In conclusion I express the hope that young Australia, instead of reproducing the names of all the counties, towns, hamlets, mountains, lakes, and rivers of Europe and Asia, will preserve the names which the aborigines of Australia gave to the distinctive features of their ancient home.

More recently, in 1986, in a speech at the Second National Nomenclature Conference Dorothy Tunbridge urged the importance of documenting the Indigenous placename networks, and of establishing an official body that would actively and systematically record Indigenous placenames, and also of reinstating the Indigenous placenames (Tunbridge 1987).

In the nineteenth century a few people recorded Indigenous placenames, perhaps the most extensive being the many Diyari names recorded by Reuther and Hillier in northern South Australia which Philip Jones's paper discusses. A few of these were taken into the introduced placename set, especially in remote areas. But the desire of the colonists to honour friends and benefactors and commemorate loved places proved very strong in many areas, and resulted in the loss of many Indigenous placenames. In the twentieth century, anthropologists and linguists started recording Indigenous placenames more systematically. Among them are the 1,100 names recorded by T.G.H. Strehlow for Arrernte country in Central Australia (Strehlow 1971), and the many names recorded by Norman Tindale. A table of about 50 such projects recording places and Indigenous names is given in Henderson and Nash (1997). This table excludes many of the tens of thousands of Indigenous placenames that have been documented and mapped as a result of heritage and sacred site surveys, land claims under the *Aboriginal Land Rights (Northern Territory) Act 1976*, and more recently of Native Title claims (Henderson and Nash 2002, Sutton, this volume). However, the documentation is rarely published, for practical and legal reasons, including the unresolved issue of Indigenous intellectual property. It lies in the offices of land councils and government departments.

Some of the projects mentioned by Henderson and Nash have resulted in publication of Indigenous placenames, for example Berndt et al. (1993) have nine maps with more than 300 placenames in the Lake Alexandrina–Coorong area of South Australia, and Dixon (1991) lists about 165 Yidiny placenames from north Queensland. A dictionary of 4,700 placenames officially recognised in the introduced placename system was published in 1992 (Appleton &

---

[2] The restriction 'recorded as placenames' is important, because, while many other places in the Sydney area have names from Indigenous languages, we do not know if they were Indigenous placenames for the same area.

Appleton 1992). About a third of the placenames in it came from Indigenous languages (Henderson and Nash 1997).

In 1993 a traineeship was funded at the Australian Institute of Aboriginal and Torres Strait Islander Studies (AIATSIS) for an Indigenous person to compile a national Indigenous placenames dictionary. The project proved too large for one person, but the final report provided some useful suggestions on investigating Indigenous placenames (Edwards c.1996). Another step towards documenting the Indigenous placename networks was taken in 1998 with the establishment of the National Place-names Project (since January 2000, the Australian National Placenames Survey), coordinated by one of the editors of this volume, Flavia Hodges, who helped organise the two placenames workshops that provided the impetus for this publication.

Documenting placenames takes several forms, depending in large part on how much the introduced placename system has replaced the Indigenous placename networks, that is, on how well the Indigenous placename networks have survived the invasion. In some areas, such as parts of Central and Western Australia, Arnhem Land and Cape York, the transmission of Indigenous placenames is unbroken. The owners of the placenames have been able to maintain their language and their residence on or near the land concerned. In such cases anthropologists, linguists and cartographers can work with speakers to get very rich data on placenames. These include not only the exact extent and location of the named feature, but also an understanding of placenames as systems of mnemonics for identifying places, and as integral to a group's understanding of its history, culture, rights and responsibilities for land. In this volume, Peter Sutton's paper describes such work among peoples living on Cape York, and Patrick McConvell does so for Gurindji people of the Victoria River area.

Another aspect of documentation of existing Indigenous placename networks concerns the use of placenames in societies. Basso (1996) shows the importance of placenames in everyday conversation in American Apache society, as mnemonics for moral tales, and Walsh (1997) and Merlan (2001) have taken this up for northern Australian societies. In this volume, Franca Tamisari discusses the culturally shared notions and images that placenames evoke for several Yirritja and Dhuwa groups in Arnhem Land, how the names both embody and provoke discussion of people's relationships with the land, and how this all relates to the words used for talking about naming. Similar themes are taken up by Baker (this volume) for the Ngalakgan, and by Wilkins (this volume) for the Arrernte of Central Australia. Wilkins also highlights the significance of vocabulary for talking about place and country.

In Arnhem Land, Central Australia and Cape York we can still see the unbroken transmission of the placename networks, and of the meanings associated with them, and we can see the diversity of toponymic practice. However, all Indigenous placename networks are under threat, and, when memories are fading, it is particularly urgent that the networks should be at least recorded.

Figure 1: Alhalkere. He is there, all painted up trying to get up to the top of the rise.
Photo: C. Macdonald

An example that illustrates what is likely to be lost, and thus the urgency for recording information on placenames, comes from spectacular 'break-away' country in traditional Lower Southern Arrernte country. The site *Alhalkere*, the Arandic word for 'nosepeg', is not named on maps. It is in the general location of Eternity Dam which is to be found on the 1:100 000 Alinerta sheet, which is part of the 1:250 000 Dalhousie map sheet in South Australia on the Northern Territory border (594850E 7064650N, Dalhousie SG 53–11). It is a piece of 'break-away', a conical hill that has split away from the main rocky range. It represents the 'erotic old man **Thudnungkurla**', and is the final site for a long, now almost forgotten myth of travel through Kuyani, Arabana and Lower Southern Arrernte country. After many nuptial adventures — always involving two women — this Ancestor arrives at a women's site on the Macumba. He is involved with 'a multitude' of women and is so worn out that he just manages to crawl over the sandhills until he comes to the hills near Eternity Dam. He knows the end is near and puts on his ceremonial gear including his nosepeg. As he tries to climb to the top he dies. He has remained there forever, a warning to all, as the strange conical outcrop with what looks like a headdress. The story makes this desolate area come to life and it is a pity that the place — like so many other important sites — should remain nameless.

In those areas where Aborigines suffered most dispossession, the transmission of the Indigenous placename network has been broken, and sometimes the only record of that network is in the overlap with European placenames. This is the case in most of the capital cities. Thus in Adelaide, the last person who spoke Kaurna as a first language died early in the twentieth century, and old Indigenous placenames survive only in placenames adopted by the Europeans, such as *Patawalonga*, a creek, and in those names recorded in the nineteenth century but not adopted as official placenames. Documenting placenames in such areas of long-established dispossession requires historical and philological reconstruction, based on archival study as well as geographical interpretation, as Amery's paper shows. For the reconstruction of a vanished Indigenous placename network it is also essential to have good documentation of existing, relatively intact, Indigenous placename networks in similar environments, so that, with due consideration of the diversity of Indigenous toponymic practices, extrapolation of naming principles can be made. Ryan's paper shows how important such cultural considerations are in reconstructing the meanings and significance of placenames.

As the introduced placename system is gradually superimposed on the Indigenous placename networks, there are transitional periods, often resulting in loss of names for features, as Harvey (1999) shows. In the Bardi dictionary (Aklif 1999) Gedda Aklif included well over 300 Bardi placenames. She has explained (Aklif pers. comm. 2000) that she did so at the request of an old Bardi man who was worried that younger people were losing the placenames, and that this hampered their ability to describe where they had been. They sometimes could not name the reefs where they had been fishing.

During the superimposition, Europeans sometimes incorporate Indigenous placenames into the introduced system, and the reverse also happens: Indigenous people incorporate introduced placenames into the Indigenous networks, as Donaldson's and Sutton's papers discuss. So, last century Kaurna people used placenames like *Adelaide*, while Europeans took over *Yurre idla* from Kaurna for Uraidla in the Adelaide Hills. But there is one difference in the early incorporations. The Indigenous people by and large took over the European placenames for places in their area, and used them with much the same reference as the Europeans used them. These places may have already existed and had names, such as hills and creeks. Or the names may have been names of newly created settlements or habitation structures. Thus Wilkins (this volume) notes Arrernte people calling town camps by the names of adjacent structures, e.g. *Trucking Yards camp*.

But the Europeans often took over the Indigenous placenames without much consciousness of their meanings or referents, as Ryan's and Monaghan's papers show. In South Australia, *Uraidla* (discussed in Amery's and Amery and Williams' papers) is a good example. It means 'two ears' and was used by the Kaurna people to refer to 'Mount Lofty and the adjoining point' (Teichelmann and Schürmann 1840). However, the Europeans used it as the name of a town nearby. Europeans were not always conscious of whether the name was a placename, and also assigned Indigenous placenames from one area to another place, as McConvell's and Ryan's papers show. This could be to commemorate a place, which is not uncommon with house and farm names. Or it could be via another commemoration; apparently the place *Tarcoola* in South Australia is named after a horse that won the Melbourne Cup. They also used words from Indigenous languages as placenames. For example, in the 1930s the goldfields near Tennant Creek were named the *Warramunga* goldfields after the Warumungu, the traditional owners of the country, a practice known in English toponymy (Cameron 1996). This name was then taken up in the 1960s as the name of a Tennant Creek discussion club, the *Warramunga Club*,[3] which then led to references to the 'Warramunga ladies', none of whom were Warumungu women.

The extreme end of such incorporation is found in newly coined names, whose form is intended to evoke the sounds of Indigenous languages. A nineteenth century example: the Adelaide suburb *Glenunga* which consists of the Scots 'glen', and the Kaurna locative ending *nga*. More recent examples: a house called *Didjabringabeeralong* and another called

---

[3]   The club appears to have been disbanded in about 1981, just before land claims made residents of Tennant Creek more aware of the Warumungu as traditional owners of the country. Files relating to the Warrumunga Club are held in TF140, archives of the Tennant Creek branch of the National Trust.

Yankaponga part way between the towns of Yankalilla and Myponga[4] (and Amery provides further examples). The lack of respect for Indigenous naming practices led to the naming in 1956 of a new Aboriginal community on Alyawarr/Kaytetye country, *Warrabri*, an invented name combining the names of two groups who had been moved there: **Warumungu** (then spelled *Warramunga*), **Warlpiri** (then spelled *Walbiri*), with perhaps the echo of New South Wales placenames like *Narrabri* and *Boggabri*.[5]

.   As time goes by, Indigenous Australians modify their strategies for naming places as Merlan (2001) indicates. They may adopt introduced placename strategies. Thus the *Nakkondi/Look* exhibition of photographs of Indigenous Australians (held at the State Library of South Australia, April 2000) contained a photo of two men standing in a garden in front of some roses. The caption was:

> Timothy and Ian, Yankalilla 1999. This is my land. I call the place Narangana, Narungga
> of my mother, and Kaurna this land of my father.

The name is unusual both because it is a blend and because it is a commemorative name, neither of which are common strategies in Indigenous Australian placenames. A different example: an Aboriginal outstation near a **Miyilpurnuru** 'crow' site in Tennant Creek has been called *Crow Downs* (Ruby Frank pers. comm. to Jane Simpson, 28 May 2000). *Downs* is a topographic descriptor which links the outstation explicitly to local stations such as *Brunette Downs* and *Rockhampton Downs*. But Warumungu placenames in general do not have topographic descriptors as their second element. Again, documentation of these kinds of changes that take place in Indigenous placename networks under the influence of the introduced network will be crucial for interpreting old records when reconstructing vanished Indigenous networks. The difficulties of interpreting placenames once transmission has been broken or severely diminished are shown in Amery's, Hercus's, Ryan's and Schebeck's papers.

Tunbridge not only stressed the importance of documenting the Indigenous placename network. She also emphasised the importance of assigning Indigenous placenames to otherwise unnamed features. Attempts to incorporate Indigenous placenames officially occurred as early as 1840 in South Australia, when Governor Gawler expressed 'the wish … to keep to the native names of localities' (see Amery and Williams, this volume), and recurred at the 1884 International Provincial Geographical Conference in Melbourne which advocated using Indigenous placenames as official policy (Henderson & Nash 1997). Tunbridge also advocated reinstating Indigenous placenames. This practice had happened as early as 1840 when, at Governor Gawler's request, the name *Field's River* for a river south of Adelaide was changed to what a settler, revealingly, described as 'its proper name of Onkaparinga' (Hawker 1975:41).[6]

---

[4]   Simpson saw the first in 2000 in Mittagong, New South Wales, and the second in 1999.

[5]   The name has since been changed to *Ali Curung*, an attempt to render the Alyawarr *Alekarenge* 'dog-associated', a reference to the Dog Dreaming of the country.

[6]   Whether 'Onkaparinga' was the proper name for the whole river is doubtful. Support comes from Teichelmann and Schürmann (1840) who give *Ngangkiparringga* as the whole river Onkaparinga. Teichelmann (1857) refers to the River *Ngangki*, as well as *Ngangkiparri* 'the lady river'. However, Meyer (1843) gives three names, including *Ngangkiparingga*, for 'Horse-Shoe' (probably the bend in the river close to the sea).

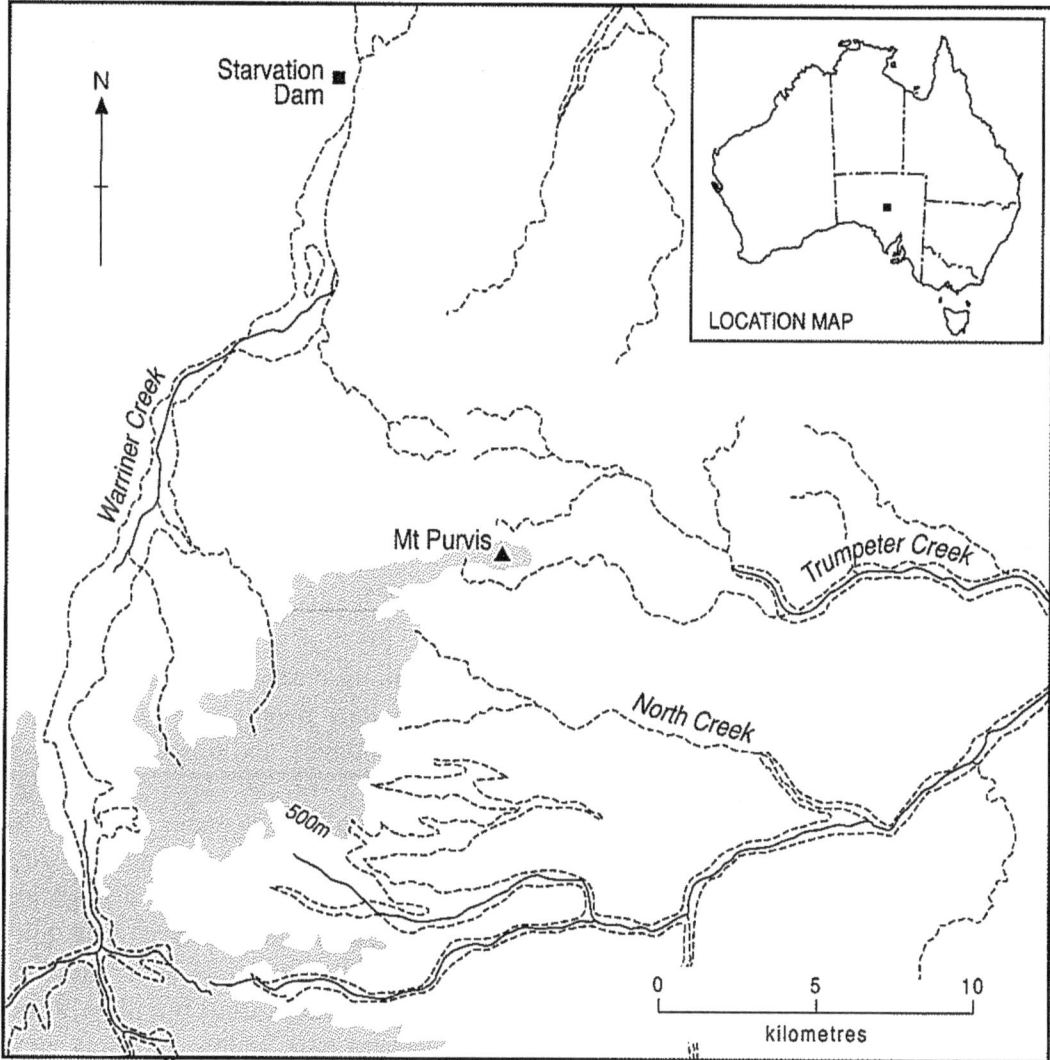

Map 1: Mount Purvis and surrounding area

Figure 2: Mount Purvis 'the Old Woman Busted' *Photo: C. Macdonald*

Like Gawler, others have wanted to restore Indigenous names, sometimes noting the discourtesy of imposing introduced names when Indigenous people are still using a rich system of names for places (Tunbridge 1987; Sutton, this volume), and sometimes arguing for the desirability of enlivening the landscape by introducing further Aboriginal names on ordinary maps. For example, ***Ulyurla Palthiyangunga*** 'the Old Woman Busted', Mount Purvis, is a well-marked place on the Billa Kalina map sheet in what was traditional Arabana country (53J 0588100E 6742900N, Billa Kalina SH53–7). The main site is a huge erosion gully on the northern face of the east–west ridge that makes up Mount Purvis. The gully has cut back to the top of the ridge about 300 metres west of the trig. In its upper reaches it is about 150 metres wide with near-vertical slopes, but it discharges onto the plain through a narrow channel with a gentle slope. The rocks exposed are pale yellow and pink and they contrast with the darker brown of the surrounds. In a myth and song cycle, again now mostly forgotten, the Ancestor **Thunpila** carries the decaying body of his dead wife in order to bury her, and at this place the body more or less explodes and 'all her innards fall out'. A letter from 'Moodloowardoo' to the *Adelaide Register* laments of this place and its name that:

> Yet all this interesting if primitive piece of geological tradition, with its excellent moral inculcating abstemiousness, is lost under a small triangle with a dot in the centre, branded on our plans Mount Purvis, in memory no doubt of some estimable but probably prosaic gentleman. (cited in Cleland 1952 from Cockburn 1908:66)

In remote areas, such as the Pitjantjatjara lands in northern South Australia, assignment and reinstatement of Indigenous placenames are relatively uncontroversial. In areas that have been settled by Europeans for long periods, assignment of Indigenous placenames may have sentimental cachet and even market value. Thus, new vineyards in southern South Australia and Victoria sometimes have names from Aboriginal languages, although the languages have not been spoken as first languages since early in the twentieth century. One might speculate that using a name from an Aboriginal language to name one's property gives the illusion of a long-term relation with a place, an unbroken chain right back to the original owners, and perhaps also the illusion of a connection with a lost age of simple living (which would link with ideas of the virtues of 'traditional' wine-making). Reid's paper in this volume discusses some of the linguistics issues involved in assigning new names.

In such areas, while assignment of names to features without names may be easy, reinstatement of names can be controversial, as the attempt to restore names in and around the Grampians National Park showed (Clark and Harridine 1990). However, rather than replacement of introduced names, dual naming policies are gradually being adopted. For instance, as Amery and William's paper discusses, Kaurna people have been working towards dual naming for some parts of the City of Adelaide, in conjunction with the City Council and the South Australian Geographic Names Board.

But in areas where both Europeans and Aborigines live, Indigenous placenames are often ignored. Thus in 1999 the town of Tennant Creek in the Northern Territory had about 50 per cent Aboriginal population. No streets or parks or public spaces were named after Aborigines, or had names in the local language, Warumungu. Only organisations primarily relating to Aborigines and town camps had Warumungu names. As for Alice Springs, Petrick (1996) lists about 550 placenames (mostly street names). About 30 have their origin in an Aboriginal language, and the majority of these are not original Arrernte placenames, but rather Arrernte words for flora and fauna applied to streets.

Since Tunbridge's paper, the State and Territory nomenclature authorities have been working towards policies on assignment, reinstatement and dual naming, in consultation with Indigenous communities. Their draft policies are given in this volume. William Watt of the South Australian Geographic Names Board has worked with Tunbridge and Adnyamathanha people to assign Adnyamathanha names to various features in the Flinders Ranges, and with Pitjantjatjara people and the linguist Cliff Goddard to assign names in the Anangu Pitjantjatjara lands in the north-west of South Australia. They have also been able to work with Ngarrindjeri people and older records to assign Indigenous names to features without existing official names along the Coorong. A further official step has been the use of the recognition of Indigenous placenames as an indicator for the state of Aboriginal languages, as part of national state of the environment reports (Henderson & Nash 1997).

Reinstating Indigenous placenames is seen as a mark of respect for the Indigenous peoples of Australia. In some areas, however, too few of the Indigenous placenames survive for this to be possible, as Reid notes for Armidale. In such cases people have attempted to remedy this by using words from the Indigenous languages as placenames, sometimes commemoratively, as in Amery's proposal to name places in Adelaide after prominent Kaurna people, and sometimes by taking words that sound good to both Europeans and local Aborigines, as Reid discusses for new official names in Armidale City taken from the Anewan language. Reid's paper also addresses the difficulties of pronunciation involved in incorporating words from Indigenous languages into the introduced placename set.

Reinstatement and assignment of names (placenames or otherwise) from Indigenous languages raise difficult and sensitive issues. Essentially the issue is this: if Indigenous names are brought into the official introduced placename system, they are in the public domain. This provides public recognition of the names, and thus of prior occupation of the country by Aborigines and Torres Strait Islanders. But once the names are in the public domain as placenames, then, as Reid puts it, they become a commodity. They can be used, say, in business names, without permission being granted by the Indigenous owners of the placenames. Some Kaurna people, for example, dislike the idea that a Kaurna placename, such as *Karrawirraparri*, could then be used in a business name, say the *Karrawirraparri* Greeting Card Company, and perhaps create favourable feelings among potential clients, without Kaurna people being consulted. Williams raises this issue in the Amery and Williams paper.

The intellectual property involved in using words or placenames from Indigenous languages is illustrated in a dispute, unresolved at the time of writing, about a new vineyard called *Koppamurra* in the south-east of South Australia (Altman 2000). There has been a telephone exchange and a station in the area called *Koppamurra* since the early twentieth century, and it has been recorded on maps since the 1940s. The form of the name suggests that it is probably derived from an Indigenous language. But the vineyard owners wanted to trademark *Koppamurra* and argued that 'any geographic use of the name would be deceptive and confusing and place its trademark at risk'. The fact that the word is likely to be a word of an Aboriginal language did not prevent the vineyard from being able to trademark it.

So, sensitive consultation is needed with communities before any Indigenous placenames are taken over into the introduced placename network. While Indigenous placenames can be reinstated to some extent, Indigenous placename networks are a different matter. Reinstating an Indigenous placename network amounts to reinstating a way of looking at the land, together with a set of associated responsibilities, and would probably only be possible on areas of land such as the Anangu Pitjantjatjara lands which Aborigines hold title to.

In the rest of the paper, we want first to outline briefly, important differences between the introduced placename system, and the set of Indigenous placename networks; and second to discuss strategies for forming placenames used in a number of Australian Indigenous languages.

First we discuss four differences between Indigenous and introduced placename systems. Then we mention three strategies for creating introduced placenames which are rarely found in Australian Indigenous placenames. Finally, we discuss the meaning and interpretation of analysable Indigenous placenames, focusing on southern and central Australia.

2   FOUR DIFFERENCES BETWEEN INDIGENOUS AND INTRODUCED PLACENAMES

The two systems of placenames differ in a number of ways, for example, whether the placenames form networks, how they act as mnemonics, what uses are made of the land, and what counts as a significant feature.

2.1 System versus set of networks

First, before 1788, Australia had no centralised government and no means of coordinating a single set of placenames. Each group had its network of placenames which linked in with its neighbours' networks, most noticeably at the boundary areas, but not only there. The result was that one place might have several names. It is also the case that different groups will have different strategies for forming placenames.

After 1788, a single official set of introduced placenames was codified. The idea was that this set of placenames would be known to and used by everyone who needed to use them. Of course, in particular areas settlers developed their own networks of unofficial introduced placenames — names for paddocks, dams and bores on farms, for example. But, in general, the single official network of introduced placenames contrasts with the many local networks of Indigenous placenames. There is also contrast in ownership. The official set of introduced names is not owned. Everyone who wants to can have access to it, through maps, gazetteers and the like. But the Indigenous placename networks are very often owned. One family may have the exclusive right to impart information about particular places, including their names. Some placenames may be powerful and may be secret or sacred, not for public distribution, as is clearly demonstrated in Tamisari's paper.

2.2 Local mnemonics versus mnemotechnics

Second, all placenames are shorthand labels given to significant geographic features for the purposes of finding them again, referring to them, and passing on the knowledge of the place to other people.

In the past, in areas where people have lived for a long time, without major disruption, placenames have developed organically as local mnemonics to refer to places. Thus, for example, the English placename *Taplow* comes from 'Taeppa's burial-mound' (Cameron 1996). We can easily imagine that soon after the mound was built, people started referring to it as in 'Take a right at Taeppa's burial-mound', and that over time when a settlement grew up in the area, people could become quite unaware that the name even referred to a burial mound rather than to the settlement.

While the organic development of placenames as local mnemonics in England is partly comparable to the Indigenous placenames networks of Australia, the English local mnemonics do not form a coherent mnemonic system in the way that Indigenous Australian placenames in many areas do. The invention of writing and of mapping meant that English placenames could be stored for long periods and sent long distances. The name *Taplow* would start to be used by people who had never been to the area, and never seen the burial mound. Writing and mapping reduced the need to have names that were memorable as having meanings that related directly to a particular place. The relation between the sense of the name and the referent of the name could be quite arbitrary. Instead, names could be used for other purposes, say, to commemorate a person, *Adelaide*, or to commemorate a much-loved place, *New England*.

In this respect the Indigenous placename sets of England differ from the Indigenous placenames networks of Australia. Indigenous Australians by and large did not have long-term placename storage devices like paper maps, with the possible exception of the Diyari toas mentioned in Jones's paper in this volume. But they had huge numbers of placenames to remember. Peter Sutton (pers. comm.) calculates that some Wik people on Cape York know several thousand placenames.

The effort of remembering not only the names, but also where the places are that they name, seems enormous to us. We are used to refreshing our memories from written material. People without such aids develop mnemotechnics, systematic mnemonic systems. In such systems, the mnemonic value of the name is important, but it must also be easily memorable as part of a system, as for example in the mnemotechnics of classical and mediaeval Europe (Yates 1969), a sequence of things to be memorised was associated with something else which had an easily graspable order (the seating plan at a feast, a room, the floorplan of a palace, a theatre).

Many groups in central and southern Australia devised similar but more elaborate mnemotechnics discussed in Strehlow (1971). To remember where a place is, and what it is like, the placename is associated with the travels of Ancestral Beings. Warumungu ancestral women go to the place *Wiitin* and leave a coolamon. That is visible now as a waterhole. **Wiitin** means 'coolamon' in Warumungu. They go east to another place, *Manaji*, where they dig bush potatoes. **Manaji** means 'bush potato' in Warumungu. Tamisari's paper contains several excellent examples of the association of topographic features with evocative ancestral activity — yellow ochre as ancestral faeces, for example. These travels and actions are represented in story, song, sand drawings, body painting and dance, and, as Wilkins (this volume) points out, the placenames are an essential part of the stories. In fact, a placename can then act as a reverse mnemonic — while it acts as an aide-memoire for locating a place, it can also act as an aide-memoire for events happening at that place, for the story (Basso 1996). That is, such placenames have a non-arbitrary relation to the place, as Merlan (2001) elaborates on for Jawoyn placenames. Moreover, as Merlan also points out, if a placename concerns actions of Ancestral Beings, those actions may also be carried on by humans at that place. Thus, circumcision was carried out at a Jawoyn place whose name invokes the story of an Ancestral Being carrying out circumcision at that place.

The sequencing of placenames according to the travels of an Ancestral Being is undoubtedly important for remembering a sequence of placenames, and thus the location of those places. In this sense Indigenous placenames can be described as a network. They are linked by more than the fact that they are names given to places (Wilkins, this volume). By knowing the story one knows something about the place, and about its location vis-à-vis other places. Places are connected by the story.

However, the dreaming track mnemonic is only one strategy used. Not all places are well connected in the network; some places may just stand on their own. They may have a story that is not linked to other places. And in fact not all areas of Indigenous Australia use the long sequences of names associated with travels as the main mnemonic technique, as Sutton (this volume) notes. Moreover, not all names are easily analysable (Walsh, this volume). The names mentioned above are clearly analysable, and are obviously easy to interpret as mnemonics. But, in fact, different Indigenous placename networks vary dramatically as to how interpretable the names are. For example, Luise Hercus has found most of the Arabana placenames are interpretable. But Sutton (this volume) observes that in the Wik region more than 50 per cent may be unanalysable. There are many reasons for this: the linguist's lack of knowledge, the loss of the language, or, as Sutton describes, the people may have switched to using another language but have kept the placenames in the earlier language. Another reason may be that places have several names. Some names may be esoteric, so that only a few people know the connection between the meaning of the name and the place that it refers to. Walsh's paper discusses the consequences of this.

Instead, other methods for preserving and transmitting the knowledge of places are used. Nash (in prep.) has observed the importance of toponymic gossip, talking about places, retelling stories of trips, as ways of keeping memory of places alive. Sutton (pers. comm.) notes the importance of competitive placename calling; he describes men competing with each other in naming places. Related to this is Aklif's (pers. comm.) observation that her Bardi teacher indicated that a person who can name a place has a special connection to that place and when a person cannot name a place anymore that connection weakens.

2.3 Uses made of the land

A third important difference between Indigenous placename networks and introduced placename systems has to do with the different uses made of the land. Australian Aborigines for the most part did not build permanent structures, such as houses, bridges and roads. This has a number of consequences for placenames. It emphasises the importance of systematic mnemonics. One can find out the name of the place *Taplow* by asking the people who live there, 'What do you call this place'? But in the unusual situation when hunter–gatherers are travelling and come to a waterhole whose name they do not know, they cannot expect to find someone living there who will tell them the name of the place. Rather, they will have to rely on being able to describe the place later to someone who might know it, and learning from them the name of the place.

The lack of permanent structures also has consequences for placenames. English placenames have been classified by Cameron (1996) into two main types: 'habitative', 'topographical' and a third, lesser type 'names of tribes'. The 'habitative' placenames reveal the changing habitation patterns of sedentary lifestyle, as well as the change of dominant languages — the -*cester* of Roman camps, the -*by* and -*ton* of towns, and the bridges, the markets, the mines, the fields.

INDIGENOUS	INTRODUCED
Kaurna: *wodli* 'house, hut'	*Fairfield*
Warkowodliwodli Klemzig (Warko?)	*Civic Centre*
Tambawodli Emigration Square (*tamba* 'plain')	and many indigenous to the UK, e.g. names ending in -*ham*, -*by*, -*thorp*, -*ton*, -*cester*.
Piltawodli the Native Location (*pilta* ?possum) (Amery and Williams, this volume)	

By contrast, Indigenous placenames in central and southern Australia have far fewer habitation structures to record. In fact, it is exceedingly difficult to find Indigenous placenames that make explicit reference to habitation. That is, we do not find names of the form 'X Camp'. Some Indigenous placename networks, however, do allow placenames to combine with generic uses of words meaning 'camp/place' such as the Arrernte use of **pmere** 'camp, country, place' discussed by Wilkins (this volume). Probably related to the lack of placenames involving habitation structures is the lack of placenames for paths or trails, other than the idea that a route is that taken by an Ancestral Being.

However, people clearly do live in the country, camp at places, and occupy them. McConvell's paper raises the intriguing question as to whether the extension of names of sites to areas has to do with the way people use the area.

2.4 What counts as a significant feature

A fourth important difference between Indigenous and introduced placename systems is what counts as a significant feature to be named. As we said earlier, placenames are labels given to significant features for the purpose of referring to them in discussion with others. What count as significant features depends on what impression the feature makes on a person seeing it, or in the case of some water features, on the person hearing it. Thus we might expect a large red mountain rising out of a plain to have a name. And we might expect water features to have names, as Donaldson (this volume) shows for the Ngiyampaa.

But we should not expect a one-to-one match between the kinds of topographic features usually given names in the Indigenous British placename set, and those given names in Indigenous Australian placename networks. Thus, in the Warumungu land claim, it emerged that a prominent hill called by the Europeans 'White Hill' was called by the name of the surrounding area, *Parakujjurr* 'cut-two', which includes two rockholes, the specific referents of the placename. Similar differences in the extent of the area referred to by Indigenous and introduced placenames are discussed by McConvell for the Gurindji. Another difference is in the naming of creeks. As Tunbridge (1987) noted, the Adnyamathanha name parts of creeks, not the whole creek.[7] This is true in many parts of Central Australia as well. However, as Peter Sutton (pers. comm.) notes, in the Wik area of Cape York, generic terms for words like 'creek' are prefixed to names, and seem to denote the whole creek.

3   THREE STRATEGIES FOR CREATING INTRODUCED PLACENAMES RARELY FOUND IN AUSTRALIAN INDIGENOUS PLACENAMES

Both the Indigenous placenames networks of Australia and the Indigenous placename system of England developed slowly. In this respect they are radically different from the introduced placenames network of Australia. The invasion and settlement of Australia took place so rapidly that in many areas there was no time for local mnemonics to develop. The administration of the country required rapid labelling of the country so that people could communicate with each other and be sure they were referring to the same place. Particular people, explorers and surveyors, were given the task of bestowing names on places. They used different strategies for designing the names. Sometimes surveyors were given instructions, which often included trying to find Indigenous names for localities, as Monaghan discusses. But a major strategy used was commemoration.

3.1 Commemoration strategies

Explorers who had to pay the bills named places after wealthy sponsors, for example *Glen Elder, Joanna Springs*. Many introduced placenames commemorate people or places by using the name of the person, such as *Adelaide*, or of their rank, *Kingston*, or the name of the place, such as *Richmond*. Sometimes a qualifier is added, such as 'New' in *New South Wales* and *New England*. We have rarely encountered such commemoration strategies in Indigenous placenames networks (McConvell, this volume). We have come across places bearing the personal name of an Ancestral Being (which may happen to be the name of a living person).

An example of a person giving their name to a site within the framework of traditional naming is that of the **Tyimama** sandhill, on the north side of the Spring Creek floodout in the Witjira National Park in Lower Southern Arrernte country. In the **Urumbula** myth and song cycle the Cat Ancestor brings handsome young women with him all the way from Port Augusta and picks up more on his journey north. He leaves behind the more elderly ones and they are represented in the landscape by hills or sandhills. One high and crimson sandhill is called **Tyimama**, and there is a **Urumbula** verse naming it, probably composed around the turn of the century. Apparently the verse and the site belong to a Dalhousie woman who had taken on the European name of 'Jemima'.

---

[7]   Possibly the European propensity for naming the whole river comes from river transport — travelling and trading up and down rivers probably assist in naming whole rivers rather than just places along the river.

But much more commonly we have seen the name of a place used as a personal name. This is akin to the way many English surnames are derived from places, and to the way that Australians in farming communities may refer to people by the name of their property. 'I heard Woolcunda on the radio yesterday.' It relates closely to ownership. Knowing the name, the story, the songs and perhaps dances and designs associated with a place is an important part of owning the place; displaying that knowledge can be a way of 'performing' one's ownership of that place. One kind of display of ownership is by bearing the name of the place. For example, Ronald Berndt's Yaraldi teacher, Albert Karloan, had a name *Djinbatinyeri*, derived from the place *Djinbatung* (Berndt et al. 1993:148).[8] Among the Warumungu, Simpson has heard people addressing the senior owner of the Snake and Star Dreaming using the name **Jajinyarra**, an important place on the dreaming track. This intimate relationship between personal names and placenames is elaborated on in Tamisari's paper describing the Yolngu practices in Arnhem Land.

3.2 Topographic descriptors

Second, many Indigenous placenames of the United Kingdom include topographic descriptors, as Baker's paper points out. They illustrate a classification of the landscape, and are found in British placenames from Celtic times onwards: *Strathspey*, valley of the Spey river, *Strathclyde*, valley of the Clyde river, *Inverness*, mouth of the Ness river, *Inveresk*, mouth of the Esk river. Many introduced placenames also include topographic descriptors: *Mount Remarkable, Lake Eyre, Point Pearce, Coffin Bay, Second Creek, River Murray* and so on. Such an explicit classification is helpful for newcomers to an area, who are reading a map and wanting to associate the explorer's names with what they see (but see Carter 1988:110 on the discrepancies between what an English explorer might call a 'lake' and what an English reader might imagine a lake to be).

We have found little trace of such classification in the Indigenous placenames we have looked at in Central and Southern Australia (not one of 689 distinct Warumungu, Warlpiri and Wakaya placenames contain a clear topographic descriptor). Donaldson (this volume) makes a similar observation for the Ngiyampaa of New South Wales, and Baker (this volume) shows that this is also true in the middle and lower Roper River area of the Top End of the Northern Territory. Arrernte (Wilkins, this volume; Harold Koch pers. comm. 1999) has a few names involving topographic descriptors such as '**pwerte** 'hill' and, rarely, X **kwatye** 'water'. However, absence of compound names including topographic descriptors is not true of all Indigenous placename networks. There are some notable exceptions:

(i)   In areas where generic classifiers are used, such as the Wik languages of Cape York, Peter Sutton (this volume) has noted the use of generic classifiers like 'creek' prefixed to a name, and given as the name of a creek.

(ii) In a number of networks in South Australia, there are compound placenames using the word for 'water', **kawi, kapi** often reduced to **-awi** (Hercus and Potezny 1999), and in Kaurna, the word *parri* or *pari* 'river',[9] as in *Warriparri*, 'Sturt Creek' discussed by

---

[8]   It was formed by dropping the ending *-ung* (probably a locative ending) and adding the 'belonging to' morpheme *-inyeri*.

[9]   The spelling used for Kaurna is that of Teichelmann and Schürmann (1840). In fact, *parri* almost certainly had a retroflex continuant.

Amery. However, the *parri* forms are not entirely convincing, because, while the Europeans applied them to the whole of the river, wanting as they did a name for the whole of the river, we do not know if this is what the Adelaide Plains people did.

(iii) Perhaps the most striking exception is provided by Schebeck's paper in this volume. He notes that almost a third of the 483 Flinders Ranges placenames he looked at are compound names including topographic descriptors. While the **awi** and **vari** (*parri*) forms mentioned above predominate, he lists 19 other topographic terms used. John McEntee points out (pers. comm. to Jane Simpson, 10 May 2000) that some, such as **Yurntu Vuri** 'sun stone/hill', involve older words for 'stone', while others, such as some of the many with **vambata** 'hill' in their names, may indicate more recent European influence in placename construction.

Environmental classifications are used in placenames. For example, in the networks we have investigated, we do find topographical descriptors, but as the main element, e.g. *Wulpayi* 'creek' (Warlpiri), *Alinjirri* 'floodout' (Warumungu). We have also come across placenames referring to ochres or pipeclay resources as the main element: *Palanmirrijangu* 'white clay-having' (Warumungu), *Wapanjara* 'red ochre' (Warumungu), *Karrku* 'red ochre' (Warlpiri), *Kartjiparnta* 'pipeclay-having' (Warlpiri), and compare Baker's Marra *Mayngu* 'red ochre'. Hercus's paper discusses problems of whether these are actual placenames or descriptions of the topography, a point taken up in Schebeck's paper.

But the major environmental classification is dominant vegetation. Thus, in the Warumungu land claim, of more than 680 distinct placenames, more than 80 clearly involve names of vegetation. Baker (this volume) notes similar properties in the Roper area, and a considerable number of the Yidiny placenames for which Dixon (1991) gives meanings refer to vegetation.

VEGETATION

Kaurna	*Korra weera, yerta* and *perre*	Adelaide, and the Torrens. From *karra* 'red gum', *wirra* 'forest'.
Yaraldi	*Muwantjangali, Muwandjingal*	dense pine forest ?from Nebuluwar to East Wellington (Berndt et al.1993:14, 313–314). Compare Ramindjeri *mowantye* 'pinewood' (Meyer 1843).
Warumungu	*Purrurtu*	'coolibah' — name of group's country.
Yidiny	*Jalngganji*	*jalnggan* 'milky pine' + **-ji** 'with' (Dixon 1991).

This emphasis on vegetation probably reflects hunter–gather attitudes towards land too. If land is not being cleared for farming, then the vegetation, ti-tree, pine or grasses are a fairly permanent part of the landscape, suitable for a mnemonic.

A possible interpretation of the lack of topographic classifiers in many Indigenous placename networks is, as Anna Wierzbicka (pers. comm.) suggests, that in these networks placenames have a default head 'place, area, country', just as in the introduced network, the default reading of an introduced name without a topographic descriptor, such as *Adelaide*, probably is 'place of habitation'. Wilkins (this volume) explores in detail the use in Arrernte of a generic **pmere** with placenames while Alpher (this volume) notes the widespread use of a generic **pin** 'home-place, country' occurring as a part of Yir-Yoront tract-names (but omissible), and Simpson has noted in Warumungu narratives the use of **manu** 'country' appearing as a separate generic word together with placenames.

3.3 Relative location

A third difference is that neither Indigenous placenames in Britain nor introduced placenames in Australia show the kinds of systematic mnemonics described above for Indigenous Australian placenames, which allow the creation of large networks of placenames. However, they do sometimes form smaller networks through the use of indicators of relative location. As English speakers we are used to placenames that have as part of the name a cardinal point, *North Adelaide*, and so on. Usually they refer to a settlement that is east or south of some other settlement, often larger. It makes sense to name a smaller settlement or structure by reference to a larger landmark, or else naming a part by reference to the whole: *South Australia, East Gippsland*. Cardinal points are classic markers of relative location, but there are also markers of relative altitude or upstream, like *Upper Sturt, Lower Mitcham, Naas* and *Top Naas* in the Australian Capital Territory, *Top* and *Bottom Baldknob Creek, Bottom Pigstye Swamp, Top Camp*; relative distance to a place, *Outer Harbor*, relation to some named place, *Noarlunga* and *Old Noarlunga* in South Australia, *Hartley* and *Little Hartley*, and *Back Yamma Forest* in New South Wales.

But we have not come across this system of explicitly marking relative location[10] in the relatively intact Indigenous placename networks in areas where we have worked in Central and southern Australia. This may be a result of the fact that people travelled much of the time, and of the lack of names for settlements or camps. Settlements are obvious reference points for relative location markers to develop from. Of course, in principle there is not any reason why people did not develop names for regular campsites that locate topographic features with respect to other topographic features — why not have 'east of the mountain', 'south of the river' as placenames?

However, the fact that the relatively intact Indigenous placename networks in some areas do not seem to use cardinal points means that we should be at least suspicious of placenames from Aboriginal languages that contain cardinal points. For example, in the nineteenth century the word *Cowandilla* was recorded as the name of a place in Adelaide. However, early sources have:

Kobandilla	Districts of the Adelaide tribe	Wyatt (1879)
kawandilla	'in the north'	Teichelmann (1857)

Where did this name come from? Is this name likely to be used by the traditional owners of Cowandilla to represent their country? Why would they call their countries 'in the north' respectively? North of what? We do not know the circumstances of the bestowal of these two names. Which Aborigines gave the Europeans these names? What did they think the

---

[10] There is in Central Australia a partial system of relative location in that places which have the same name are sometimes said to be linked. But it is nowhere near as extensive as the system in the United Kingdom.

Europeans were asking? What did the Europeans think the names referred to? But we do know that two important cultural brokers, Mullawirraburka and Parnatatya, who worked with the missionaries and protectors recording the language of the Adelaide Plains were men whose country was south of Adelaide. Thus, from their perspective *Cowandilla* would be in the north. If they were asked the name of the Cowandilla area, they might well have said, 'It's in the north'. However, see Amery (this volume) for an account of another Kaurna name involving a cardinal point: *Patparno, Patpungga* Rapid Bay (Wyatt 1879) 'in the south', for which we do know one instance of the use of this name when it cannot have meant 'south of where we are talking'.

This lack of relative location indicators does not hold in certain other areas, such as the Wik people of Cape York, as Sutton describes. Hence, the cardinal points in the Adelaide Plains name could be genuine names. We simply do not know enough about the toponymy practices of the Kaurna to judge.

4 SEMANTIC CONTENT OF INDIGENOUS PLACENAMES IN SOUTHERN AND CENTRAL AUSTRALIA

The meanings and interpretations of Indigenous Australian placenames are the subject of most of the papers in this volume; however, several focus on general problems of interpreting placenames. Thus Schebeck discusses what is a placename, as opposed to a description or a secondary name alluding to some aspect of the story. This relates to the problem of distinguishing between what is a proper name and what a common noun. In some languages, such as Yankunytjatjara (Goddard 1985), or Ngalakgan, Alawa and Marra (Baker, this volume), the distinction is made in the morphology; special allomorphs of case endings appear on proper names. But in others, placenames are harder to distinguish from common nouns. However, distinctive endings, such as the comitative and locative suffixes that are often found on placenames, serve a different purpose, marking out a name as belonging to a particular language. Thus Kaurna has many placenames ending in the locative **-nga** or **-(i)lla**, while its neighbour Ramindjeri has placenames ending with a locative **-ng**: *Ramong* and *Wirramulla* both refer to *Encounter Bay* and have the same root. Genitives can have the same function: the Warlpiri place *Yankirri-kirlangu* 'emu-belonging' is translated closely in Warumungu as *Karnanganja-kari*. Indeed, if names are interpretable and are transparent mnemonics, then they are likely to be translated by neighbouring languages. Thus the same feature may have two different names, both with the same sense, but with different forms. An example from Hercus's fieldwork: 'the Fire-track', i.e. the Macumba River, *Makampa/URingka* (from **maka-wimpa** in Arabana, **uRa-ingka** in Arrernte). This translatability reduces the ability to track movements of people and languages over time through placenames, as Harvey (1999) shows for the Top End, and as Sutton notes in his paper. But, as Walsh's paper points out, Indigenous placename networks vary as to their analysability. Merlan (2001) says that a considerable number of Jawoyn places had associated stories but they had names that were hard to relate to the story. Alpher's and Baker's papers show that in some areas there is plenty of evidence for archaisms being retained in placenames. Likewise Sutton discusses ancient placenames in Cape York whose forms conflict with the phonological structure of the currently dominant local languages. He considers the question of whether stable land–language relations are likely to lead to more or less analysable placenames.

Here we summarise our experience in interpreting central and southern Australian Aboriginal placenames. We divide them roughly into four types (excluding those that linguists have been unable to analyse), along a continuum.

Semantic Content of Analysable Indigenous Placenames in Southern and Central Australia

descriptions of ----> topography and environment	description refers to Ancestral Beings and environment simultaneously *literally*	----> *figuratively*	reference to Ancestral Beings
Parnttarr-kujjurr 'two cracks' **Karlukarlu** 'boulders' **Purnungkurr** 'swamp'	**Kijjiparraji** 'white ghostgum' (the *Mungamunga* women turn into ghost gums)	**Para-kujjurr** 'two body-entrances' = two rock holes **Wiitin** 'dish' = depression in creek	**Karnkka** 'moon' **Kartti kujjurr** 'two men' **Maliki-kirlangu** 'dog belonging' **Ngurru pakinyi** 'nose pierced'

(these are Warumungu placenames)

In the first column are names that transparently describe the topography and environment, discussed in this paper, and most strikingly in Schebeck's paper. In the centre, these often occur with suffixes meaning 'having', 'associated with', 'belonging'; thus Warlpiri *Ngalyipi-parnta*, 'having *ngalyipi* (snakevine)'. Ash's paper discusses the use of the comitative **-baraay** in Gamilaraay, and Donaldson does the same for Ngiyampaa **-puwan**. Sometimes these are no longer productive in the modern language; thus Warumungu has a suffix **-riji** or **-yiji** found in a few words **lirrppiriji** 'claw-having = goanna', and in a number of placenames, *Manajiriji* 'bush.potato-riji'. (It is, of course, possible that these names designate environment and mythology simultaneously. The bush potatoes are there because ancestral women planted them.)

On the right are names that appear to refer only to aspects of events on an Ancestral Being's journeying, such as throwing a boomerang, urinating, making fire, standing up a spear-thrower and so on. Names such as the Warumungu *Ngurru pakinyi* '(someone) pierced a nose' belong here. Tamisari's paper discusses these. Merlan (2001) provides interesting Jawoyn examples in which the placename is an ideophone for an action of the Ancestral Being, as the place *Gurngurnbam* where **gurngurn** denotes the thudding of a kangaroo. In between are names which refer simultaneously to the story and to the landscape. This may be literally (as in a place named by a tree which the Ancestral Beings did something with) or figuratively, such as Warumungu *Jalkaji* for a spear-thrower which represents a tall upright rock. These names sometimes have variant forms, thus Warumungu *Ngurrujarrpijara* 'nose-enterer', and *Ngurruparramanyi* 'nose-sat.down' refer to the same place. Both are allusions to an action happening at the place that is the source of the placename. Baker's paper gives further examples.

Probably most names fall into the middle two kinds. That is, if a name appears to be purely topographical or environmental, it may be that we have lost the associated story. And if a name appears to be purely mythological, it may be shorthand for what an ancestor did at that place. Thus, indirectly, names such as *Ngurru pakinyi* 'nose pierced' also refer to features of the landscape, but the metaphor is not as transparent as it is with *Jalkaji*.

Different placename networks have different hierarchies of places, (McConvell, this volume). Some include both named large areas, 'countries' or 'tracts' (see Alpher's paper for this term), and named sites within those countries, as Baker discusses, while others do not. Most of our observations cover names for sites, rather than names for countries. But we note that Warumungu countries include names whose senses are at both ends of the continuum, *Purrurtu* 'coolibah', the name for the country belonging to the holders of **Jalajirrppa** 'white cockatoo' dreaming, and *Warupunju*, an Arandic word for 'fire', the name for the country belonging to the holders of **Warlukun** 'fire' dreaming.

A final point on form. Placenames evoking a story associated with a place are sometimes expressed as a verb or phrase, or, as Wilkins (this volume) observes, as a headless relative clause, and examples are provided in many papers in this volume, as well as in Merlan (2001). This is an unusual feature of Australian Indigenous placenames compared with introduced placenames (although it is found in other Indigenous placename systems, e.g. Kari's (1989) discussion of Alaskan Athabaskan placenames). Languages vary as to what percentage of names are expressed this way; Schebeck (this volume) notes that only 14 of the 483 Adnyamathanha placenames he examined were of this form. However, in the Roper area Baker notes that there are a large number of such placenames.

In Arabana some of these are simply verb forms, often marked for the past tense; sometimes the verbs are preceded by a noun subject or object:

Pakalta	'he is digging for somebody else', Mount Arthur, Arabana (Present benefactive)
Ngampayiwalhuku	'(they) came to get a grinding stone'
Kudna-tyura-apukanha	'they (the ancestral Emus) had diarrhoea long ago', Rockwater Hill in Arabana country, where there are lots of green stones
Thidna-pakanha	'digging with his foot', the Tidnabucca waterhole on the Macumba, also another site near Coward Springs
Wati–warakanha	'he blocked the track', hill near Anna Creek in Arabana country, the reference is to the Ancestral Turkey who tried to 'perish' the Initiands

RELATIVE CLAUSE

Wabma tharkarnayangu	'(where) the Snakes stood up', Wommaturkaenamana Bore, in Arabana country
Yunga kurdalayangu	'(where) his waterbag fell down', reference to an Ancestor who got such a fright that he dropped the waterbag he had just filled; alternative name for Burraburrina Well
Pilparu palthiyangunha	'(where it, i.e. the whole camp) was split by Lightning', Blanket Waterhole near Belt Bay, Lake Eyre
Tyalpiyangunha	'(where they, the Initiands whom the Turkey was trying to perish) cooled down', Coolibah Yard near the Lower Neales, close to Lake Eyre.

Such names are common in some areas, uncommon in others. The syntax of these is well worth further investigation; for example, Merlan (2001) notes the rarity of transitive subjects compared with intransitive subjects and transitive objects in Jawoyn placenames.

CONCLUSION

We have tried to show that Indigenous placename networks in many areas can be viewed as systematic mnemonics. Within the networks we have suggested that the hunter–gatherer way of life is reflected in the lack of habitation names, and in the lack of placenames formed using relative location indicators such as cardinal points. We have shown also that there are differences in what counts as a significant feature to be named. In terms of the actual names used, we have claimed that commemoration strategies characteristic of introduced placename systems are rarely if ever found.

REFERENCES

Aklif, Gedda, 1999, *Ardiyooloon Bardi Ngaanka: One Arm Point Bardi dictionary*. Hall's Creek: Kimberley Language Resource Centre.

Altman, Carol, 2000, What's in a name: a vineyard or a dot on the map? *The Australian*. 10 February 2000, p.5.

Appleton, Richard and Barbara Appleton, 1992, *The Cambridge Dictionary of Australian Places*. Cambridge, UK: Cambridge University Press.

Basso, Keith H., 1996, *Wisdom Sits in Places: landscape and language among the Western Apache*. Albuquerque: University of New Mexico Press.

Berndt, Ronald M. and Catherine H. Berndt, with John E. Stanton, 1993, *A World That Was: the Yaraldi of the Murray River and the Lakes, South Australia*. Melbourne: Melbourne University Press at the Miegunyah Press.

Cameron, Kenneth, 1996, *English Place Names*. London: B.T. Batsford.

Carter, Paul, 1988, *The Road to Botany Bay*. London: Faber and Faber.

Clark, Ian D. and L.L. Harridine, 1990, *The Restoration of Jardwadjali and Djab Wurrung Names for Rock Art Sites and Landscape Features in and around the Grampians National Park: a submission to the Victorian Place Names Committee*. Paper presented at Koorie Tourism Unit, Victorian Tourism Commission, Melbourne.

Cleland, John Burton, 1952, Meteoric crater — native legend. *The South Australian Naturalist* 27(2):20.

Cockburn, Rodney, 1908, *Nomenclature of South Australia*. Adelaide: W.K. Thomas & Co.

Dixon, R.M.W., 1991, *Words of Our Country: stories, place names and vocabulary in Yidiny, the Aboriginal language of the Cairns–Yarrabah region*. St Lucia: University of Queensland Press.

Edwards, Dale, c.1996, *National Indigenous Place Names Dictionary Project: Final report*. 49pp. Canberra: Australian Institute of Aboriginal and Torres Strait Islander Studies.

In association with the Geographical Names Board of New South Wales and the Department of Employment, Education, Training and Youth Affairs.

Goddard, Cliff, 1985, *A Grammar of Yankunytjatjara*. Alice Springs: Institute for Aboriginal Development.

Harvey, Mark, 1999, Place names and land–language associations in the Western Top End. *Australian Journal of Linguistics* 19(2):161–195.

Hawker, James C., 1975, *Early Experiences in South Australia*. Adelaide: Libraries Board of South Australia.

Henderson, John and David Nash, 1997, Culture and Heritage: indigenous languages. Australia: State of the environment technical papers series (natural and cultural heritage). Department of the Environment, Canberra.

Henderson, John and David Nash, eds. 2002 *Language in Native Title*. Native Title Research Series, Canberra: Aboriginal Studies Press.

Hercus, Luise A. and Vlad Potezny, 1999, 'Finch' versus 'Finch-water': a study of Aboriginal placenames in South Australia. *Records of the South Australian Museum* 31(2):165–181.

Kari, James, 1989, Some principles of Alaskan Athabaskan toponymic knowledge. In M.R. Kay and H.M. Hoenigswald, eds, *General and Amerindian Ethnolinguistics: in remembrance of Stanley Newman*, 129–149. New York: Mouton de Gruyter.

Map sheets:

SA Department of Mines and Energy, n.d. Alinerta 6045 1:100 000 Map sheet. Adelaide.

Joint Operations Graphic, 1984, Billa Kalina, South Australia 1:250 000. Series 1501, SH 53–7, edition 1. Compiled March 1984.

Division of National Mapping, 1981, Dalhousie, South Australia. 1:250 000 Map sheet. SG 53–11, edition 1.

Merlan, Francesca, 2001, Form and context in Jawoyn place-names. In J. Simpson, D. Nash, M. Laughren, P. Austin and B. Alpher, eds, *Forty Years On: Ken Hale and Australian languages*, 367–383. Canberra: Pacific Linguistics.

Meyer, Heinrich A.E., 1843, *Vocabulary of the Language Spoken by the Aborigines of the Southern and Eastern Portions of the Settled Districts of South Australia*. Adelaide: James Allen.

Nash, David (in prep.), Ethnocartography: understanding central Australian geographic literacy. Draft, 2 October 1998, 30pp. Presented to Australian Anthropological Society Annual Conference, 2 October 1998.

Petrick, Jose, 1996, *The History of Alice Springs through Landmarks and Street Names*. Alice Springs, the author.

Stephens, Edward, 1889, The aborigines of Australia. *Journal of the Proceedings of the Royal Society of New South Wales* 23:476–503.

Strehlow, Theodore G.H., 1971, *Songs of Central Australia*. Sydney: Angus and Robertson.

Teichelmann, Christian Gottlob, 1857, Dictionary of the Adelaide dialect. MS. (with double columns). 99pp. Held in South African Public Library.

Teichelmann, Christian Gottlob and Clamor Wilhelm Schürmann, 1840, *Outlines of a Grammar, Vocabulary, and Phraseology, of the Aboriginal Language of South Australia, Spoken by the Natives in and for Some Distance around Adelaide.* Adelaide: published by the authors, at the native location.

Troy, Jakelin, 1994, *The Sydney Language.* Canberra: the author.

Tunbridge, Dorothy, 1987, Aboriginal place names [edited address delivered to the Second National Nomenclature Conference (now the Committee for Geographic Names in Australia), 21 August 1986]. *Australian Aboriginal Studies* 2:2–13.

Walsh, Michael, 1997, The land still speaks? Language and landscape in Aboriginal Australia. In D. Rose and A. Clarke, eds, *Tracing Knowledge in North Australian Landscapes: studies in indigenous and settler ecological knowledge systems*, 105–119. Darwin: North Australia Research Unit.

Wyatt, William, 1879, Vocabulary of the Adelaide and Encounter Bay tribes, with a few words of that of Rapid Bay. In J.D. Woods, ed., *The Native Tribes of South Australia*, 169–182. Adelaide: E.S. Wigg & Son.

Yates, Frances A., 1969, *The Art of Memory.* Harmondsworth: Penguin Books.

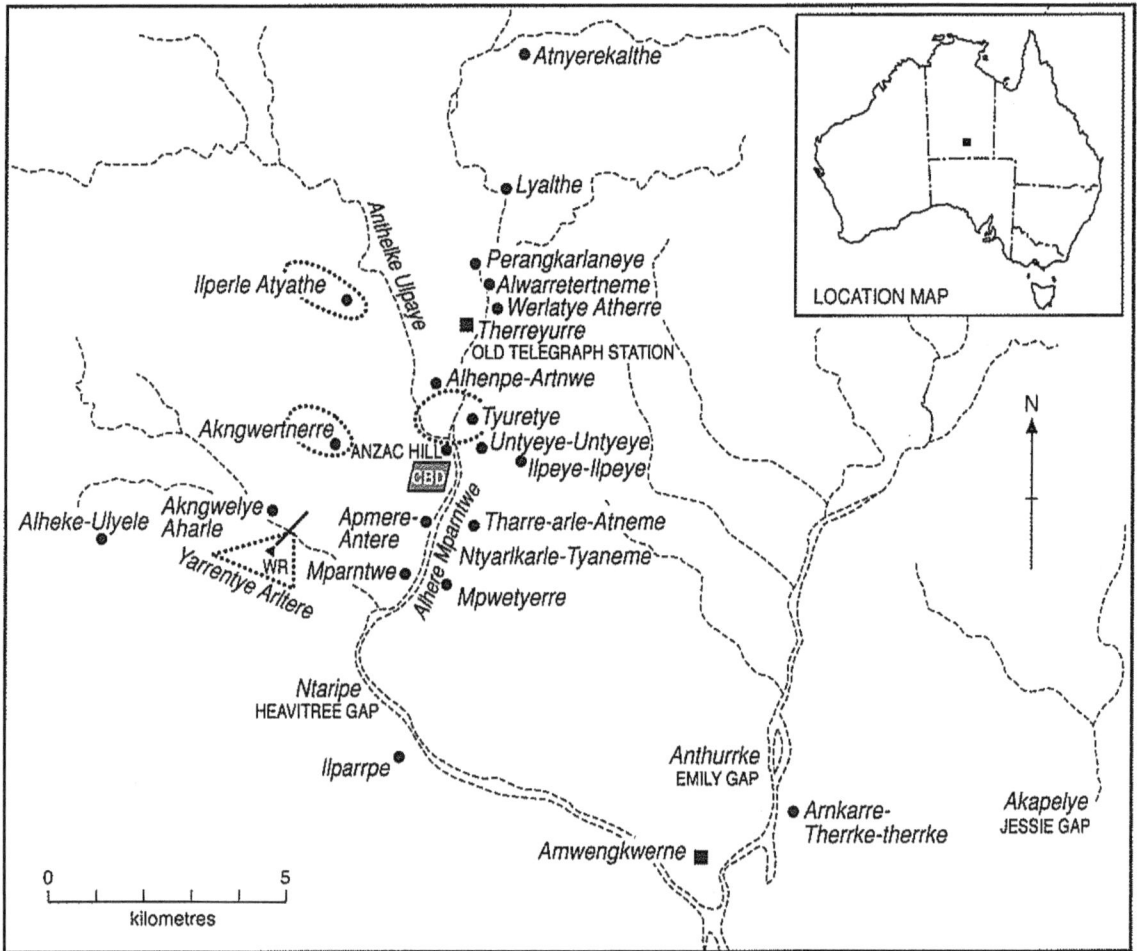

Map 1: Apmere Mparntwe 'Alice Springs'
     Places named (and pointed to) by a senior Arrernte man, Wenten Rubuntja.

2 THE CONCEPT OF PLACE AMONG THE ARRERNTE[1]

David P. Wilkins

Now as to the concept of space, it seems that this was preceded by the psychologically simpler concept of place. Place is first of all a (small) portion of the earth's surface identified by a name. (Einstein (1953:xiii) in Jammer (1954))

1 PLACE

It is necessary to preface the discussion of *place* in Mparntwe Arrernte with a justification as to why it falls within the domain of entities rather than the domain of space.[2] Einstein's quote echoes a popular view, within some semantic approaches, that fundamental to the domain of

---

[1] This paper was written in the mid-1980s, based on a study of about 150 placenames, and discussions around them. At the time, I was still formally affiliated with the Yipirinya School. My deepest thanks go to the Mparntwe Arrernte members of the Yipirinya community with whom I worked closely in the 1980s. This paper was originally to appear as a chapter of a dissertation with quite a different organisation than the one I eventually produced (Wilkins 1989). Still, it had a relatively wide circulation. Many people gave comments on an earlier draft, and I apologise that I have failed to respond adequately to all the comments. For comments and discussion that they may not even remember giving, I am indebted to Avery Andrews, Edith Bavin, Alan Dench, Gavan Breen, Nick Evans, Cliff Goddard, John Henderson, Robert Hoogenraad, Harold Koch, Bill McGregor, David Nash, Jane Simpson and Anna Wierzbicka.

This paper had been relegated to a pile of papers I thought would never see the light of day because I had decided I would never do the necessary work of updating and correcting that was required. However, the editors asked to include it in this volume, even without the necessary updating. It is published here with minor modifications, updating of some references (in square brackets), the addition of a map, and a definition of **pmere** (which also appears in Wilkins 2000). While there are some details in the paper which, 15 years on, I do not totally subscribe to, I still stand behind all the major points.

Note that the spellings used are those of the mid-1980s. We have not updated the spellings of the Arrernte words to reflect the changes in Henderson and Dobson (1994), according to which **pmere** would appear as 'apmere', **pwerte** as 'apwerte', and **ntherrtye** as 'antherrtye'.

[2] This paper was originally written to be a chapter in a semantically organised language description of Mparntwe Arrernte. One major section was on language pertaining to entities in the world. Other sections included Space, Time, Causation, and so on. It was particularly relevant in the structure of the description that the discussion of 'place' (namely **pmere**) appeared in the 'entities' rather than the 'space' section.

L. Hercus, F. Hodges and J. Simpson, eds, *The Land is a Map: placenames of Indigenous origin in Australia*, 24–41. Canberra: Pandanus Books in association with Pacific Linguistics, 2002.

25

space is the notion of *place*. The most radical position is that *place* is a universal semantic primitive through which location and other spatial relations are to be explicated (cf. Wierzbicka 1980). Such a claim would suggest that some form of lexical or morphological equivalent to English 'place' should exist in all languages and that this necessarily underlies the concepts of space and location. This position, however, runs into difficulties as we shall see.[3]

A fundamental error surrounding the concept of *place* within semantic theory had been the failure of semanticists to recognise that the English word 'place' has a meaning range which covers two distinct semantic notions. One notion is that of *'place as entity'* and the other is that of *'place as one of the two arguments of a locational/spatial predication'*. *Places* in the second sense are generated into existence whenever such spatial predications (as represented by adpositions, spatial cases, or other morphological forms) apply. Such *places* are created out of any type of real-world entity and they exist as places as long as the predication holds true for them. *Places* in the first sense are a subtype of entity and exist, like other entities, regardless of such predications. An 'arm' is not a *place as entity* and, as a consequence, one cannot say: '*Arms are places'. However, an 'arm' can be a *place as spatial relation*. For example, when a fly sits on someone's arm, one can say (in English): 'His arm is the place where the fly is'. 'Towns', on the other hand, are an example of *places as entity* and one is able to say 'Towns are places'. Of course, *places as entities*, like other entities, may occur as *places as spatial relation*.

This conflation of senses is not by any means a universal. Many Australian languages, for instance, possess a lexical item which designates *place as entity* but not *place as spatial relation*. In these languages this second notion tends to be conveyed (or, more precisely, generated) by the use of spatial cases such as locative. This clearly brings the status of 'place' as primitive into question for these languages.

The failure to distinguish these two notions has given rise to some misleading semantic claims. Lyons (1977:693), for example, states outright that 'places are not entities'. This is not

---

[3] Abbreviations used in this paper include:

A	subject of transitive verb	MpA	Mparntwe Arrernte
ABL	ablative	NomNEG	nominal negator
AFTER	'after'-ative	NPP	non-past progressive
ALL	allative	PASS	'do verb action while moving past'
Anm	Anmatyerre	PC	past completive
AS.WELL	as well	pl	plural
DAT	dative	POSS	possessive
dist	distal	REL	relative clause marker
DS	different subject	REVERS	'do verb action back to/while going back'
ERG	ergative	S	subject of intransitive verb
GO&DO	'go to a place and do verb action'	sg	singular
IMP	imperative	SS	same subject
INCH	inchoative	VbNEG	verbal negator
LOC	locative	1	first person
mid	mid-distant	3	third person

Morphemes are separated by hyphens: 1pl-POSS. Portmanteau morphemes are joined together: 1sgPOSS. Asterisk * indicates an unacceptable form.

Words of Indigenous languages are in italics, with the exception of the emphasised **pmere**, the topic of the paper. Placenames in Indigenous languages are in italic bold.

true unless we restrict the use of 'place' to *place as spatial relation*. If we do this, his following statements are consistent with the position taken here.

> As places are not entities, so entities are not places; but in so far as they occupy space, entities may serve to identify the spaces that they occupy ... What must be emphasised is that in all such instances we are relating to a place. But we refer to the place indirectly in terms of the entity that it contains; and this is tantamount to treating the entity as a property of place. (Lyons 1977:693)

This paper, then, deals primarily with *place as entity*. The discussion begins by looking at the semantic range, and applications, of the Mparntwe Arrernte lexical item that encompasses this particular notion of *place*, **pmere**. Following this is a closer examination of the Mparntwe Arrernte conception of geographical place and principles of placenaming. The discussion demonstrates how the particular conception of *place as entity* is reflected by the linguistic constructs and the lexical choices that are used to talk about and describe places and how, conversely, one needs to understand this largely culture-specific conception if one is to model how and why native speakers use the grammatical forms that they do when talking about *place*.

1.1 Pmere — camp, country, place

While the majority of Australian languages do not possess any lexical item whose central meaning corresponds to English 'place', they usually have a key lexeme, with very broad semantic application, that translates some senses of that word. **Pmere** is the corresponding Mparntwe Arrernte form and, as we shall see, its central reference to land and place, as related to other entities, makes it a very significant item of vocabulary. Its meaning range is typical for Australian languages and Mparntwe Arrernte speakers most often translate it as 'camp' or 'country' rather than 'place'.[4] As such, **pmere** may designate a 'camp' in the sense of *an area containing several families and distinctive groups of people (single men, single women, visitors, and so on) all sharing some major resources but living in close proximity within a number of separately organised shelters* (example (1)). Alternatively **pmere** may refer to a camp in the sense of *a specific family or group dwelling place; more specifically the actual shelter shared by that group*. This latter case, often translated as 'home' or even 'house', may be viewed as the smaller 'camp' within a 'camp' (example (2)).

(1)   ***Pmere*** *Stevens mape-kenhe, Macmillan mape-kenhe,*
     camp   Stevens group-POSS   Macmillan group-POSS

     ante Rice mape-kenhe yanhe-le   neme,
     and Rice group-POSS there(mid)-(LOC) be-NPP

     ikngerre-thayete-le.
     east-side-LOC
     'The Stevens', Macmillan's and Rice's camp is there, on the east side (of town).'

---

[4]   Dixon (1980:105) discusses in some detail the range of the Dyirbal noun *mija*, often glossed by bilingual informants as 'camp' or 'hut'. Similarly, Myers' (1976:158ff) discussion of Pintupi *ngurra* shows it to have a range comparable to **pmere**. Moreover, he proposes that the Pintupi view of reality is based mainly on three closely related concepts, one which is encoded by *ngurra*.

(2)   ***Pmere***   *tyenhe*   *yanhe*   *tne-me.*
      shelter   1sgPOSS   there (mid)   stand-NPP
      'My house (shelter) is (stands) there.'

When used in the sense of 'country' or 'land' **pmere** refers to a large area perceived as having rough boundaries which may or may not be presently inhabited but for which Dreamtime stories are known and for which responsibility and care of the land rest with a certain person, family or group (example (3)). In this sense it has overlapping reference with *ahelhe* 'land, ground, dirt'.

(3)   ***Pmere***   *nwerne-kenhe,*   ***Pmere***   ***Mparntwe,***
      country   1pl-POSS   place   Mparntwe

      Undoolya-*nge*   *antekerre*   *inte-me,*   *Arlpere-kenhe*
      Undoolya-*ABL*   south   lie-NPP   Warlpiri-POSS

      pmere   *kenhe*   *altule*   *inte-me.*
      country   but   west   lie-NPP
      'Our country, Mparntwe, is (lies) south of Undoolya, but Warlpiri country is (lies) west.'

Not surprisingly the three different applications of **pmere** discussed so far are each associated with a different one of the three existential–positional verbs [Wilkins 1989:5.1.3.1]. Shelters 'stand' (*tne-* (2)), camps 'sit' (*ne-* (1)) and country 'lies' (*inte-* (3)). Moreover, there is a hierarchy of possible inclusion of one referent of **pmere** within another which follows this same order (Figure 1).

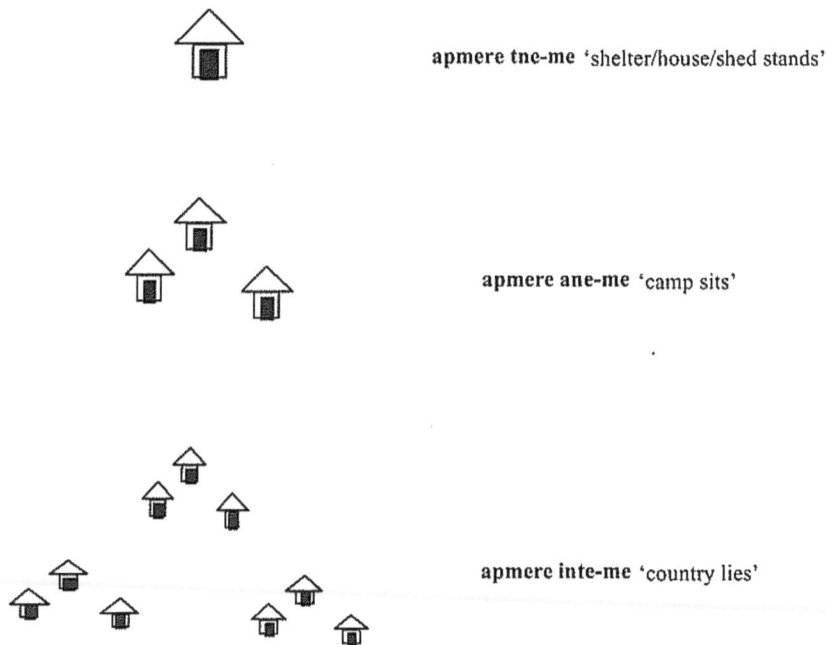

apmere tne-me 'shelter/house/shed stands'

apmere ane-me 'camp sits'

apmere inte-me 'country lies'

Figure 1: The inclusive nature of the referents of **pmere**

Thus far **pmere** has been described largely in terms of its association with people, and while this is perhaps its most common association, it is by no means part of the core meaning. An ants' nest may, for instance, be described as *yerre-kenhe* **pmere** — 'ant-POSS dwelling place', and the place within a spear thrower where a spear end is fixed and where the spear lies can be described as *irrtyarte-ke* **pmere** — 'spear-DAT place' — 'place for spear'. This latter phrase is a good example of when it is appropriate to translate **pmere** with 'place'. These and earlier examples, however, show that core to the sense of **pmere** is that it is *a place where, habitually, someone/something lives/exists, or could potentially live/exist*. So, unlike the general, abstracted sense 'place' can have in English, **pmere** is strongly identified as a place defined by its association with another entity or entities. Note further that, while various nominals may take the locative case, the location thereby designated is not necessarily called **pmere**. *Amwelte-le* 'arm LOC' — 'on the arm', for example, specifies a location which in English could be called a 'place' but, in Mparntwe Arrernte, could not be called a **pmere**. In short, an 'arm' is not *a place as entity* and is, therefore, not a **pmere**.

1.1.1 Pmere *as generic/classifier*

Pmere has been discussed in its role as a simple designator, but it can also function as a generic term, or classifier, for 'geographic place' — a function which is not shared by the corresponding word in many other Australian languages.[5] As such, it enters into a generic-specific construction [Wilkins 1989:3.4; Wilkins 2000] of which three subtypes may be identified. The distinction of subtypes is based on the nature of the *specific* element in the constructions.

Firstly, when using a *placename* it is common to precede it with the generic **pmere**. So, while one can say:

(4)   *Ayenge* **Sydney**-*werne* *lhe-ke.*
       1sgS   Sydney-ALL   go-PC
       'I went to Sydney.'

it is just as common to say:

(5)   *Ayenge* **Pmere Sydney**-*werne lhe-ke.*
       'I went to Sydney.'

In the second subtype (examples (6) and (7)) the specific element which follows the generic **pmere** is the name of the *totem for a place*. In some cases a construction of this sort can provide the actual name of a place. In all cases it can be used as an alternate designation for a place. One must rely on context to specify which place, among a number of places of the same totem, is being referred to. Out of context it is assumed that the place designated is *the*, or one of *the*, major centre(s) for that totem.

---

[5] Yankunytjatjara (Goddard 1983), for instance, possesses a small range of generic/classifiers but *ngurra* 'camp, place' is not among them. By contrast, Yidiny (Dixon 1977) *bulumbu* appears to share this function with **pmere**.

(6) *Re*   **Pmere Ntyarlke**   *anper-irre-nhe-nke.*
    3sgs   place   Ntyarlke   pass-INCH-PASS-PC
    'He went past the ***Ntyarlke caterpillar*** place.' (Which in context is the place known
    as ***Uletherrke -Mount Ziel.***)

(7) **'Pmere Mparntwe'**   *ante*   **'Pmere Yeperenye'**   *nyente*   *ile-me.*
    place Mparntwe (Alice Springs)   and   place   Yeperenye   one   tell-NPP
    ''*Mparntwe* (Alice Springs)' and '***Place of the Yeperenye***' mean the same thing.' (In
    other words the two phrases refer to the same place.)

It is uncommon for the totem name to appear on its own without **pmere** to designate a place,
but it is not unattested.

In the final subtype, a nominal referring to a *type of place* may provide the specific element
for such a construction. This may be a place that occurs naturally as part of the landscape:

(8) *Artwe-le*   *tywerrenge*   **pmere**   *inteye*   *kwene-ke*   *arrern-irtne-ke.*
    man-ERG   sacred.objects   place   cave   inside-DAT   put-REVERS-PC
    'The man put the sacred objects back in the cave.'

(9) *Itne*   *ngentye*   *are-tye-lhe-ke*   **pmere**   *urrkale-ke.*
    3plA   soakage   see-(GO&DO)-PC   place   mulga.country-DAT
    'They came upon a soakage in mulga country.'

or it may be a place people make for a specific purpose:

(10) **Pmere** *arnkentye-le*   *artwe*   *tyenhe*   *neme.*
    place   single.men's.camp-LOC   man   1sgPOSS   live-NPP
    'My (initiated) son lives in the single men's quarters.' (A way of saying: My son is
    single.)

or it may be a type of place invested with special significance:

(11) *Mwantye*   *lhe-Ø.*   *Nwerne*   **pmere**   *meke-meke-ke*   *itw-irre-me.*
    carefully   go-IMP   1plS   place   sacred.site-DAT   near-INCH-NPP
    'Go carefully. We're approaching a sacred site.'

We can see from the examples of the three subtypes above that the generic use of **pmere**
refers to *places located at a specific point on the earth's surface.* Moreover, for the generic to
be used with a specific term, the place designated must be *of social, functional or religious
significance to people.* This significance is inherent in placenames and totem names, but may
or may not be inherent in 'type of place' nominals. When the specific term is not, by itself,
necessarily a significant place, then context must allow the place to be interpretable as
significant to people, for **pmere** to be used with it in a generic-specific construction. A cave,
for instance, is a type of **pmere** when it houses sacred objects (example (8)) but it may not be
seen as such otherwise. Another example which demonstrates this point involves **pwerte**
'rock; hill'. One could not say, for instance:

(12)　　*Arrwe*　　　　**pmere**　　　　*pwerte-le*　　　*neme.*
　　　　rock.wallaby　　place　　　　hills-LOC　　live.NPP
　　　　'*Rock wallabies live in the hills.*'

But one could say:

(13)　　*The*　　*arrwe*　　　　*tyerre-ke*　**pmere**　*pwerte-ke.*
　　　　1sgA　rock.wallaby　shoot-PC　place　　hills-DAT
　　　　'I shot the wallaby in the hills.'

especially where the 'hills' referred to are known to the addressee. This is consistent with the meaning of **pmere** discussed previously, although the generic use has a slightly narrower (more 'socio-centric') sense.

1.2 Named places and placenames

1.2.1 *Conceptions of geographic place and linguistic consequences*

Many writers have noted, and elucidated, the special significance of the land and landscape for the Aboriginal Australians (Bell 1983, Berndt 1976, Elkin 1938, Spencer & Gillen 1927, Strehlow 1971). Strehlow (1970:135) comments that:

> For the Aboriginal Central Australian the totemic landscape formed a firm basis for religion, for social order, and for established authority itself.

It is not strange, therefore, to find that the concept of geographic place is very different from the one held by Anglo-Western culture and that this has certain linguistic consequences. This has already been touched on, in part, in the discussion of **pmere** (1.1), but here the issue will be examined more closely in a discussion of named places — an important subclass of those places to which the term **pmere**, in both its designating and generic functions, is applicable.

Traditionally all named topographical places are important sites, or areas, which were created by the activities of totemic ancestors during the Dreamtime. They are the track, or imprints, which evidence the continuing reality of the Dreamtime. The subject matter of many Dreamtime stories is, therefore, an account of how individual sites came into being.

Rarely, if ever, in traditional stories or paintings is a place depicted on its own without depicting some connection with other places. There are at least four types of connection that can exist between places.

Firstly, places may be located along the same path of travel of one Dreamtime ancestor, or group of ancestors. Each place therefore represents different scenes from the complete story of that ancestor, and thereby shares the same totem. Dreaming tracks have been known to extend very long distances through the country of a number of different groups, and it is interesting to note that texts will employ special linguistic mechanisms to indicate the movement of an ancestor from the area of one group to that of another. Apart from a simple overt reference to the change, which is not always present, linguistic markers representative of the difference between the two groups may be used. Thus if a totem moves from Western Arrernte country, or Pitjantjatjara country, into Mparntwe Arrernte country, a song in the first dialect/language may be sung in the early part of the text and, when the change of area occurs, a song in the second dialect/language will be used. Similarly, reported conversation may first contain lexical items, or even full sentences, from the one group and later will use ones from the second group. Most interestingly (from the point of view of what sort of morphemes can diffuse from one area to the next) it should be noted that in the narrative itself particles, and even suffixes, of roughly corresponding meaning in the two dialects or languages will be substituted for one another to signal the area change. In the following example from a Dog Dreaming text in which the ancestor moves from Mparntwe country into Anmatyerre (Ti Tree) country, the signal of the change is the switch from using the Mparntwe Arrernte allative form, *-werne*, to the Anmatyerre allative form, *-werle*.

(14) *Re*   *lhe-me-le,*   *lhe-me-le*   *pmere*
     3sgS   go-NPP-SS   go-NPP-SS   country

    arrpenhe-werne.   *Pmere-k-irre-me-le*   *re*
    other-ALL(MpA)   Country-DAT-INCH-NPP-SS   3sg

    inte-ke.   *Ingweleme*   *kem-irre-me-le*
    lie-PC   Morning   'get-up'-INCH-NPP-SS

    aweth-anteye   *lhe-ke.*   *Lhe-me* *anteme* *pmere*
    again-AS.WELL   go-PC   Go-NPP   now   country

    *kngerre-**werle**,*   *pmere*   *kwatye-rle*
    big-ALL (Anm)   country water-REL

    *ne-me-rle-**werle**.*
    be-NPP-REL-ALL (Anm.)
    'He travelled and travelled to another place and when he got there he camped. When he got up in the morning he went off again. Now he's going to an important place (in Anmatyerre country), to a place where there's water (in Anmatyerre country).'

This demonstrates not only a link of place to place, but also a link of people to place, and of language to place.

Secondly, the same totemic association may be attributed to different places (for example, they may both be caterpillar sites, or honey-ant sites). This does not entail that the places be connected by a Dreamtime path. Where this association exists, the places are seen to have a similar nature and something that affects one place will then affect the others, as well as affecting the totem for the place and the people associated with the place and the totem. Thus, though removed in space, two places of the same totem may be treated as parts of the same whole and so become subject to the grammatical rules pertaining to parts and wholes [Wilkins 1989]. For instance, with respect to switch reference [Wilkins 1988, 1989], where one place is the subject of one clause and another place is the subject of a linked clause, *same subject* marking (*-mele*) may be used on the dependent clause to indicate the unified nature the two places are perceived to have via their shared totemic affiliation (see example (15a)). *Different subject* marking (*-rlenge*) could also be used with respect to the same two places, but in this case the construction would focus on the fact that they are different places which are physically distant from one another in geographic space (see example (15b)). Thus, one can emphasise either unity or separation, and this is consistent with the treatment of other separable parts of the same whole [see Wilkins 1988 for more examples].

(15) a. *Pmere nhakwe kurn-irre-**mele**, pmere nhenhe kurn-irre-ke.*
site that(dist) bad-INCH-SS site this bad-INCH-PC
'When that place became defiled, this (related) place (also) became defiled.'
[Same-subject (SS) marking emphasises these are two places united by the same totemic affiliation.]

(15) b. *Pmere nhakwe kurn-irre-**rlenge**, pmere nhenhe kurn-irre-ke.*
site that(dist) bad-INCH-DS site this bad-INCH-PC
'When that place became defiled, this (other) place (also) became defiled.'
[Different-subject (DS) marking emphasises these two places are distinct and distantly separated entities (and so backgrounds the fact that they share a totemic affiliation).]

Note also that one place may have more than one totemic association since it may be the meeting point of two or more ancestors, or else it may be a place where the tracks of different totems cross over. Thus, a single named (socially important) place may manifest a network of relations (and totemic connectedness), and so, depending on speaker, or context, one relation rather than another may be given emphasis.

The third relation involves places being linked by virtue of the patrilineal pair association attributed to them. Strehlow (1970:110) stresses the importance of the *nyenhenge* section areas[6] (*nye* = 'father' -*nhenge* = kin DYADIC) as far as leadership and authority are concerned. Arrernte kinship divides the eight skin classes into four *nyenhenge*, or patrilineal pairs [cf. Wilkins 1989:1.2.4.1]: *Kngwarraye–Peltharre*; *Penangke–Pengarte*; *Perrwerle–Kemarre*, and *(A)ngale–Mpetyane*. Each named place is associated with one of these four pairs and because of this classification one can talk about places using the language of kinship. One can, for example, attribute a kin-relationship term to the association between two places depending on how the two *nyenhenge* pairs of those places are related to each other. Thus if two places have the same affiliation, they can be described as *kenhenge therre*[7] 'two brothers together', since two brothers, like the two places, have the same skin classification (see Green, et al., 1984, sect. 2, p.15). If one place is designated *Kngwarraye–Peltharre* and another place *Penangke–Pengarte*, then the two places can be referred to as *ipmenhe(nhe)nge therre*, 'two mother's mothers together'. In the kinship system *Peltharre* and *Pengarte* call each other *ipmenhe* 'mother's mother'. There is also evidence that people can refer to places by a kin term, the choice of the kin term being dependent on the person's own individual skin classification and the *nyenhenge* classification of the place [for an elaboration of this point, see Wilkins 1993].

It would appear that the totemic association(s) of a place can be totally cross-cut by the *nyenhenge* association. This means that two places of the same totem need not be of the same *nyenhenge* section.

Finally, places can be associated by geographical proximity. They can, for example, be geographically contiguous areas or they may be places contained within a single totemic area. An area may contain several sites of different totemic affiliation but be considered the main domain of a single totemic ancestor. Usually such an area contains a site of major importance which is often the site where a totemic ancestor emerged from the land or went back into the land.

---

[6] Strehlow's spelling *njinana*. The term literally means 'father-together'.

[7] From *kake* 'elder brother' reduced to *ke-* and -*nhenge* 'kin DYADIC' followed by *therre* 'two'.

It should be obvious from the preceding discussion that for Mparntwe Arrernte people a topographic place is not conceived as simply an isolatable point in space. Instead, a named place is a point within a network of relations and it is these relations that give it definition. These relations are not only, or even mainly, with other places, but also with people and things through kinship and totemic affiliation. The language reflects these associations at the levels of lexicon, grammar and discourse.

An Mparntwe Arrernte person's view of a place is necessarily subjective since s/he is related to it along parameters of varying intimacy. S/he may share the place's totemic affiliation — it may in fact be his/her conception site; or s/he may share its patrilineal pair association; or s/he may be related to people responsible for the land, and so on. All these relations will determine one's varying obligations to a place and constrain his or her behaviour towards that place, including how, or whether, s/he talks about a place. More specifically, this also determines the actual lexical choices and grammatical constructions the individual can use when talking about a place.

One further point which emphasises this conception of place, and the importance of the Dreamtime in such a conception, has to do with definitions given by Mparntwe Arrernte speakers for the phrase **pmere ulerenye**.[8] This phrase is usually glossed as 'a strange place', but in further elaborating its meaning one speaker said:

(16)   *Altyerr-iperre   ne-tyekenhe,   'pmere   ulerenye'.*
         Dream-AFTER   be-VbNEG   pmere   ulerenye

         Altyerre   Urrperle-kwenye.
         Dream   Aboriginal (black)-NomNEG
         "'**Pmere ulerenye**' is (a place that's) not from the Dreamtime. It's not an Aboriginal Dreaming.'

Another speaker explained **pmere ulerenye** by saying 'nobody knows the stories for that place'. Examples of places that were considered **pmere ulerenye** were places like Sydney and Melbourne. The phrase designates a place that lies outside the network of relations outlined above, a place to which an Mparntwe Arrernte person cannot calculate their relation, and so does not know how to behave with respect to it, because it is not part of the world that is organised by the Dreamtime.

2 PLACENAMING

> F.J. Egli in his *Nomina geographica* (Leipzig, 1893, 2nd ed.) has demonstrated that geographical names, being an expression of the mental character of each people and each period, reflect their cultural life and the line of development belonging to each cultural area. To this statement should be added, that the form of each language limits the range of terms to be coined. (Franz Boas, 1934)

While it is probably a universal that all languages contain placenames, the way placenames are structured, and how placenames are treated grammatically, may differ from one language to another. For instance, some languages may draw names from a specific lexical class of proper names in which each name has no function apart from its naming function, while

---

[8]   In this phrase **pmere** is in generic function and *ulerenye* is a 'type of place' nominal (2.1).

others construct names from the general stock of lexemes and grammatical structures of the language and tailor the name specifically to the place designated. English leans more towards the former type (cf. names like Sydney, Canberra and Melbourne) while Mparntwe Arrernte leans, as we shall see, towards the latter type.

An apparent consequence of this distinction is that while all proper names are singular definite referring expressions, and hence have a denotational function, names that arise in the second fashion may have, as in Mparntwe Arrernte, a strong connotational function. Another possible distinction between languages concerns whether placenames are formally distinguished from other noun phrases in the language or whether they are only distinguished on semantic grounds and otherwise 'cannot be distinguished, in terms of their internal grammatical structure, from noun phrases constructed according to the productive grammatical rules of the language' (Lyons 1977:640). Yankunytjatjara (Goddard 1983) is an instance of the former type; proper names have a special citational ending (*-nya*) and certain case endings, such as the locative, differ in form according to whether the noun phrase is a proper name or not. Mparntwe Arrernte is more of the latter type with the only formal distinguishing feature between placenames and other noun phrases being the ability for a placename to be preceded by the generic **pmere**.

Having encountered some of the salient features of the Mparntwe Arrernte conception of geographic place above, it should not be surprising to find that traditional placenames are not arbitrarily associated with the places they designate. While it may, at times, be difficult to analyse the meaning of a placename, older speakers state categorically that *all* placenames have a meaning and that this meaning relates to the account of the totemic ancestors that created the place and the actions they performed. In fact, a large number of placenames are analysable and there are a number of ways they can be seen to make reference to a place's Dreamtime origin. The various structures and lexical items used in placenaming are discussed below.

2.1 Analysable placenames

Headless relatives: A name may make actual reference to the significant event that happened at the place as the following examples show.

(17) **Ntyarlke-rle**   **Tyane-me**
       Ntyarlke-caterpillar-REL   'to-cross-over'-NPP
       'The place where the *Ntyarlke* (totem) crosses over.'

This is the name of a place in Alice Springs where the sacred *Ntyarlke* caterpillar crosses over the Todd River on its travels from the west.[9]

(18) *Aherlke-ke-rle*
       'to-dawn'-PC-REL
       'The place where the sun rose.'

Aherlkekerle is a site on the east side of Alice Springs where the *Ntyarlke* caterpillars, who were travelling in the middle of the night from the west, decided to rest until sunrise before doing anything else.

---

[9]   This is the place that Spencer and Gillen (1927:97, 98) spelt *Unjailga i-danuma*.

(19) *Itn-ake-rre-rle*   *Atnarnpe-ke*
3plS-(same-patrimoiety-dif.-generations)-pl-REL   descend-PC
'The place where they (the group of fathers and sons) descended.'

This is the place where a group of men, all of the same patrimoiety (fathers and sons together), who were travelling along the MacDonnell Ranges, descended from the ranges and thereby created a gap that exists there.[10]

A significant feature of the three placenames above is the presence of the suffix *-rle*, the relative clause formative [cf. Wilkins 1989:8.1.1.18, 10.1.3], which commonly attaches to the first element of a relative clause. Furthermore, all three can be seen to be reduced clauses containing a verb and, where appropriate, the subject of the verb.[11] These placenames can thus be analysed as a sort of headless relative, where the head is presumably something like 'the place where ...' In a sense the head can be provided for such constructs since, as I mentioned earlier (see §1.1.1), all placenames may be preceded by the generic for places **pmere** — for example, *Pmere Ntyarlkerle-Tyaneme*.

Naming by totem: Previously (see §1.1.1) it was observed that generic-specific constructs with **pmere** and a totemic name may be used as the name of a place. Thus one of the names for Jessie Gap is *Pmere Kapelye*, where *kapelye* is the name for a kind of lizard which is but one of three totems for that place. In this particular case it is more common, but not totally necessary, to refer to the place by the full phrase including **pmere**. However, a site along the young boy Dreaming track, *Werre Therre* (boy two) — 'the place of the two boys', is more commonly referred to without being preceded by **pmere**. Whether **pmere** is, or is not, commonly used as part of the name appears to be dictated, at least in part, by presently unexplored principles of syllable count and euphony.

Naming by particular totemic part: Another way of naming a place is where a body part of a totemic Being is referred to and hence calls up the whole. Emily Gap, an extremely important site for the caterpillar totem, is named *Pmere Nthurrke*, where *nthurrke* designates 'the black entrails of a caterpillar'. In many such cases the body part referred to is meant to conjure up, for those who know the story of the place, the particular Dreamtime event that occurred there. *Werlatye Therre* 'Two Breasts', for instance, is a significant women's site along the *Women Dancing* Dreaming track and it was at this place that the women were attacked by young uninitiated men and had their breasts cut off.

Part–whole constructions may also form the name of a place. In such a case we have the totem name followed by the part name. This often occurs in cases where the part is so generally attributed to a range of different things that it would not uniquely call up the image of the totem on its own, unlike *nthurrke* and *werlatye* which are fairly restricted to certain entities (i.e. caterpillars and women respectively). As an example, Spencer and Gillen (1927) identified the place *Kngwelye Artepe*[12] (dog back) 'Dog's Back', a ridge that is part of the dog

---

[10] Spencer and Gillen (1927:97) record this as *Innagurra-nambugga*.

[11] The verb *Aherlke-* in example (18) is a verb that takes no subject.

[12] Their spelling is *Gnoilya Teppa*.

Dreaming track. This last example is a case where there is a geographic surface instantiation of the totem and the placename clearly mirrors how the place looks.

Naming by description: The final type of placename that we will discuss involves a completely descriptive term which designates a prominent feature of the landscape that embodies the totem. For example, **Apere Therre** (red-gums two) refers to a site where two gums stand and into which some of the Dreamtime caterpillars of Alice Springs entered and then came out as butterflies. This appears to be the least common of naming practices although it should be mentioned that several placenames, which otherwise fall into the categories already discussed, do contain topographic terms like *atwatye* 'gap', *pwerte* 'hill', *ntherrtye* 'range', *inteye* 'cave', and so on (e.g. *Uyenpere Atwatye* 'Spearbush Gap').

A place is not restricted to having just one of these name types. There is evidence that a place can be referred to both by a prominent feature and by a name depicting an event. Moreover, as mentioned in §1.1.1 it seems that all places can be designated by their totem name. The question of whether all places do, or did, have more than one name, and therefore what significance might attach to each name type, remains to be investigated.

2.2 Placenames that are not immediately analysable

Having discussed names that can be analysed, we must now consider, briefly, placenames that do not appear to be analysable; in terms of Mparntwe Arrernte at any rate. Strehlow (1970:70ff) has shown how phonemic changes were intentionally employed 'in ritual sphere in order to make even ordinary words less easily intelligible to those who had not yet been admitted to full knowledge'. He goes on to comment that if these conscious phonemic changes were taken into account then 'many of the proper names of the supernatural personages and sacred sites found in the myths and traditions would become meaningful'. One example that he provides involves the name for Jay Creek — *Iwepetheke*.[13] This is traditionally explained as meaning 'the web covered it' and while *iwepe* means 'web', Strehlow claims that *(a)theke* is a deformation of the past tense of the verb 'to cover', which in Mparntwe Arrernte is *arte-ke* and in Western Arrernte is *ita-ke*.[14] Thus a lamino-dental stop has replaced an apical stop. So, phonemic confusion is likely to be one complicating factor in the analysis of placenames.

Another complicating factor appears to relate to the main origin of the totemic ancestor of a place. The larger Mparntwe area is caterpillar country and the places within that area which belong to one of the three caterpillar dreamings — *Ntyarlke*, *Yeperenye*, and *Utnerrengatye* — by and large have names that are analysable in terms of Mparntwe Arrernte. Other totemic Beings whose origin or main place is outside the Mparntwe area have places within that area attributed to them. The names of these places are often not rendered in Mparntwe Arrernte, but instead reflect the linguistic area of the totem's main affiliation. Thus, a dog Dreaming site in the Alice Springs town region is called **Arengke-rle**. *Arengke* is an Eastern Arrernte or Anmatyerre word meaning 'dog', not the Mparntwe Arrernte word (*kngwelye*); the track of the dog Dreaming enters the Mpartntwe area from the north from out of the region in which *arengke* is used as the common term. Another place, this time of the Devil Dog totem, is

---

[13] Strehlow's spelling is *Iwuptataka*.

[14] Strehlow's *etaka*.

named *Yarrentye Arltere* (devil white), or 'the White Devil'. *Arltere* is the Alyawarr word for 'white' and it is from this linguistic area that the Devil Dog entered into the region of Alice Springs.[15] These names are recognition of the fact that the totem of the named place has its main association with another area and, further, helps to keep track of the actual origin and direction of travel of the Dreaming line of that totem. One can but speculate as to the effect such conventions would have on the transmission of new vocabulary from one group to another.

One further feature of Mparntwe Arrernte placenaming which should be discussed is the use of the name designating a significant site for the larger area containing the site. This larger area may contain other less significant sites (and their corresponding placenames). So, as mentioned previously, *Mparntwe* specifically designates the most important site of the Yeperenye totem and this site is located within the township of Alice Springs. But, this name also designates the whole of the traditional area of the Mparntwe Arrernte speakers, which contains as one part of it the township of Alice Springs and hence the more specific site. Places within such large areas may themselves exhibit the same feature of narrow and wide reference on a smaller scale.

It is interesting to notice that the English names attributed to the town camps also reflect a similar feature of naming. Many camps get their English name from a prominent English-named place which they are close to. Thus the camp that is near an old-age home called 'Old Timers' is itself called 'Old Timers'; the camp that is behind the drive-in is 'Drive-in Camp' (*Nthepe*); and the camp near the place where the railway trucking yards used to be is called 'Trucking Yards Camp' (*Nyewente*). Here an English name for a prominent place can be seen to designate a nearby Aboriginal place as well — naming by contiguity.

From the above discussion we can see that while the origin of individual places is detailed in Dreamtime stories, the names of places themselves can be seen as a sort of mnemonic by which stories can be remembered and kept straight. In short, placenames themselves keep track of totems and events. A placename in Mparntwe Arrernte, as elsewhere in Arrernte country, is more than a simple referring expression; it is a compressed image linked to and recalling a larger Dreamtime history. In other words, placenames call up semantic information that one does not have access to without access to the cultural norms. An understanding of this aspect of the Mparntwe Arrernte language and culture is essential for understanding the important role of placenames in song, incantations and texts, and appreciating the subtleties that lie therein.

CONCLUSION

It has been demonstrated here that in Mparntwe Arrernte the conception of *place as entity* designated by the term **pmere** is a complex one. The evidence supports Goddard's (1986:9) criticism of the use of 'place' as a universal semantic primitive in which he states that: 'Despite its general appeal, there may be no general or ideal sense of "place" that can be translated simply into other languages.' **Pmere** is invested with important cultural significance and is inextricably linked with other social entities such as people and totems. Such denotational and connotational complexity clearly precludes its use as a semantic primitive despite its being the closest lexical equivalent to English 'place'. Moreover, while all **pmere** may be 'places', not all 'places' are **pmere** — *place as entity* should not be

---

[15] The Mparntwe Arrernte word for 'white' is *mperlkere*.

confused with location as system. It may well be that the general applicability of the word 'place' in English to designate any location is the exception rather than the rule among the world's languages. This is of course an empirical question which bears investigation.

We may conclude by reiterating that the cultural perceptions of *place as entity* are clearly reflected in language, from the level of morphology to the level of discourse. The existence of a generic term for socially-significant geographic places; the ability for two places to be treated as parts of the same whole even when separated in space; the applicability of kin terminology to geographical places; the significance of placenames as compressed images and mnemonic devices; and the importance of place in story and song all give evidence of the Mparntwe Arrernte conception of place and of how and why people talk, or fail to talk, about places the way they do. Finally, it should be clear that an understanding of the Dreamtime, and the philosophy embodied in it, is a *sine qua non* for understanding the concept of **pmere** 'place, camp, country, home'.

APPENDIX: DEFINING PMERE

This definition was published in Wilkins [2000], and represents an attempt to define the meaning of **pmere** in Natural Semantic Metalanguage. In the 1980s, when I wrote this paper, I was deeply involved in several discussions surrounding Anna Wierzbicka's work with and on semantic primitives (e.g. Wierzbicka 1980). Two of the main issues were: What are the correct semantic primitives? and How does one identify and describe cultural key words? More specifically, there was an issue as to whether her proposed primitive 'world' was actually suited to the semantic description of 'place' and 'time'. The semantic primitives research later became the Natural Semantic Metalanguage approach. I subscribe to some of the main tenets. Mainly, words in other languages can be rigorously defined using structured definitions formulated as a series of propositions using simple language. I see these definitions as rendering knowledge structures associated with the use and interpretation of a word in a given speech community. While most words give some insight into the socio-ideology of a group, cultural key terms like **pmere** provide especially rich insights into the cultural community.

pmere

Places of different kinds which are
> thought of as being important to people.

Thinking of such places,
> people could say these things about them

these are places where someone [people or Ancestral Beings]
> could live at for some time.

They can be of different sizes,
> some being big enough for only one person or Ancestral Being
> some being big enough for many people

or Ancestral Beings to live together

some being so big that they contain within them other places of this kind.

Such places have proper names of their own (like people).

Some places of this kind are classified as kin

and are said to be related to other places,

and to certain people,

in the same way people are related to one another.

All such places came into being because of

actions of Ancestral Beings during the Dreamtime.

Smaller places of this kind are associated

with one or a few individual Ancestral Beings,

and through these individual ancestors are associated

with other places and with certain people.

Such places are of varying degree of importance and power.

Some places are so important and powerful

that only certain people can go to them, and

if someone else goes to them,

they may become sick and die.

Some places can be visited by anyone at any time.

All places of this kind are meant to be associated with people who are responsible for looking after them,

and who must do the things that keep the place and

the Ancestral Beings that live there healthy.

Every person is born with a close association to a few different places.

REFERENCES

Bell, Diane, 1983, *Daughters of the Dreaming*. Melbourne: McPhee Gribble.

Berndt, Ronald M., 1976, Territoriality and the problem of demarcating sociocultural space. In N. Peterson, ed., *Tribes and Boundaries in Australia*, 133–161. (Social Anthropology Series No. 10). Canberra: Australian Institute of Aboriginal Studies.

Boas, Franz, 1934, On geographical names of the Kwakiutl. In Dell Hymes, ed., *Language in Culture and Society*, 1964:171–181. New York: Harper & Row.

Dixon, R.M.W., 1977, *A Grammar of Yidiny*. Cambridge: Cambridge University Press.

— 1980, *The Languages of Australia*. Cambridge: Cambridge University Press.

Elkin, Adolphus P., 1938, *The Australian Aborigines*. Sydney: Angus & Robertson.

Goddard, Cliff, 1983, A semantically-oriented grammar of Yankunytjatjara. PhD thesis, The Australian National University, Canberra.

— 1986, Wild ideas about natural semantic metalanguage. Paper presented at the Semantics Workshop, Australian Linguistics Society Annual Conference, Adelaide.

Green, Jenny, Rod Hagen, J. Spierings and Elspeth Young, 1984, Ti Tree land claim: claim book. Unpublished report. Alice Springs: Central Land Council.

Henderson, John and Veronica Dobson, 1994, *Eastern and Central Arrernte to English Dictionary*. Alice Springs: Institute for Aboriginal Development.

Jammer, M., 1954, *Concepts of Space*. Cambridge, Massachussetts: Harvard University Press.

Lyons, John, 1977, *Semantics*. Cambridge, UK: Cambridge University Press.

Myers, Fred, 1976, To have and to hold: a study of persistence and change in Pintupi social life. PhD thesis, Bryn Mawr College.

Spencer, Baldwin and F.J. Gillen, 1927, *The Arunta: a study of a stone age people*. London: MacMillan and Co.

Strehlow, T.G.H., 1970, Geography and the totemic landscape in Central Australia: a functional study. In Ronald M. Berndt, ed., *Australian Aboriginal Anthropology*, 92–140. Perth: University of Western Australia Press.

— 1971, *Songs of Central Australia*. Sydney: Angus & Robertson.

Wierzbicka, Anna, 1980, *Lingua mentalis*. Sydney: Academic Press.

Wilkins, David P., 1988, Switch-reference in Mparntwe Arrernte (Aranda): form, function, and problems of identity. In Peter Austin, ed., *Complex Sentence Constructions in Australian Languages*, 141–176. Amsterdam: John Benjamins.

— 1989, Mparntwe Arrernte (Aranda): studies in the structure and semantics of grammar. PhD thesis. The Australian National University, Canberra.

— 1993, Linguistic evidence in support of a holistic approach to traditional ecological knowledge. In N. Williams and G. Baines, eds, *Traditional Ecological Knowledge: wisdom for sustainable development*, 71–93. Canberra: CRES.

— 2000, Ants, ancestors and medicine: a semantic and pragmatic account of classifier constructions in Arrernte (Central Australia). In Gunter Senft, ed., *Systems of Nominal Classification*, 147–216. (Language, culture, and cognition 4). Cambridge, UK/New York: Cambridge University Press.

3 TRANSPARENCY VERSUS OPACITY IN ABORIGINAL PLACENAMES

Michael Walsh

1 THE BASIC QUESTION: WHAT PROPORTION OF PLACENAMES OF INDIGENOUS ORIGIN ARE TRANSPARENT — IN SOME SENSE?[1]

This question arose because of the discrepancy I had noticed between a relatively high proportion of transparency of about 66 per cent reported by Dixon (1991:125ff.) for the north Queensland language, Yidiny, compared to the relatively low proportion of transparency of about 20 per cent observed by me in my own work in the Darwin area. This led me to wonder whether there was some kind of norm in the proportion of transparency to opacity in the Aboriginal[2] placenames of a particular area. In turn this led me to send a query to Peter Sutton and, as I might have expected, he gave me a lengthy and thoughtful reply which indicated that the basic question I had put concealed a range of complications. That reply from Peter Sutton has found its way into his paper in this volume.

Otherwise I have received advice that the pattern across Australia is far from straightforward. Patrick McConvell (pers. comm.) reports that about 80 per cent of placenames in the Victoria River Downs area of the Northern Territory are transparent. Margaret Sharpe (pers. comm.) indicates that most placenames in the Bundjalung area (north-east New South Wales) are transparent but in the Alawa area (Northern Territory) few placenames are transparent. Franca Tamisari (pers. comm.) reports for north-east Arnhem Land that placenames vary in transparency according to a person's age and knowledge.

Comments of this kind suggest that 'transparency' is not a simple notion when applied to Aboriginal placenames. We therefore need to consider a number of types of transparency.

---

[1] This paper was delivered in Canberra at the Placenames of Indigenous Origin: An Interdisciplinary Workshop in association with Australex, the Australian National Placenames Survey and the Australian Language Research Centre, 31 October 1999. It has taken into account comments provided by participants at that workshop.

[2] In this paper I prefer 'Aboriginal' to 'Indigenous' because the latter term can also refer to Torres Strait Islanders. My own experience and knowledge overwhelmingly relate to Aboriginal people rather than Torres Strait Islanders. It may be that the observation made here about Aboriginal placenames can also be applied to Torres Strait Islander placenames but I make no claims in that regard.

L. Hercus, F. Hodges and J. Simpson, eds, *The Land is a Map: placenames of Indigenous origin in Australia*, 43–49. Canberra: Pandanus Books in association with Pacific Linguistics, 2002.

2   TOWARDS A TYPOLOGY OF TRANSPARENCY

2.1   Clarity — focus on the semantic content of the form of the placename

The first category deals with the semantic content of the *form* of the placename. A placename like Camelot or Troy may carry with it a good deal of evocative baggage but this is not evident from the actual forms 'Camelot' and 'Troy'. In the sense intended here a placename like Newtown is rather obviously relatable to the two forms 'new' and 'town'. At first blush a placename like Ashfield might similarly be related to two forms, 'ash' and 'field'. There is a difference between the two placenames in terms of my own knowledge. While the purported bipartite structure of Newtown seems reasonable enough, it also happens to be supported by historical accounts. The other placename, Ashfield, might — for all I know — be an anthroponym, named in order to commemorate someone called Ashfield. While the personal name Ashfield might ultimately be shown to derive from two parts, 'ash' and 'field', that would not be the primary basis by which the place was named. It is a relatively simple matter to check the etymologies for placenames like these from the City of Sydney but it is more likely to be quite a difficult matter for placenames in Aboriginal Australia. Simply asking someone — whether in Sydney or the outback — is no guarantee of an accurate answer. In my experience people who have lived in a place like Newtown for years might not be aware of the structure of the name until it is pointed out to them. For a name like Whitfield, people may guess and happen to be right (in terms of known history); they may recognise part of the name, -field; or they may be unable to say anything about it at all. Here I should point out that there may be differences for Aboriginal and other Australians in expectations about knowledge. An Anglo-Australian may be able to analyse 'new town' but might not expect to know anything more about why it was called Newtown. However, with many (potentially) transparent Aboriginal placenames, speakers expect that someone will be able to analyse the semantic content of the name, perhaps by relating the semantic content to the topography of the place.

Persisting with placenames from Sydney, Maraylya is for me quite opaque in terms of the form of the placename; while for Matraville I can recognise -ville as a recurring partial found elsewhere in Erskineville or Wentworthville. However, it has been pointed out to me that even if Maraylya is opaque, it is connotatively autochthonous. 'Maraylya does carry the baggage of sounding autochthonous, unlike say Cricklethorpe which though unanalysable by most of us will always sound English. This is connotative clarity, or perhaps ethnic provenance clarity — the cultural group of origin is suggested, if not confirmed' (Sutton 2000). In sum we have a number of points along a scale of clarity that would include:

very clear	Newtown = new town
apparently clear	Ashfield = ash + field/?Ashfield [anthroponym]
partly clear	Matraville = ?Matra + ville
unclear	Maraylya

Clearly degrees of clarity will depend on such factors as the level of knowledge about the semantic content of the form of the placename.

2.2 Recognition — the extent to which the semantic content of a placename is recognisable

The second category is concerned with the extent to which the semantic content of a placename is recognisable. Notice that I have not restricted myself to the semantic content of the form in this definition. So the recognition of semantic content is potentially broader, relating not just to the form but also to the associations (for places like Camelot and Troy) that are more or less familiar to people. In the arena of Aboriginal placenames one potential line of cleavage in recognisability is that between Indigenous people and analysts. Needless to say the sets do not have to be mutually exclusive. Another contrast is non-analysts vs analysts and this has already been discussed above. A non-analyst might claim that Nightcliff, the suburb of Darwin, is somehow relatable to 'night' and 'cliff'. However, it is known by analysts to be Knight's Cliff originally associated with a certain Mr Knight.

In the Aboriginal sphere native-speaker knowledge of course will vary considerably and folk etymologies may creep in. It may happen that a linguist or other analyst's deductions are in sharp contrast with native-speaker knowledge. And of course a linguist or other analyst's deductions will sometimes be wrong.

The two categories so far described are familiar enough in non-Aboriginal settings in Australia, but the next two categories have features that are particularly relevant to Aboriginal settings.

2.3 Candour — the extent to which (past) ownership of placenames is acknowledged

In Aboriginal Australia placenames are owned. A particular name will sit in a particular territory which is owned by a particular Aboriginal group. Languages are also owned by a particular group (see Walsh, 2002) and this adds a complication to Aboriginal placenames. A given place would usually be owned by only one Aboriginal group, but some places have names in two or more languages; so that while the place may be owned by just one group, the names may be owned by more than one group. The category of candour covers the extent to which (past) ownership of placenames is acknowledged.

Let us consider a particular place with three placenames. From the perspective of one group, X, the place has three names from three groups (X, Y & Z): X group owns the place and one of its names, while the other two groups each have a name in their own language for that place — for which they do not claim ownership. Meanwhile group Y claims ownership of the place, says it has two names for the place in its own language and acknowledges a third name from the neighbouring group Z. In this scenario group X has no place at all. It would be an example of group Y showing a relative lack of candour concerning ownership of that place by group X. This scenario might occur in a situation where group X in the past was generally acknowledged as the owner of the place in question, but now there are differing viewpoints which will depend on the perspective of the group. The newcomers, group Y, may want to emphasise their ownership of the place in question rather than acknowledge ownership of some kind by another group. Clearly this process is not completed overnight; so there may be a period in which there is considerable debate about the 'true' owners, and problems in the translatability of names can result in the loss of history of ownership (Hercus & Simpson, this volume). Degrees of candour will affect the relative transparency of a placename for different groups (see also Harvey 1999, esp. 192–194).

2.4 Openness — the extent to which constituents of a placename package are open to the wider public

Finally it is necessary to consider the extent to which information about places in Aboriginal Australia is open to the wider public. I know of a placename in the Darwin area which has become closed in recent years to the extent that most of the 'placename package' (described more fully in §3) is no longer open — including the form. That is to say, for at least some Aboriginal people it is inappropriate to use the placename either in spoken discourse or in its written form. There is also information associated with this placename that should not be referred to in a public setting. Clearly this raises problems for the study of this placename from an outsider's perspective. One point of interest for the study of placenames in Aboriginal Australia is how people refer to a place for which the public use of the 'real' name is proscribed. It would also be interesting to know how widely this phenomenon applies. Among the reasons a placename is proscribed is because it is the personal name of the speaker, or of someone who has passed away, or because it is the place where the speaker was initiated. For example, the English pronunciation of Banka Banka was proscribed after the death of a man called Banka Johnny. The station was then called 'Station' or 'Jinarrinji' after a site near the main Aboriginal camp on that station (Simpson 2000).

3   PLACENAME PACKAGE

In the previous section I adopted the rather inelegant expression 'placename package': here it should be unpacked. It seems to me that relevant information about a placename should include at least the following:

> phonological form/graphic form
> location/feature
> 'story'
> semantic content
> ownership

In the previous section I indicated that parts of a 'placename package' could be more or less open (to a wider public). One indication was the proscription of the phonological form/graphic form of a placename. There can be instances where these two forms interact with each other in such a way as to reduce the relative transparency of a placename. One example can be found in a placename somewhat to the west of Darwin which has appeared in graphic form both as Inyarrany and Iynarrayn. In each instance a digraph (ny, yn) is used to represent a palatal nasal. For some Aboriginal people the phonological form has remained constant, while others have adopted a spelling pronunciation so that the third syllable has become two syllables which rhyme with 'Annie'. For those Aboriginal people who know the name only by ear, the spelling pronunciation can be confusing or even scarcely recognisable. The second spelling, Iynarrayn, is more likely to force a three-syllable pronunciation, but the pronunciation still may not correspond with the phonological form that was usual in the past.[3]

---

[3] This example arose in a land claim setting where much discussion hinges on placenames. For some further examples of phonetic confusion, see Walsh (1999:168–172).

About location I will be brief. Some placenames can refer to some feature that will establish its location for those in the know: Schebeck (this volume) following Hercus refers to these as 'intermediate names'. Some placenames have narrow scope and wider scope so that the precise location will depend on access to the scope intended in a particular usage. In either case there are degrees of transparency about the precise location to be associated with a particular placename. Associated with a placename's location is the extent to which a topographical feature is referred to in the placename, either implicitly or explicitly. Sutton (2000) observes: 'In the Wik area many placenames explicitly contain a reference to a feature e.g. point-, creek-, well-.' It would be useful to investigate the frequency of appearance of topographic descriptors and their range. We might, for instance, anticipate explicit references to 'point' on the coast but the distribution of such explicit references is yet to be studied in any detail. Also worthy of research attention is the width of the reference: does a particular placename refer to a mountain or to a mountain and the surrounding plain?

The next two components of the placename package, 'story'[4] and semantic content, can overlap but it is nevertheless useful to keep them separate. The 'story' can be an account of a figure from the 'Dreamtime' who carried out certain activities at or near the place in question. There may be varying accounts and there may be a desire on the part of Aboriginal people to restrict a 'story'. Harold Koch (pers. comm.) has referred to the central Australian placename Iwupataka (aka Jay Creek) (Walsh, this volume), for which Aboriginal people deliberately promoted a false meaning in order to protect the restricted story that is referred to by this oral name (Strehlow 1971:71) (and see Wilkins, this volume). Semantic content of a placename potentially includes all the associations that may apply to that place. These may include events like a fight or a burial that have taken place in historical times. Either of these would constitute a story about the place but not *the* 'story'.

The notion of ownership has already been briefly canvassed in §2 under the heading of 'candour'.

4 MULTIPLE NAMES FOR A PLACE

In Aboriginal Australia it is relatively common for a given place to have multiple names. Multiple naming can also be seen in places outside Aboriginal Australia. A city in Turkey is currently known as Istanbul but at one time was known as Constantinople and at another time as Byzantium and each name has its own associations. Here I simply want to provide an overview of some of the multiple naming practices and draw attention to the implications for transparency.

The simplest case is one place having two names. Such doublets can be intralectal or crosslectal. For intralectal doublets where there are two names for the one place in the same lect,[5] both placenames may be opaque, both transparent, or one opaque and one transparent. In fact the situation can be more complex than this, as exemplified by Sutton (this volume). The same applies to crosslectal doublets where the two names for the one place come from different lects. An intriguing case is provided by Laughren (2000) of a single name within a lect referring to more than one place.

---

[4]   I use the quote marks around the word 'story' because there are connotations about this word which are not appropriate to accounts of 'Dreamtime' events. Such accounts are not just-so stories for Aboriginal people and certainly they are not fictional.

[5]   I use the term 'lect' so as to avoid buying into issues concerning dialect, variety and language.

The Nyirrpi story as I know it goes something like this (McConvell, this volume):

The original Nyirrpi name referred to a soakage on Waite Creek where a group of people decided to set up an outstation from Yuendumu in the mid-1970s. For some time people camped out there near that soakage (Nyirrpi is Desert Oak in Luritja and Pintupi, not in Warlpiri by the way). There was a water shortage problem and so water used to be trucked out there from Yuendumu in a water cistern.

The people who go around putting down bores to look for potable water in good supply that merited putting in a bore, eventually found water further south on Waite Creek close to a soakage called Jirtirlparnta. Eventually a bore was put down there and people moved near the bore but kept calling the place Nyirrpi as by that time they were known as the Nyirrpi mob, especially by the White authorities who just called the area Nyirrpi. Between when the people were at the 'real' Nyirrpi and their movement to Jirtirlparnta (alias new Nyirrpi > [present] Nyirrpi or Nyirripi officially) they were camped at another soakage called Walyka. This was also to the north of Jirtirlparnta.

Now people refer to the original Nyirrpi as 'Old Nyirrpi' and to original Jirtirlparnta as 'Nyirrpi' (See McConvell, this volume).

Some places will have more than two names. One example from my own research concerns a place near Darwin called King's Table. Here we have a toponymic triplet with a Larrakia name, a Wagaitj name, and a Kungarakayn name. The question then arises as to which name is used by which people. In my experience (biased by strong association with Larrakia and Wagaitj people) the most commonly used name is the Wagaitj one although many/?most people would acknowledge the place as being owned by the Larrakia. I suspect the Kungarakayn would use the Wagaitj name or the Kungarakayn name but I could only guess about the likelihood of one over the other in a given situation.

It is unclear (to me at least) how multiple naming works and what its function is. We have some case studies (like Schebeck, Sutton and Tamisari, this volume) but we are yet to gain a comprehensive picture for Aboriginal Australia. There are numerous potential questions that might arise. Is there an upper bound on the number of names a particular place might have? As lects decline in usage, are placenames more or less vulnerable? When a lect becomes a lingua franca and expands beyond its original territory, to what extent will places in the new territory gain names from the lingua franca?[6]

5 TEMPORAL DIMENSION

Finally,[7] and briefly, one must consider the robustness of placenames over time. Often enough the time depth for the first known recording of an Aboriginal placename will be quite shallow. For the area near Darwin, Basedow (1906) carried out relatively early recording of placenames in 1905. Checking Basedow's recordings with Indigenous people in 1979 I was able to match a significant number of placenames but some were unrecognisable. Basedow had recorded 47 names of which about 34 are clearly recognisable. Other placenames I recorded in 1979 in this area are no longer recognisable (see Brandl, Haritos & Walsh 1979:60–67). This is just one instance of placenames becoming less transparent because of the temporal dimension.

---

[6]   I am grateful to Pat McConvell for raising the issue of lingua francas.

[7]   Jane Simpson and Peter Sutton have read and provided valuable comments on a draft of this paper. I am most grateful to them and also to David Nash who responded to a number of queries.

REFERENCES

Basedow, H., 1906, Anthropological notes on the western coastal tribes of the N.T. and S.A. *Royal Society of South Australia Transactions*, vol. 31.

Brandl, Maria, Adrienne Haritos and Michael Walsh, 1979, *Kenbi Land claim to Vacant Crown Land in the Cox Peninsula Bynoe Harbour and Port Patterson Areas of the Northern Territory of Australia by the Northern Land Council on behalf of the Traditional Owners*. Darwin: Northern Land Council.

Dixon, R.M.W., ed., 1991, *Words of Our Country: stories, place names and vocabulary in Yidiny, the Aboriginal language of Cairns–Yarrabah region*. St Lucia, Qld: University of Queensland Press.

Harvey, M., 1999, Place names and land–language associations in the western Top End. *Australian Journal of Linguistics* 19(2):161–195.

Hercus, Luise and Jane Simpson, this volume, Indigenous placenames: an introduction.

Laughren, Mary, 2000, Email to Michael Walsh, 17 May 2000.

Schebeck, Bernard, this volume, Some remarks on placenames in the Flinders Ranges.

Simpson, Jane, 2000, Email to Michael Walsh, 31 May 2000.

Strehlow, T.G.H., 1971, *Songs of Central Australia*. Sydney: Angus & Robertson.

Sutton, Peter, 2000, Email to Michael Walsh, 29 May 2000.

—— this volume, Placenames of the Wik region, Cape York Peninsula.

Tamisari, Franca, this volume, Names and naming: speaking forms into place.

Walsh, Michael, 1999, Interpreting for the transcript: problems in recording Aboriginal land claim proceedings in northern Australia. *Forensic Linguistics* 6(1):161–195.

Walsh, Michael, 2002, Language ownership: a key issue for Native Title. In John Henderson and David Nash, eds, *Linguistic Issues in Native Title Claims* Canberra: Native Title Research Series, Aboriginal Studies Press, 230–244.

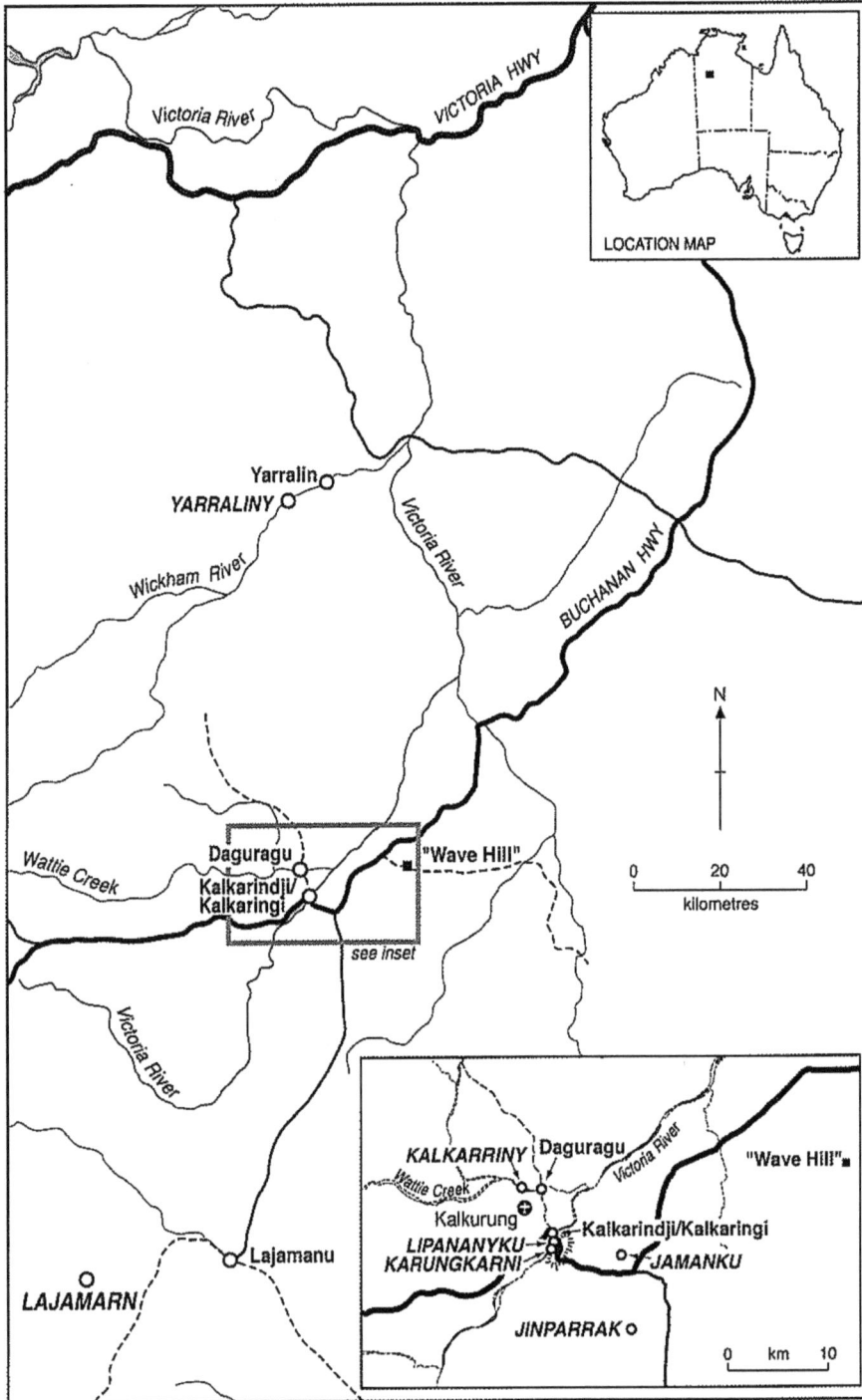

Map 1: Approximate locations of Indigenous and official placenames
in the Victoria River District, Northern Territory

4 CHANGING PLACES: EUROPEAN AND ABORIGINAL STYLES

Patrick McConvell

1 ENGLISH PLACENAMES: TAKING THE NAME OF YOUR HOME WITH YOU WHEN YOU MOVE

Gurindji people had special cause to note European-Australians' fondness for taking placenames with them when they moved, because this is what happened to the name of the most important cattle station in the region, where most Gurindji people came to live — Wave Hill. Buchanan (1935:71) records why Sam 'Greenhide' Croker gave this name to the camp the pioneer pastoralists made on the Victoria River in 1883:

> Greenhide Sam, struck by the sharp undulations of the plateau, suggested the name of Wave Hill, by which it has been known ever since ...

The land formation is clearly visible from the bridge over the Victoria River to this day; the wave-like shapes along the hills beside the river are very striking and possibly unique in the region. This area is called Lipananyku and Karungkarni by the Gurindji after the two large waterholes in the Victoria River.

The first Wave Hill station was established a kilometre or so from there. A police station, and later a 'welfare' settlement, were established close to the station homestead also bearing the name Wave Hill until the 1970s. Gurindji found this transfer of the name Wave Hill relatively understandable, as the places so named were all within the same general area.

Much more puzzling to them though were the subsequent transfers of the name Wave Hill to new station homesteads quite far away from the original location. First, following the 1924 flood which washed away the original station at Malalyi-malalyi, Wave Hill was moved to higher ground to a new site further from the river at *Jinparrak*, which was occupied until after the 1966 strike at Wave Hill, when the station — with its name 'Wave Hill' — was moved even further away from the river to the 'new station' at *Jamanku,* some 30 kilometres from the original 'Wave Hill' and nowhere near any undulating hills. This was originally the Number One stockcamp site of Wave Hill, and was generally known as 'Number One' by Aboriginal people in the 1970s to 1980s.

For the Gurindji this behaviour with placenames is emblematic of a deep difference between Europeans and Aborigines. A sentiment often voiced by Aboriginal people is that Whitefellas are always changing their law, whereas Blackfella law stays the same. For

L. Hercus, F. Hodges and J. Simpson, eds, *The Land is a Map: placenames of Indigenous origin in Australia*, 50–61. Canberra: Pandanus Books in association with Pacific Linguistics, 2002.

Gurindji the difference in treatment of placenames is an aspect of this difference. Since placenames are part of the Law (*yumi*) assigned by Dreamings (*mangaya, puwarraj*) to a specific place, mere human beings cannot lift those names up and drop them in other places. Of course, naming is not part of the important 'law' of Whitefellas in the same way; it is considered to be of symbolic significance, which often means of lesser importance.

In the light of this behaviour, it seems paradoxical to Gurindji that the Aboriginal people have been labelled 'nomadic' and their relationships to place seen as variable and transient. Rather they see European-Australians as the true nomads who shift from place to place taking names with them.

While many Gurindji may not be aware of this, the custom of transporting names to new locations and particularly 'old country' names to the colonies is an established part of European culture.

2 TRANSPORTING ABORIGINAL NAMES IN THE EUROPEAN DOMAIN

2.1 New centres of influence: the airstrip

As Europeans appropriate Aboriginal placenames as their own, or acquire power and influence over how such names are used in areas where Aboriginal people still use their own system, the tendency for such names to be shifted from place to place also increases.

The changing names in the Wave Hill area provide further examples of this. During the 1970s there was a move to 'Aboriginalise' the names of former missions and welfare settlements. In the case of Wave Hill Settlement this came about at the same time as the area was actually de-Aboriginalised. In 1974 an area was cut out from the Daguragu pastoral lease to form an open township which came to be dominated by White commercial interests. This move, of which the local Gurindji people were not aware at the time, also prevented this area from being claimed under the Land Rights Act (1976) when the rest of Daguragu was successfully claimed in 1981.

For a few years the 'township' was known as Libanangu which was as close as people who wrote the name could get to the traditional name of the Wave Hill area — Lipananyku. However, following the upgrading of the airstrip, which was between Libanangu and Daguragu, it was decided (by whom is not clear) that the airstrip should have a new name; and soon after the township itself was assigned that name (without consultation with Aboriginal people as far as I know). The airstrip was named Kalkurung but variations, with a *-ji* ending unknown among the Gurindji, also circulated at the time. The town is now known as either Kalkaringi or (since 1986, officially) Kalkarindji. This was named after the waterhole on Wattie Creek downstream from Daguragu which was closest to the airstrip — Kalkarriny. As usual White people had problems with the palatal nasal 'ny' at the end of the word. Somebody heard it as ng and transcribed it like that (the spelling Kalkaring can also be found ephemerally at that time); someone else I think must have had an inkling of the actual sound and known that this was transcribed *-nj* by some earlier linguists: hence Kalkarinj. I think people then pronounced it 'Kalkarindge'. What happened next was that someone added a vowel *-i* to this word to make it easier on the tongue. I assume this was an English-speaking European, although it may have been done with the cooperation of Warlpiri residents of the township, whose language does not favour consonant-final words (although adding *-u*, *-pa* or *-ku* is their usual way of dealing with this, not adding *-ji*).

No Gurindji person would have dreamed of applying the name Kalkarriny to Lipananyku, the township, since they are quite different places on different rivers. The central meaning of many Gurindji names is a waterhole, and while that name can be and often is applied to a small creek running into a main river at the waterhole, applying the same name to waterholes on different creeks because they are relatively close to each other seems impossible. However, it is worth drawing a distinction between the process involved here and the earlier one of transferring the name Wave Hill to different places as people moved. The extension of Kalkarriny, the airstrip (or airport, as it was then more grandly called by some), to the township is one of extending a major name of a place to satellite places in its vicinity. This is not completely alien to the Aboriginal practice, as will be discussed below, although basic rules were broken, as noted above. What is alien is the principles that decide how such extensions are made: in this case from an airstrip to a town — indeed a town that has been established in one form or another for more than one hundred years under the name Wave Hill, and presumably for at least hundreds under the name Lipananyku. Actually extension of a name of an airstrip to a town (already named) is fairly strange even in European-Australian terms. Imagine renaming Sydney Mascot.

2.2 The influence of historical Aboriginal immigrants in shifting names

The establishment of Hooker Creek settlement, later called Lajamanu, provides another example of transfer of names from one place to another, at least according to Gurindji and Wanyjirra people. The settlement was established primarily because Vestey's cattle stations wanted a labour pool (Berndt & Berndt 1987; McConvell 1989). The process of setting up the settlement seems to have been a comedy of errors, from its initial siting in a place different from that surveyed, to the several attempts to bring reluctant Warlpiri there from the southern Tanami to a place on Gurindji/Kartangarurru land, followed by the departure of many of them back to their own country. When the fashion of naming settlements with versions of Aboriginal names came into vogue in the late 1970s, the name Lajamanu was provided by the Warlpiri residents. This is the Warlpiri version of the name Lajamarn, which actually refers to a place 30–40 kilometres west of where the settlement was. The traditional owners and custodians of sites associated with the *Wampana* (hare wallaby) Dreaming that passes through Lajamarn were disgruntled that the name had been appropriated and misapplied. They regarded this as yet another slight on them by the Warlpiri and Europeans, in collusion, in a long line beginning with their failure to consult with them over the siting of the settlement. This then is yet another type of transfer of place by either a mistake or perhaps manipulation by Aborigines from elsewhere with the help of European authorities, who were either unaware of the other story about the name, or ignored it.

2.3 Substituting the name of a planned destination for the eventual site in a move

Finally in this set of examples from the Victoria River District, there is the case of Yarralin. When the Aboriginal people on Victoria River Downs (VRD) station went on strike in 1972, following the example of the Wave Hill strikers, they went to live at Daguragu. During their exile at Daguragu they made plans to return to an area they could use for living not too far from the station, again on the pattern of the establishment of Daguragu by the Wave Hill strikers (Rose 1991:229–30; 1992:22). The place they chose was the waterhole Yarraliny on the Wickham River. However, the strikers were being assisted by Jack Doolan, then a Department of Aboriginal Affairs officer, who came up with a different plan (perhaps after talking to the VRD management) of occupying the old Gordon Creek homestead area, known as '6-mile', called *Wangkurlarni*. Doolan's account of this (1977:111) suppresses the disagreement over the site, which was quite acrimonious at the time, and claims that the Aboriginal people call Gordon Creek, Yarralin. This is what happened when they did move back to '6-mile' in 1973: it was officially called Yarralin, but it took many people who knew better many years to use that name, and some never did. This name transfer was a result of a political 'fudging' process in which the name of an original intended goal of a move is substituted for the actual goal — like calling the Caribbean the West Indies. Here European intervention and motives seem to be paramount, but most of the Aboriginal people involved played along, probably judging in this case that getting some land back was more important than arguing over names. An additional factor could be that the real 'Yarraliny' was only 4–5 kilometres away from the 'new' Yarralin and on the same river, thus one could be considered to be within the sphere of influence of the other.

2.4 Summary

The survey above could be continued, but it is sufficient to show, I think, that Aboriginal placenames assigned by official 'Whitefella' processes are much more likely to be the wrong names according to Aboriginal law than the right ones. A variety of motivations including some Aboriginal interventions and plain mistake lie behind these results, but a key issue is the Whitefella's conviction that humans can arbitrarily alter names, a belief vigorously denied by most Aboriginal people.

3 EXTENSION AND CHANGE OF PLACENAMES BY ABORIGINAL PEOPLE

This is not to say that Aboriginal people themselves do not change or shift placenames. Changing of placenames becomes necessary because of death taboo, for instance, where a name is similar to that of a recently dead person. The focus in this paper, however, is on taking the name of one place and applying it to another place.

My hypothesis here is that this arises overwhelmingly because of extension of one name, which is at the centre of a sphere of influence, to other places that are considered in some sense part of, or attached to, that sphere. This kind of extension of a 'big name' to a wider associated zone has been noted frequently in Aboriginal Australia, for example among the Yolngu of north-east Arnhem Land (Morphy 1984:26; Keen 1995:509). For various reasons the extensions might come to be considered more central than the original centre over time.

The 'sphere of influence' and 'centrality' within that sphere are social concepts rather than purely geographical, whether the change or extension is explicitly mediated through a social group associated with a place or not. The nature of such spheres is changing as the configurations of Aboriginal life change, but it is not clear that Aboriginal people are adopting European patterns of transporting names to any significant extent. Below we shall look at some patterns of extension of names to determine how far they fit with the hypothesis above.

3.1 'Big places' and waterholes as centres of spheres of influence

A common occurrence in Gurindji country is for the following group of places to have the same name:

(a) a waterhole on a main creek or river;

(b) a creek flowing into that waterhole or nearby;

(c) a large hill near the waterhole;

(d) the general area around the waterhole.

Such places may be called *ngurra jangkarni* which could be glossed 'important place', but also 'big camp' and the 'sphere of influence' around the place tended to be associated in people's minds with the immediate foraging range around a large camp. Thus groupings of people are associated with name extent.

The whole question of the size of the 'sphere of influence' around a place came to be hotly debated when sacred sites legislation and more particularly the NT Land Rights Act placed great legal weight on the existence of a 'site' on the land. The case of *Julama* was a test case which followed on the Daguragu land claim as a result of a repeat claim by the Gurindji (McConvell & Hagen 1981; Toohey 1982).

Justice Toohey (1982:22) had excluded a small area in the south-west of the claimed area on the grounds that it was part of a clan estate which was largely outside the claim and that there was not a named site within the area. It was successfully argued that the sphere of influence of named sites extended across the area in question, which was mainly part of a river bed. The Gurindji were bemused, some amused and some annoyed, about the apparent obtuseness of *kartiya* (White people) in not recognising something blindingly obvious to them, that this part of the river links to and is part of the same owned area as other parts where names are more centrally located. The word *ngurra* 'place' can be translated 'site', and can be used to mean a 'camp' in a specific and confined space, but also means a wider 'country'.

3.2 'Runs' and the influence of social groups on placename extension

There is variation in different parts of Australia as to whether local group countries or 'clan estates' have names. In the Victoria River District they generally do not, although the tendency for them to be named increases as one moves further north. Occasionally, however, a particular named feature so dominates a local group area that the two names are used virtually synonymously. This is most evident in the case of distinct named ranges of hills such as *Martpirlin* in southern Daguragu station, or *Yunurr* north-west of Mistake Creek.

Among the Gurindji and Malngin country the association between a person or group is marked by linguistic forms. The suffix *-ngarna* has the meaning 'denizen, inhabitant of' but in conjunction with such country names it designates a more specific type of relationship in Aboriginal law (*yumi*).

In all cases I know of such a name being generally recognised, the person is both a traditional owner by descent and a person who lived most of their life in the area and/or consistently camped and foraged there.

Martpirlin-ngarna and *Yunurr-ngarna*, for instance, were interpreted as meaning particular individuals, traditional owners who had also spent a lot of time in the area concerned. In the latter case, when a man formerly known as *Yunurrngarna* died, an older woman, also a traditional owner who knew the area well, came to be called by this name.

The suffix *-mawu* is similarly used with areal placenames to signify a group of people associated with the area, for example *Marpirlin-mawu, Yunurr-mawu*. Once again traditional ownership (through descent) is often implied but long-term residence is probably the most important factor.

Where there is no dominant named feature, in the sense of a large range of hills occupying most of the area or a creek running right through it, a name of a single important site may come to stand for the whole area in a metonymic way, in expressions like the above. However, this usage does not generally extend to using such placenames alone to signify the whole area.

The use of terms derived from areal extensions to describe territories connected to groups of people opens up the possibility that the social group and its range are as significant as environmental features in determining the way a placename can be extended.

Similar patterns are found in wide areas of Australia. In a large area of eastern Queensland local groups are known by a term which has a suffix *-barra*. Among the Girramay and Jirrbal, the group and the area are known by such a term. (In fact, the suffix is a regular productive one in this language meaning 'connected with' — although whether this indicates that this is the centre of diffusion of this suffix in its toponymic/ethnonymic sense is unsure.) The root to which this suffix is attached can be either of two types: (a) a dominant environmental type usually either a soil type or a dominant plant species; or (b) an important placename, for example:

(a)
Marrany-barra 'Yellow stringy-bark — connected'

Jagurru-barra 'Fan palm sp. — connected'

(b)
Yinyja-yinyja-barra 'Mount Smoko — connected'

Girjal-barra 'Mount Bronco (?) — connected' (The English placename for the mountain where the Storm story originates is given in Dixon and Koch (1996:311) but I could not elicit or confirm this name among Girramay people or from maps.)

These terms are called 'family runs' in local Aboriginal English and are associated both with traditional ownership and residence in, and use of, the area. The names of type (b) may have a more restricted use applying to smaller groups and countries and a wider usage encompassing a number of smaller local groups and countries which may have their own *-barra* names. The important factor in whether this two-tier extension of the placename is possible seems to be the relationship between the groups involved: if they are considered to be part of a larger whole then the placename of one can achieve broader application.

3.3 Innovations in names and apparent shift

The above cases of application of extension of placenames to wider areas on the basis of the range of connected social groups do not constitute cases of actual shift of names from one place or another. However, it is possible that an appearance of such a shift might be brought about if, after an extension of a name, the original referent of the name were to be assigned a different name.

Innovation of placenames is not as common a feature of Australian Aboriginal culture as it is of other cultures around the world (cf. Basso 1997), although processes of discovering new signs of Dreaming action, which may lead to new placenames, have received some more attention recently (Povinelli 1993:149–150; Merlan 1998:216ff.). Most Australian Aboriginal people tend to deny that placenames are assigned by humans, but attribute them to the activities of the Dreamings. However, some names clearly have been changed in recent times: a few places take their name from where a person of that name was initiated in living memory; or from the special bereavement name of someone who died at a place.

Moreover some names are recent types of coinages which can be distinguished in their form from older placenames which are either unanalysable in terms of the present language or use archaic morphology. Among the Ngarinyman and Gurindji, the newer names by contrast tend to be of the types:

Plant species + jarung 'having' e.g. Jalwarr-jarung 'Having fig trees' or

Dreaming name + LOCATIVE e.g. Jantura-rla 'Turkey –LOC' 'Turkey dreaming'

Contrast Dreaming + LOC + rni in older placenames

While it would be stated by local Aboriginal people that the Dreaming put the fig trees there, or the turkey had travelled that way, the placename itself may not be in the song for the place and there may not be such insistence that the name came from the ancient time. People regard such names as more descriptive of the dreaming action in the place than direct from the Dreaming's voice, which is the case with many other placenames.

3.4 Repeated names in different countries

Another pattern which looks something like shifting or copying of names to new locations is the phenomenon of the same placenames being found in different countries. In some cases placenames associated with a particular Dreaming Story are repeated, in two or more local group countries, either along the same Dreaming Track, or where the same story is enacted at different places. This is not a question of humans taking the names to different places, but of the Dreamings repeating the same words and actions at different places in their travels, according to local people. Some of these examples may involve succession processes, whereby the stretch of the songline belonging to one group is extended using the reference point of the same placename.

3.5 Modern placename shift and new spheres of influence

As the conditions of Aboriginal life change it is inevitable that there will be change in what motivates the assignment and transfer of names. People do not tend (at least at present) to spend as much time hunting and gathering in inaccessible ranges of hills, but prefer to go where motor vehicles can take them. It is uncertain then if, for instance, anyone will merit the name *Yunurrngarna* or other similar epithets in future. There will be traditional owners, but not those who additionally spend a lot of time in the particular area they own, if it happens to be inaccessible by vehicle. It is possible, however, that the intensive foraging formerly needed to validate such naming will be dropped as a requirement. If, for instance, a community is set up on the outskirts of these hills and referred to as *Yunurr*, the leading person in this community could be called *Yunurrngarna,* and the central reference of the name transferred to the settlement area, not the range.

This residential aspect of Aboriginal life has changed markedly over the last 30 years thanks to the 'outstation' or 'homelands' movement. The Gurindji and Malngin people, who spent time in their traditional country and were given names like *Yunurrngarna* as a result, lived most of the year on cattle stations in Aboriginal camps near the homesteads or stockcamps, and were able to live and forage on their country only by luck during stockwork or during the 'walkabout' stand-down in the Wet season (McConvell 1989). Other groups less involved in the pastoral industry, like the Warlpiri, were initially herded into large government settlements or missions, but in the 1970s these began to break up and people set up 'outstations' in more isolated areas.

Mary Laughren (pers. comm.) reports the following regarding the Warlpiri placename *Nyirrpi:*

> When people moved out to Nyirrpi — a soakage on Waite Creek — in the 70s to form an outstation from Yuendumu they found that there wasn't a sufficient amount of water there to warrant the building of a bore. For a while, when Nyirrpi soakage dried up they were camped around another soakage called Walyka (= cold to touch). Later when sufficient water was found and a bore installed on the banks of Waite Creek near yet another soakage called Jirtirl-parnta, people moved there for good and starting calling it Nyirrpi — as this name had become that used by Whites and indeed others to refer to this group — 'the Nyirrpi mob'. It is the gazetted name for the place and most young people probably do not know they are actually living at Jirtirlparnta, not the original Nyirrpi. For a while people used the term 'New Nyirrpi' but then dropped that modifier and it became Nyirrpi.

Nowadays Nyirrpi has many houses and other facilities and is 'like Alice Springs' according to residents, far removed from when I first saw it in 1975 — a couple of tin sheds with an uncertain water supply. See Walsh (this volume).

This is reminiscent of the case of Yarralin above, where the original proposed destination name was carried over to the actual site where people settled. However, while White involvement is evident in this case, the major factor appears to be the identification of the social group as the 'Nyirrpi mob' by other Aboriginal people as well as by Whites.

In this sense the transfer is perhaps parallel to the extension of a term for a central site to an area occupied by the group, as discussed above as an old traditional practice. The nature of the social group is naturally somewhat different as it is formed round a modern 'outstation' movement; but the processes and principles of the transfer of names, based on a group interest centring around a main place, seems to be at least as much due to Aboriginal tradition in this case as to European influence.

3.6 Naming after placenames

Identification of individual Aboriginal people with places, through both the places and people being named from the words of the Dreaming's song, is a key part of Aboriginal culture in many regions. Within the Aboriginal domain in its modern manifestations, placenames also become the source of names for other things, such as buildings and institutions. In Darwin, for instance, Nungalinya College is named after the important Larrakia site Nungalinya 'Old Man

Rock'. When I was at Strelley people named the community newspaper Mikurrunya after the nearby range of hills. In both these cases, however, the institution was only a short distance from the place that provided the name — maybe 3 kilometres in the case of Nungalinya; and about 5 kilometres in the case of the distance from Mikurrunya to the community and school. The institutions would fall within the same 'sphere of influence' geographically.

The process started by these developments can lead to a shift in the reference of a placename. Aboriginal people in Darwin are generally aware of the original meaning of Nungalinya, but the combination of the use of an English term for the original site and the application of the old term to a new location could easily lead to a change. If you asked most people in Darwin 'Where's Nungalinya?' they would direct you to the college.

4 IMPLICATIONS FOR HISTORY

I wrote about European-Australians' influence and interference in the question of where Aboriginal placenames were assigned at the start of this paper. These examples were drawn from places where Aboriginal people are in the majority and at least nominally could have wielded some power. In this area of the Northern Territory most of the Aboriginal names that eventually made it onto the Whitefella 'map' (albeit perhaps in the wrong places) could do so because of the increase in interest in applying Aboriginal names to Aboriginal communities in the 1970s. In fact in the VRD there are still very few Aboriginal placenames on the maps — nearly all of the names remain English, despite hundreds of Indigenous placenames being known throughout the region. When this region was settled, very few Indigenous placenames were used. There are therefore no examples of European settler treatment of Indigenous names in the area in the early days.

This is somewhat unlike a number of areas in southern Australia where there are many towns and other features with Aboriginal names. Those names were appropriated early on by European Australians and used by them as they wished, with little or perhaps no consultation with the Aboriginal people who remained in the areas. Given what we know of the European culture of placenames, it would seem likely that they would shift them from the original location to one more centred on European social groups and 'spheres of influence', and from then on from one homestead or township to another as they moved. Rob Amery has told me of a number of such examples in the Adelaide area, and it is likely that there are many others not so well documented. Such a history could well obscure the early Indigenous location of the place, but early maps and records might reveal at least a good part of the story. In some cases it may be that the existence of Aboriginal groups attached to named places and areas may have constrained such complete absorption of the names into the Whitefella placename culture, at least for a while.

CONCLUSIONS

This paper began by pointing out a great divergence between how Aboriginal and non-Aboriginal people treat placenames, at least as perceived from the Aboriginal side. Whitefellas transfer names from one place to another as they move; such behaviour is 'not law' in the Aboriginal domain. Names belong to certain places forever and human beings who tamper with this law display a dangerous level of hubris.

However, at another level, the fundamental process involved in the extension of names from central locations to other parts of a zone shows a definite similarity across the two cultures. In both cases, the zone across which a name can be moved is defined by the social networks licensed, and the law and ideology that define human/land connections in the respective societies. As David Nash reminded me, the movement of the name 'Wave Hill' from one homestead location to another, while perplexing to local Aboriginal people, is perfectly in accord with the relevant zone in Anglo-Australian understanding and law — the cattle station 'property'. Indeed this maintenance of one name for the central point in such a property is a legal requirement, which would entail changes in numerous official records if altered.

On a larger scale, the transfer of names from one part of Australia to another, and even from Britain to Australia, is a symbolic gesture which makes a statement about continuity — even in a vague sense — of some (usually Anglo-Celtic) network of settlers.

Traditionally oriented Aboriginal people are following the same patterns of extension of naming, in a certain sense, but not the same types of zones or social networks. The ability to move is highly restricted by a set of geographical constraints and customary rules about zones. Now, however, all groups have become enmeshed in European-Australian social and legal structures and have been affected by the patterns of movement imposed on them. These include the concentration in settlements during most of the twentieth century, followed by the possibility of moving back to homelands in the last 30 years, an option which for most Indigenous people is highly restricted by the availability of land not already claimed by Whites, and other practical factors.

Contradictions will surely arise over naming practice due to these changes. Say an Indigenous group owns or leases a cattle station, for instance, and wishes to move its station headquarters to a different location (as was mooted for a while at Daguragu). Would the group have to retain the station name in defiance of Aboriginal tradition, or adopt the new (traditionally correct) name of the new location in defiance of European-Australian pastoral practice?

Other examples in this paper concern the fact that people ended up in different outstation locations from where they had originally planned, for a variety of practical, and often European-imposed, reasons. Here the application of the 'wrong' name to the eventual location is disputed by many, but usually eventually accepted by most. The reason is that the people setting up the outstation have become a socially defined group of 'settlers' by their focus on the place, even before they move. Those going to X are referred to as the 'X mob', and this naming of the social group is then transferred to their settling place, whether or not that was strictly X previously, as long as it is in some sense in the same zone as X. This process of a shift of placename from one location to another nearby, mediated by association of the name with a social group or band, is one that probably has parallels in traditional life, although the context has changed, and is still changing.

REFERENCES

Basso, Keith, 1997, *Wisdom Sits in Places: landscape and language among the Western Apache.* Albuquerque: University of New Mexico Press.

Berndt, Ronald M., ed., 1977, *Aborigines and Change: Australia in '70's.* Canberra: AIAS.

Berndt, Ronald M. and Catherine H. Berndt, 1987, *End of an Era: Aboriginal labour in the Northern Territory.* Canberra: AIAS.

Buchanan, Gordon, 1935, *Packhorse and Waterhole.* Sydney: Angus and Robertson.

Dixon, R.M.W. and Grace Koch, 1996, *Dyirbal Song Poetry: the oral literature of an Australian rainforest people.* St Lucia: University of Queensland Press.

Doolan, Jack, 1977, Walk-off (and later return) of various Aboriginal groups from cattle stations: Victoria River District, Northern Territory. In Berndt, ed., 1977:114–123.

Keen, Ian, 1995, *Knowledge and Secrecy in an Aboriginal Religion.* Oxford: Clarendon Press.

McConvell, Patrick, 1989, Workers and camp-mob: indigenous articulations in the cattle industry. Unpublished MS.

McConvell, Patrick and Rod Hagen, 1981, *Land Claim by Gurindji to Daguragu Station: anthropological report.* Alice Springs: Central Land Council.

Merlan, Francesca, 1998, *Caging the Rainbow: place, politics and Aborigines in a north Australian town.* Honolulu: University of Hawai'i Press.

Morphy, Howard, 1984, *Journey to the Crocodile's Nest.* Canberra: AIAS.

Povinelli, Elizabeth R., 1993, *Labor's Lot: the power, history and culture of Aboriginal action.* Chicago: University of Chicago Press.

Rose, Deborah Bird, 1991, *Hidden Histories: black stories from Victoria River Downs, Humbert River and Wave Hill Stations.* Canberra: Aboriginal Studies Press.

— 1992, *Dingo Makes Us Human: life and land in an Australian Aboriginal culture.* Cambridge: Cambridge University Press.

Toohey, J., 1982, *Land Claim by Gurindji to Daguragu Station.* Report by the Aboriginal Land Commissioner. Canberra: AGPS.

5   IS IT REALLY A PLACENAME?

Luise Hercus

There are few topics as challenging as the study of Australian placenames. Their formation varies from region to region, they may be analysable or not, they may refer to the actions of Ancestors, they may be descriptive, or 'indirectly' descriptive: an Ancestor is said to have noticed some particular feature and named the place accordingly. That feature may or may not be permanent. Placenames may be single morphemes or consist of a whole sentence, they may show archaic features. In any case they are unpredictable: we can never guess what a place was called. We can also never be sure we are right about a placename unless there is clear evidence stemming from people who have traditional information on the topic. In the absence of such evidence we have to admit we are only guessing. While, for instance, within the bounds of grammatical rules one can know with reasonable surety how to make certain verb forms, with placenames there are no such bounds; the only limitation is the extent of human imagination. That is why, in those cases where there is clear evidence, they can tell us so much about how people thought about their country.

This paper is intended to illustrate this from two specific angles, the use of generic terms as names, and the use of apparently 'silly' names.[1]

1 GENERIC TERMS[2]

If we look at general dictionaries of the past, rather than just placenames, there are many instances of the use of general terms where one might have expected a specific term. This happens sometimes even in the most brilliant and valuable documents from the past. Thus Stone's important vocabulary of Wembawemba (1911) has the following entries:

---

[1] In general, Aboriginal words transcribed from modern recordings are in bold typeface; Aboriginal words quoted in early literature are in italic typeset and Aboriginal words transcribed from modern recordings and used as a placename are in italics, bold typeface.

[2] Italics have been used for words that have been recorded in the recent past on tape, inverted commas are used for data from written records.

L. Hercus, F. Hodges and J. Simpson, eds, *The Land is a Map: placenames of Indigenous origin in Australia*, 63–72. Canberra: Pandanus Books in association with Pacific Linguistics, 2002.

fire sticks (saw)	*wannup*

when it is clear from all the other evidence of Wembawemba that **wanap** is simply the general word for 'fire, firewood', and does not necessarily have the specific meaning of 'fire sticks'.

net peg	*gunneneuk*

when it is clear that **kani** is the general term for any large stick or pole, and **kani-ny-uk**, i.e. **kani** followed by a linking consonant and the third person possessive marker, means simply 'his/her/its pole' and has no direct association with a net.

necklace (crayfish claw)	*mannunyeuk*

when **manya** is the general term for 'hand', and **manya-ny-uk**, that is **manya** followed by a linking consonant and the third person possessive marker, means simply 'his/her/its hand' and has no association as such with a crayfish and still less with a necklace.

In the case of placenames the situation is a little different: a general term may in fact BE the actual specific term. There are just a few obvious examples of this:

For the Paakantyi people of the Darling River, the Darling river was **Paaka** 'The River', the only one that mattered: it had no other name.

The most spectacular example is that of the Diamantina/Warburton. Anyone who has flown over the southern Simpson Desert will have seen the deep channel of this river as it winds towards Lake Eyre. North of Clifton Hills in the far north-east of South Australia a smaller channel comes out of the Diamantina westwards, twists and bends and ultimately rejoins the main channel before it reaches Lake Eyre. For hundreds of miles to the north there are only sandhills and no sign of any river or creek, nothing, until one reaches the lowest flood-outs of the Hale and Plenty rivers. So for the Simpson Desert Wangkangurru people the Diamantina/Warburton was THE Creek, *Karla*. It had no other name. The smaller ana-branch that wound out of it was THE Little Creek, *Karla-kupa*, the Kallakoopah. In both these cases a general term doubled as a specific term. An interesting point is that in the case of the Kallakoopah this analysis is no longer quite true: the adjective **kupa** 'small' is the standard term for 'small' in the closely related Arabana language, but Wangkangurru people say **nyara** for 'small'. As this name and several fixed locutions show, Wangkangurru people must have said **kupa** once upon a time. So the name Kallakoopah represents an archaism and also shows how a general and completely analysable name can be on the way to becoming at least partly unanalysable and therefore uncontroversially a name.

Figure 1:'THE Creek', Karla, winding through the southern Simpson Desert.
Photo: L. Hercus

Figure 2:'THE Little Creek', the Kallakoopah, winding through the Simpson Desert.
Photo: L. Hercus

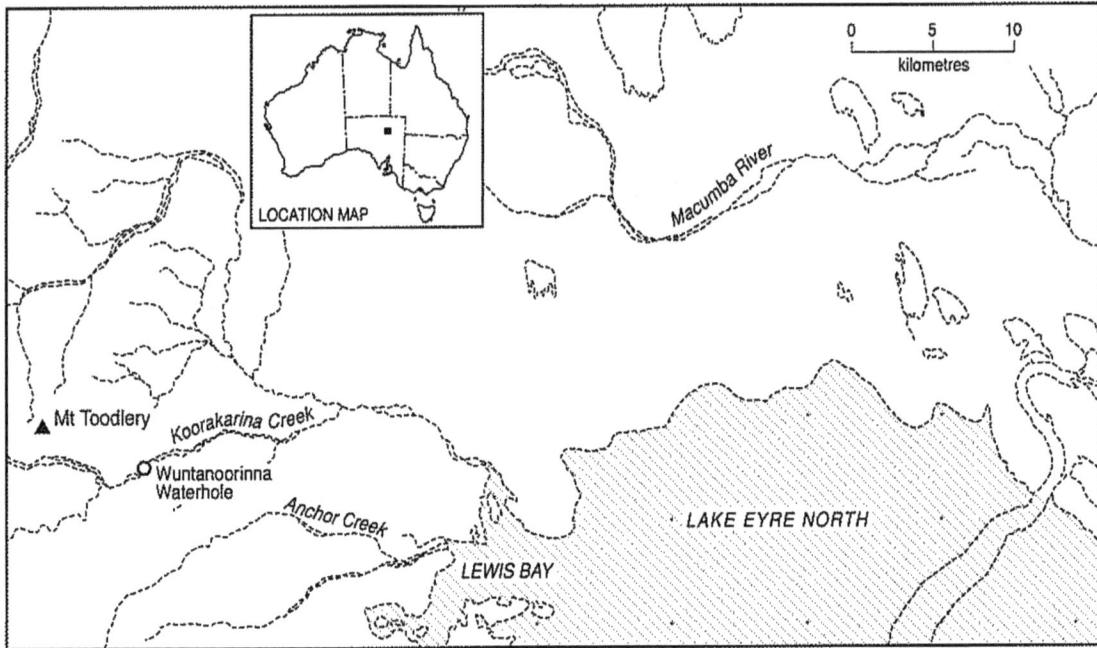

Map 1: Mount Toodlery and surrounding area

There can be other cases where a general term is used because there *is* no specific term: the feature in question may seem to Europeans as meriting a specific name, but not in Aboriginal eyes. Thus about 15 kilometres to the west of the north-westernmost corner of Lake Eyre the maps show Mount Toodlery. This is not a prominent peak, it is nothing but an elevated rocky tableland, and the highest point of it has a trig. It is the sort of place where it is absolutely impossible to drive, and I would say even to walk, just an elevated vast expanse of rocks and boulders. So presumably the Arabana people, whom the surveyors must have asked about it, called it **thurliri** 'rough stony country'.[3] There was a name for an elevated part of the tableland that was closest to the Wuntanoorina waterhole, it was *Unthu-nyurinha kadnha*: it was where the Initiands were kept in the Two Men History associated with Wuntanoorina *Unthu-nyurinha* waterhole, and the name refers to circumcision. The part of the elevated tableland with the trig, however, did not have a specific name. So it was not surprising that some years ago, when I started talking about Mount Toodlery, nobody seemed to know where I meant. When we actually got into the vicinity of this place the Arabana speaker Arthur McLean said: 'that is **thurliri** all right, rough country' — i.e. **thurliri** was a term that could be used for *any* such area, and there are plenty such in Arabana country.[4] In these more traditional regions, this is a rare instance of such a word ending up as a specific placename. There seem to be altogether relatively few instances of general terms used as specific, both of

---

[3] The Arabana word *thurliri* 'rough tableland country' has an etymological connection with the well-known Arabana word *thurla* 'sharp stone', which has become part of archaeological vocabulary in English as 'tula adze'.

[4] To illustrate the unpredictability of placenames: Mount Robinson to the west of Mount Toodlery is an equally inhospitable big plateau and traditionally it was called *Yaltya watinha* 'track of the waterhole frogs', because the by then desperate Ancestral Waterhole Frogs travelled all the way up there in their search for the swamp by the Macumba where they ultimately came to rest.

the *Karla* type where the general term doubles as a specific name, and the *Thurliri* type where there is no specific name.

In Victoria,[5] where there was considerable language loss at a much earlier date, lists of placenames, just like the ordinary dictionaries, contain a number of what appear to be general terms identified as referring to a specific item/location. In the case of dictionaries, as in the examples from the work by Stone, quoted above, this is usually because the questioners expected there to be a special term when there was none, which involves to some extent a lack of communication. In the case of locations the question arises: are they really placenames or just general terms that have been used, perhaps again through lack of communication, in this case usually between a surveyor and Aboriginal people. A surveyor may have pointed at a creek and asked 'what do you call this?', only to get the answer 'the creek', and he may have then carefully noted and reported the term as a placename, when the specific name of the creek may have been quite different. Typical Victorian examples are:

From **yaluk**, 'creek' in Central Victoria and some western Victorian languages:

Yallock Creek	former name of the location of Koo Wee Rup, south-east of Melbourne
Yallock Creek	creek that flows into Western Port Bay (near Koo Wee Rup)
Moyne River	Port Fairy River, originally known as *Yalloak*, 'a shallow swampy stream' (Lane 1869 in Smyth 1876 vol.2:186)

From **pirr** 'creek' in Wembawemba, Perepaperepa and Djadjawurrung:
Barr Creek is the name of part of the Gunbower Creek, close to the Murray River. The name indeed corresponds to **pirr**: because of the fact that when an -r- sound follows the **i**, there was some lowering of the vowel. The spelling of the placename presumably represents this; other early spellings are *bur*, *ba*. There is also a Piccaninny Barr Creek, known in 1870 as Little Bar Creek (Contract Surveyor to Surveyor-General, 8/8/1870).[6]

From **brim**, 'spring':
Brim is a place on the Henty Highway north of Warracknabeal. The name was originally the name of a station, and it was recorded by the surveyor Chauncey and translated as 'a spring or well with water': Smyth (1876 vol.2:205). The name probably derives from the station and may not be the result of immediate communication by a surveyor with Aboriginal people.

Brim Springs is just south of Horsham, and there is also mention of a name Brim Brim for the same area. This reduplicated form is also quoted by Thornly in Smyth (1876 vol.2:60) with the meaning 'spring'.

From **buluk**, 'lake, swamp' in Djadjawurrung, Djabwurrung and southern Central Victoria:
Buloke. It is not surprising that this general term should be used for Lake Buloke near Donald in Central Victoria. It is a particularly important large lake, prolific in wildlife once upon a time, and notorious in more recent times as the place where, at dawn on the opening day of the season, shooters sometimes shot each other instead of the ducks. It can, however, be

---

[5] I am deeply indebted to Ian Clark for permission to use his database on Victorian placenames.

[6] Information courtesy of Ian Clark.

shown from data of the 1870s that there was also a specific term **Banyenong** for this lake alongside the general term 'THE lake'.

The same word is represented by the name of Lake Bolac, which is just off the Glenelg Highway, a little more than halfway between Ballarat and Hamilton. It is not as big as Lake Buloke, but is the biggest lake in the area, and must have constituted an important resource for local people. It too might well therefore have been THE Lake.

Bulluc-bulluc. This name was noted in Robinson's Journal (Robinson journal, 19 March 1840, see Clark 1998) for an area which is now part of Melbourne. The reduplication indicates that this may mean 'a lot of swamps', 'a swampy area'.

From words meaning 'head', 'hill', i.e. **purk, purp, panyul**, and *kowa*, 'high mountain':
Bork-bork. The use of the word for 'head' to mean 'hill' is widespread. The word **purk** for 'head' is given for Djabwurrung by Dawson (1881:xix *puurk*, 'head') and the general term 'Bork-bork', 'heads, mountains', is attested to refer to a forest hill north of Raglan, not far north of Beaufort (Porteous in Smyth 1876 vol.2:179). This term was also used to refer to the Pyrenees 'Range at head of Waterloo' (Surveyor-General's list of names in Smyth 1876 vol.2:194). The Pyrenees were The Hills in this area, and the use of the term for just one small part, the hill north of Raglan, may be an error.

Panyul is a mainly Central Victorian word for 'hill'. It occurs as 'Banyule', the name of a hill in the Heidelberg area of Melbourne and this has survived in the name 'Banyule Flats Reserve'.

Cowa, according to surveyors' returns, was the original name for Mount Arapiles. As this stands out all alone, away from the Grampians, it is understandable that it should be called The Mountain. The same name is attributed to the whole of the Grampians Range by Wilson (Smyth 1876:178).

From words meaning 'sand', *kolak*, **kurak**, and *maloga* (Yota-yota):
There are several placenames that simply represent the word for 'sand' in various Victorian languages. The northernmost is Koorakee in New South Wales some 18 kilometres north-east of Robinvale. This represents Mathi-mathi / Letyi-Letyi **kuraki** 'sand', and it corresponds to **kurak** in the other Kulin languages. A form *kulak* is found in Dhauhurtwurrung and the Colac language, hence the name of the township of Colac.

Korrac 'sand' is the name of a parish by the Avoca River, south-east of St Arnaud in Victoria (Tully 1997:88).

Cor.rac.cor.rac, i.e. **Kurak-kurak**, 'a lot of sand' is reported to be the name of a creek east of the Woardy Yalloak Creek: Robinson journal, 12 August 1841 (I. Clark:1998).

Torn. This word is given in Dawson's own vocabulary as 'hummocks of sand' for 'Peek whurrung', the language of the Warrnambool–Port Fairy people, and this would undoubtedly be the meaning of the placename. It is, according to Dawson, a place just west of Warrnambool which does not have a special European name.

Maloga. The name of the old mission station by the Murray meant 'sand' in Yota-yota.

There are general terms for other natural features which also occur as placenames, examples are:

Pyalong, between Kilmore and Heathcote in Central Victoria, simply means 'red gum trees'. The reduplicated form of the same word, Bael-Bael 'a lot of gum trees', is the name of a lake near Kerang. *Piel*, an old name of a salt lake in the Camperdown area (Scott in Smyth 1876:183), represents the same word.

Terrick Terrick near Pyramid Hill gets its name from Wembawemba/Perepaperepa **terik** 'gravel'; Drik Drik 'limestone' in far south-west Victoria, just north of the Lower Glenelg National Park, is probably a cognate word.

The uncertainty of placenames is truly evident here: all these placenames may well have been just general terms noted as names by mistake. We can make some tentative suggestions from the nature of some of the places. Of the names listed above Torn is the one most likely to be of the *thurliri* type, that is, a place that did not have a specific name. It is probable that Dawson's informants had been asked 'what do you call this place?' and they came up with a general rather than a specific answer 'hummocks of sand'.

2 'INTERMEDIATE' NAMES

In Arabana country on the western side of Lake Eyre there are many creeks rising from the rough ranges, and some longer rivers that traverse some distance before reaching Lake Eyre. There is no unique creek, unlike the Simpson Desert situation, and all the major creeks have Arabana proper names. These sometimes do not follow the same pattern as the English names: thus on modern maps the Pootneura Creek joins the Lora Creek and this then joins the Arkaringa Creek. In Arabana nomenclature the Pootneura Creek is the dominant creek all the way and the other creeks join it. Apart from the proper names, people sometimes referred to creeks by what might be regarded as a semi-generic or 'intermediate' term. Creeks can differ from each other by the type of trees that grow there, and so people may refer to a creek as 'red gum creek', **apira-karla**, usually the largest; 'box creek', **pitha-karla**, usually also a main creek with permanent or semi-permanent waterholes; 'gidgee creek', **urinyingka-karla**, 'red mulga creek' **amuna-karla** (only in the far north), or 'tea tree creek' **ityara karla** (only in the far south). These terms are mainly used in circumstances where the speaker thinks it is quite clear to the listener which creek is meant and he does not need to give the actual specific name. Similarly swamps are sometimes referred to by the Arabana equivalent of 'box-swamp' or 'canegrass swamp', when there is also a quite specific name. These new descriptive names are as it were familiar terms, the speaker visualises the specific place and does not need to give the specific proper name. I am aware of only one such name that has come down to us superseding a specific placename.[7] These 'intermediate names' are however represented in English translations, even when there is a well-known Aboriginal proper name, hence Box Creek (for *Ulyurla karla* 'Old Woman Creek'), and the former Box Creek siding on the old Ghan line north from Anna Creek, 'Red Mulga Creek' south of Dalhousie, and in Kuyani country, well to the south, the politically notorious 'Canegrass Swamp' (for *Piya-piyanha*) in the neighbourhood of Olympic Dam.

There are some descriptive placenames in Victoria which could once upon a time have been such intermediate terms, for example, Wycheproof, that is, *Wityi-purp* 'basket-grass hill'; Narragil Creek near Maryborough from **ngari-wil** 'having she-oaks'; Dharugill, a spring in the Heytesbury area south of Camperdown, 'a spring etc., a plant' according to Scott in Brough Smyth (1876:184) probably from *dharuk-wil*, 'having *dharuk* plants, plants with edible roots'; Wood Wood, a township near the Murray River down from Swan Hill from **wurd-wurd**, 'bulrushes'. The names in question may simply be descriptive specific names, but they may also be 'intermediate' type names.

---

[7] This is **Ityara karla** 'Tea Tree Creek', which is used by Arabana people for Stuart Creek South, in Kuyani country superseding an unknown previous Kuyani name.

3 'SILLY' NAMES

In placenames it is easy to find some that look at first sight like quite silly mistakes arising from a lack of communication:

Thus there is a small township south of St Arnaud in Victoria named Winjallok. The surveyor Philip Chauncy recorded this name (cf. R. Brough Smyth 1876 vol. 2:211). He also recorded that it meant 'where is it?' This interpretation is indeed correct. The place is in Djadjawurrung country and this language is very close to the neighbouring Wembawemba. The Wembawemba word **wintya** means 'where' (Hercus 1992). Grammatical markers can be added to this, with a linking consonant -l-. Hence, with the addition of the third person marker -**l-uk** we have **wintyaluk** 'where is he/she/it?' It looks like an obvious example of lack of communication, and as Tully (1997:92) suggests in his dictionary of Djadjawurrung: 'It appears that the informant was not sure where Chauncy meant'. Chauncy however recorded the translation: would he have recorded the sentence if he had doubts about it representing a placename?

Mount Beckworth in Central Victoria, a few kilometres west of Clunes, was once called *Nananook*. This name was recorded by A. Porteous in R. Brough Smyth (1876 vol. 2:178), and he gives the translation 'behind'. This name probably represents a word similar if not identical to Wembawemba **nyaninyuk** 'the back of his/her neck' (**nyani** 'back of neck', -**ny**-linking consonant for the third person after nouns ending in vowels, and -**uk**, third person marker). The word for 'back of neck' is used in many general contexts where the term 'back' would be used in English, hence the translation 'behind' is not surprising. Tully (1997:89) suggests 'this appears to be an example of confusion between Porteous and his informant, as *nananook* means literally behind him'. But why would Porteous, like Chauncy, knowing what the word meant, record this as a placename?

Another such placename is 'Nawalah', 'what is it?', Wilson in Smyth (1876 vol. 2:178). It is the name of a place to the north-west of Stawell. This name can be analysed as **nya wala**, 'what is it?' The interrogative base for inanimates in Werkaia and other W. Kulin languages is **nya**: in Werkaia the full form for 'what' is the reduplicated **nya-nya**, but in Djabwurrung the simple **nya** was used: the placename implies that this was also the case in Jardwadjali. The pronominal base in both Werkaia and Wembawemba is **wal**, hence **wala-ny-uk** (with the addition of the third person possessive marker) means 'he, she, it'. The third person marker was not necessarily required, one could therefore say **nya wala**, 'what is it?' This could well have been an utterance of a perplexed Elder, but was it? Why would a surveyor, knowing what the name really meant, list it as a placename?

Wategat is the name of a place just west of the Grampians: it is mentioned by Thornly in Smyth (1876 vol. 2:62) as *Wateegat* with the translation 'come, come here'. The verb **warda**-means 'to come' in neighbouring Werkaia. The second person plural imperative of an intransitive verb is not attested for Werkaia, it was -**iaty** in Wembawemba (transitive -**akaty**). The Werkaia ending might well have been -**ikaty**, with the **k** preserved as in the transitive. Wategat is in Yardwadjali country and so the name of the place actually gives us grammatical information about the second person plural imperative in Jardwadjali and probably Werkaia. In any case the name means 'You (pl) come!' Could it have been that an Elder wanted to draw the attention of a surveying party to some feature, and the word was noted as a name by mistake?

There are similar types of placenames in an area where traditions have remained alive longer, and this can throw some light on the situation. Just to give one example: in the far north-east of South Australia, south-east of Birdsville and about 25 kilometres from the

Queensland border, is a large, usually dry, lake called Etamunbanie. This represents **ita** 'that way' **manpa** 'go' + **ni** placename marker. The general meaning is 'you go that way (while I go this way)' in Yawarawarrka[8] (Breen MS). It is very tempting to think that a Yawarawarrka man, accompanying the surveyor to this rather desolate and windswept spot, got tired of going around with him and used this opportunity of saying so, more or less telling him to get lost, without the surveyor understanding what was going on and noting it down carefully as a placename. In this case, however, there are still people who know the derivation: the actual explanation of the name is quite different.

This lake, Etamunbanie Lake of the maps, is set in the middle of a wide plain: it belonged to the Mindiri Emu tradition, one of the most important traditions of the Lake Eyre Basin. An area near Etamunbanie Lake was one of the three main sites for the Mindiri ceremony for which people came from far afield. The main Mindiri dancing ground was by a tree on the edge of the lake. The myth relates how, after they had completed a ceremony there, two groups of emus split up and one said to the other **ita manpa** 'you go that way (while we go this way)'. There is a verse which contains these exact words, followed by some rude departing jibes at this second lot of emus. This is how the place got its name according to the traditions of Yawarawarrka people.[9]

It is therefore quite possible that in the Victorian instances there were similar explanations: an Ancestor, in circumstances which we do not know, may have said **wintyaluk** 'where is it?' or **nyaninyuk** 'behind him' or **nya wala**, 'what's that', or *wateegat*, 'come here!' Porteous and Chauncy and other surveyors may not have made mistakes at all. As in the case of the possible general terms given as specific placenames, we cannot know for sure, but it could well be that these names, as in the example of Lake Etamunbanie, are not due to lack of communication, but reflect the utterances of Ancestors to whom these places held a special significance.

We do not know the answers to the suggestions raised, but the study of Australian placenames is in its infancy. In Victoria, in particular, work in progress by young scholars on some of the earliest maps and on early manuscript material means that we may ultimately recapture at least some of the Aboriginal vision of the landscape that is provided by placenames.

REFERENCES

Breen, J.G., n.d., *A Yandruwantha dictionary*. MS.

Clark, I.D., 1998, *The Journals of George Augustus Robinson, Chief Protector Port Phillip Aboriginal Protectorate*. Melbourne: Heritage Matters.

Dawson, James, 1881, *Australian Aborigines*. Melbourne: George Robertson. Facsimile edition 1981. Canberra: Australian Institute of Aboriginal Studies.

Hercus, L.A., 1992, *Wembawemba Dictionary*. Canberra: the author.

---

[8] The main source of information on the Yawarawarrka comes from J.G. Breen's work on the neighbouring and closely related Yandruwantha language (Breen:MS).

[9] The information actually came from a Wangkangurru speaker who was familiar with the area.

Smyth, R. Brough, 1876, *The Aborigines of Victoria and Other Parts of Australia and Tasmania*, vol. 2. Melbourne: Government Printer.

Stone, A.C., 1911, The Aborigines of Lake Boga. *Royal Society of Victoria Proceedings* 23:433–468.

Tully, J., 1997, *Djadjawurrung Language of Central Victoria, including Place Names*. Maryborough: the author.

DOCUMENTING PLACENAMES

6 ON THE TRANSLATABILITY OF PLACENAMES IN THE WIK REGION, CAPE YORK PENINSULA[1]

Peter Sutton

1 HISTORY

In the late 1980s I obtained Australian Research Council and local Wik organisational funding to compile a database of site records from western Cape York Peninsula between the Embley and Edward Rivers, in a project based at the South Australian Museum. The main field data came from myself (mapping from 1976 onwards), David Martin (mapping from 1985) and John von Sturmer (mapping from 1969). Small amounts also came from fieldwork by Roger Cribb and Athol Chase, who mainly mapped in 1985, although Roger returned to focus more on archaeological mapping and site work later.[2]

There has also been further database compilation of mainly new (that is, post-1990) field data, especially from the inland sector, amounting to adding a further 200+ pages of supplementary site reports to the 1,000 pages already produced. This, like the longer 1990 report, is still under restricted access at the time of writing, because of the role of these documents in the ongoing Wik Native Title Determination Application before the Federal Court. The data have now been massaged into a more widely accessible form, using Access

[1] Primary acknowledgements are due to the Wik and Wik-Way peoples of western Cape York Peninsula for their long-standing commitment to collaborative efforts with scholars who have carried out fundamental ethnographic research among them, myself included. Particular thanks are also due to John von Sturmer and David Martin for sharing major cultural mapping data from which I have drawn some parts of this paper.

Funding for the research behind this paper came from the Australian Institute of Aboriginal and Torres Strait Islander Studies, Commonwealth Department of Education, University of Queensland Department of Anthropology and Sociology, Aurukun Shire Council, Aurukun Community Incorporated, Department of Aboriginal Affairs, the Australian Heritage Commission, the Australian Research Council, and the Cape York Land Council.

I thank Jane Simpson and Barry Alpher and Bruce Rigsby for helpful comments on earlier drafts, Luise Hercus for help with comparative data, Flavia Hodges for keeping me writing to a fierce deadline, and Michael Walsh for stimulating me in email conversations to think about the byzantine topic of translatability in Aboriginal placenames.

[2] At the end of that ARC project the work was produced as Sutton et al. (1990).

L. Hercus, F. Hodges and J. Simpson, eds, *The Land is a Map: placenames of Indigenous origin in Australia*, 75–86. Canberra: Pandanus Books in association with Pacific Linguistics, 2002.

for the information and a Geographic Information System for the site plots. The records have many fields per site and information is in a majority of cases sourced back to a particular set of Aboriginal consultant names and dates, notebook pages, tape tracks and film roll and shot numbers, such that more detail is available offstage. Most of the material in the database is in English, although the original data may have been in several other languages.

There is a separate database report prepared for the Wik Native Title case by John Taylor, who has been mapping since 1969 mainly in the Mitchell–Edward Rivers region, which deals with sites in the southern end of the Wik area.

Once the Wik Native Title case is over we will have to come to a decision about the future of the material. We have clan heads' permission as of 1985 to publish the 1990 version that was being drafted at that time. We may or may not have to renegotiate that permission.

I am, however, able to briefly outline the nature of this mapping research archive. On a rough count there are about 3,000 site records from approximately 120 clan estates between the Embley and Edward Rivers, and inland to just west of about Coen. Very roughly, around 1,600 of these sites have been visited and ethnographically mapped on the ground. In a number of cases this involved reconstructing specific camps or events and their personnel, remembered from times past by those who were there. We often recorded remains of material culture and routinely focused on economic resource aspects of sites, their seasonal use, camp type, plants found there or nearby, historical memories of the place, religious significance of the place, camp composition and hearth arrangements, ownership matters, restrictions on behaviour, movement pathways between sites, foraging ranges from base-camps, and so on.

Our botanical specimens are now stored at the Queensland Herbarium. These include some exhaustive collecting transects across a range of typical environmental types, plotted onto airphotos. The collecting and ethnobotany of a number of species were carried out opportunistically by most of us, but there was at least one field trip where ecologist Dermot Smyth and I concentrated more or less exclusively on the ethnobotany of the area between the Archer and Kendall Rivers, and Dermot has also done some independent work in the area.[3]

The relation of sites to estates was often focal to this research. Estate descriptions and tenure histories vary in detail from very little to a great deal. These site-based bodies of data cross-link to genealogies covering about 100 estate groups, and to land tenure studies by von Sturmer and myself.[4]

On top of all this is a large multi-volume report currently in the Federal Court, which contains the items listed below under Sutton (1997). Together with the 1990 volume, also in evidence, these reports total about 1,800 pages.

Besides these reports, there are various digitised bodies of material, including mission personal record cards which have been keyboarded, a Wik ethnobotany database by myself and Dermot Smyth, a genealogy database in Brother's Keeper, researchers' own word-processed field note material, audio tape catalogues at the Australian Institute of Aboriginal and Torres Strait Islander Studies, and so on.

Time, space and legalities do mean, however, that here I am able only to take a 'tip of the iceberg' approach.

---

[3] I have never had the funded time to write the ethnobotanical material up in detail, although an abbreviated version is sprinkled through my *Wik–Ngathan Dictionary* (1995).

[4] Von Sturmer (1978), Sutton (1978).

2 COMPLICATIONS

Placenames in the Wik region seem to offer endless complications. This becomes much clearer as one tries to go through lists of site names and, as I have tried to do, allocate them on a percentage basis to simple dichotomies such as:

'translatable'	'untranslatable'
'archaic word'	'contemporary word'
'original language'	'introduced language'
'autochthonous language'	'other Aboriginal language(s)'.

One complication, for example, lies in the fact that in different languages, and sometimes even in the same language, a site may have both an opaque (untranslatable) name and a translatable name — so does one count this as one of each or one half each for the two categories? And where a site has two or more opaque names, one each in different languages, should one count this as one case of opacity or as two, three etc.? A notable example is the small lagoon and important base camp site of *Aayk = Thuul = Thuulu* just inland from Cape Keerweer.

A sub-complication for this part of the discussion is that it is clear in some instances that the opaque name is old and endogenous and the transparent name is very probably recent and exogenous. For example, there are cases in the area between the Archer and Embley Rivers where an old opaque name in the local language (Andjingith, Adithinngithigh etc.) is matched and in some cases now replaced by a Wik-Mungkan name meaning '[So-and-so] Story Place'. Thus the Wik name is not a translation of a former site name but its semantic content depends on the pre-existing cultural content of the site itself.

For example:

Iwiken (Adithinngithigh, no translation available) = *Pach-aw* (Wik-Mungkan) '[Shooting] Star Story Place'

Puk-aw (Wik-Mungkan) 'Baby Story Place' (no original Ndrra'ngith [?] name recorded, perhaps lost)

Uk-aw (Wik-Mungkan) 'Brown Snake Story Place' (no original Andjingith name recorded, perhaps lost)

These are a bit reminiscent of those Australian–English placenames, like Emu Creek, Kangaroo Rock, The Twins, and so on, which in some cases can be shown to be actually based on the presence of the relevant Dreamings, but this is not obvious from the name itself. I have come across many examples over the years. In the case of Lake Cadibarrawirracanna or *Kardipirla Warrakanha*, 'Star Dancing Place', it is easier for most people when hearing this name to recall Slim Dusty's song about it than to reflect on Australia's original legends.[5]

Certain sites particularly associated with the former Aurukun mission's activities and thus with Europeans — a very small minority of places mostly north of Archer River — are moving towards the loss of their Indigenous names in the Wik area, and a few have already

---

[5] This lake is just east of Coober Pedy. Information and translation from Luise Hercus (pers. comm.).

got there. Among this class of places are *Waterfall, Pera Head,*[6] *Lowdown Swamp, Mr Little Crossing, Kilpatrick ~ Bill Patrick*[7] *Landing, Police Lagoon,* and *Bamboo.* Even further south, younger generations are by now unlikely in most cases to know that the *Moving Stone* (sometimes *Movie Stone, Movie Girl*) is at *Um Thunth.* Even the old name of *Thew-en* for *Cape Keerweer* (sometimes *Kep Kwiwi, Kek Piwi*) is now somewhat antiquarian. But, in general, local placenames have largely endured in the Wik region and north of it to the Embley River.

3 THE PUZZLE OF LANGUAGE SHIFT

Nevertheless I have often puzzled over why such language shifts seem to have occurred so rapidly for some placenames, in the face of the strong retention of at least one common Aboriginal language in daily usage in the region. Consider the case of *Hagen*[8] *Lagoon,* the name of an outstation that has been used over a number of years. Even old people who know the lagoon's original name of *Ochenganh-thathenh* seem to prefer the English name for the outstation. The English name refers to a mission staff member of many years ago who is otherwise little known or remembered. *Ochanganh-thathenh* means 'saw a mudshell', a far more poetic name.

This anglicisation of placenames has happened with several outstations, including for example Bullyard (*Am*), Ti Tree (*Wanke-niyeng* etc.), South Arm (*Yaaneng*), Emu Foot (*Tha' Achemp*),[9] Stony Crossing (*Othungam*) and North Kendall (*Kuchent-eypenh*). This may reflect the cultural location of such places on the cusp of local people's interaction with local government and the state, and thus the separation of the outstation name from the original site name in terms of its real denotata. That is, the outstation name is not that of a natural place so much as of an installation and its people. In the case of *Aayk,* which has operated intermittently as an outstation from 1971 to the time of writing, the outstation name has effectively become separated from the original *Aayk,* which is a small lagoon, not just semantically but spatially. When the outstation was moved away from this lagoon to be closer to a new dry-weather airstrip at *Mulpa'el-nhiin,* it became known as 'New Aayk'. And note that this is within a community where the first language of children is still a Wik language.

Wooentoent, for which there is I believe no translation but which is at least euphonious, was in the 1970s being replaced by the imaginative (?) 'Green Point'. *Pooerroeth,* known in Wik-Mungkan as *Piirrith,* gave rise to English *Peret,* the name of a cattle outstation yard and dip, in the 1950s or 60s.[10] After a cyclone the residential focus of the cattle operation shifted to the other end of the airstrip, but the buildings retained the name of Peret, until my own residence there brought about a reinstatement of the old name to the new outstation location, *Watha-nhiin.* Given this means 'white-tailed water rat sitting', its loss would have been regrettable, I thought. The name seems to have retained standing.

---

[6] A curious case in that the original name of *Malnyinyu* seems to have also been in use among the mission staff before the 1960s.

[7] From Bill Patridge Landing, see MacKenzie (1981:213).

[8] From Jimmy Hogan's Lagoon, see MacKenzie (1981:101).

[9] The original placename means literally 'foot emu', i.e. emu's foot.

[10] Not to be confused with the late Peret Arkwookerum, whose first name was a rendering of his English pronunciation of the translation of one of his major totems, Parrot.

By the early 1990s another small settlement had begun back at the other end of the airstrip, but this time *Pooerroeth* was no longer the base of the English name of the site. Two European men working for the community set it up for cattle operations and, later, as a detention centre. During this process it had become 'Cattle Camp'.

One of the most distasteful moments I ever had at Cape Keerweer, which is just south of *Pooerroeth*, was when conversing with a non-Aboriginal fisherman whose family had established itself illegally on the mud flats of the lower Kirke. He blithely referred to all the fishing spots and creeks he was using by entirely new, English, and banal names. Having not long mapped the same places and recorded their real names — and virtually none apart from the Kirke itself and Cape Keerweer had any other known names according to official maps — I wondered if I were witnessing the beginning of the end. This small settlement was later removed on orders of the local council.

Not far away are what have become increasingly known, even to the local people, as 'Big Lake' and 'Small Lake' — imaginative names again. 'Small Lake' is now the preferred name it seems, although it was earlier known even to the missionaries as *Munpun*, more accurately the name of a base camp site on its shores. More properly it is known in its own estate's language as *Uthuk Eelen* (Small Milky Way) or *Weenem Eelen* (Small Lawyer Cane). Just south of it is *Uthuk Aweyn* (Big Milky Way), also known as *Weenem Aweyn* (Big Lawyer Cane), but it is almost always now known as Big Lake.[11]

There are cases where the Wik equivalents of opaque site names are also opaque and clearly old, for example, *Thokali* (in earlier records thus, and also 'Dugally', probably an Andjingith name, no longer used) = *Thoekel* (Wik-Elkenh) = *Thukel* (Wik-Mungkan). The last of these is possibly a borrowing from Wik-Elken which has been on the coast a lot longer than Wik-Mungkan. But in fact one hears 'Love River' being used in the region more even than the lingua franca's *Thukel*.

Many site names are compounds which contain a translatable morpheme and an opaque one, and are thus perhaps old compounds, or a new compound incorporating an old word (for example, *Yagalmungkanh?*) or snatches of mythic phrase (cf. *Pulthalpempang?*), so I am not sure how to count these in terms of any attempt to quantify the semantic tendencies of Wik placenames. Should they score a half in each category or one in each?

4 WHAT DOES 'UNTRANSLATABLE' MEAN?

There is a halfway house even within this particular set, namely those site names that contain an obviously translatable element plus one that looks like a contraction or archaic version of a typical site-name verb, a class in which I include for the Wik area *-sits, -sees, -camps, -lies, -stands*. But these 'contracted' or 'archaic' descriptions are the linguist's deductions, not something offered by an informant, at least in a number of cases.

Furthermore, where we have failed to elicit an answer to a direct question about meaning and cannot find the word(s) in our dictionaries, we cannot be sure that the word was not simply a word we did not know but could have been told about, or an uncommon term such as the name for a minor species, a remembered archaism, or one of those Wik synonymous doublets that is only used in the Big Language register. We may thus be wrongly counting the

---

[11] Paired sites like this occur elsewhere, for example the *Thiikanen Eelen* and *Thiikanen Aweyn* (Small Island, Big Island) of the Love River estuary, and the *Yaal Eelen* and *Yaal Aweyn* (Small Freshwater Stream, Big Freshwater Stream) of the banks of the Kirke River estuary.

term as being simply opaque (that is, to our informants as well as to ourselves) when (to somebody) it was actually transparent at the time.

Another snare for the unwary here is that the northern Wik languages have all undergone neutralisation of unstressed vowels and final vowel loss, so there are many monosyllables, and homophones abound. Without an informant's reaction one might make the wrong translation of what appears to be a word from the language, or perhaps identify an ancient and opaque word as a modern and meaningful one. The danger of this increases with the shortness and phonological simplicity of the word, monosyllables that lack consonant clusters being dangerous above all other words. And with an informant's reaction as to translatability one would still always be worried by the possibility of folk-etymology and word-play.

5 LINGUISTIC PREHISTORY

An added difficulty is that when people have succeeded to estates formerly held by people of a different language, they have clearly neither kept all their predecessors' names for sites nor replaced them wholesale, in at least all the cases I have had time to look at. Near the coast, in the Norman–Archer River region and south to Love River, and in a case or two further south beyond there, are many ancient names which have retained their pronunciations even though they conflict with the sound systems of the dominant local languages, especially the lingua franca Wik-Mungkan.

Different sites within the same estate may thus be named in quite distinct languages, two at least. This is easily established on phonological grounds, the commonest being that names in the area north of Archer River often reflect the presence of a voiced/voiceless stop series (for example, *Waager, Thaadh, Chejedem, Wobeb*). These voiced stops are not found in northern Wik languages, only southern ones originally spoken far from this area. But they are retained in Wik-Mungkan and other northern Wik varieties when occurring in placenames north of Archer River.

Pre-stopped nasals (for example, *Mbang*, often now pronounced *Bang*), rhoticised apicals (*Katra*), consonant clusters such as /mr/ or /kl/ (for example, *Upumren, Amran, Intheklok*), all impermissible in any Wik language, also occur in this area north of the Archer. There is even some historical evidence of sibilants in these northern placenames (for example, '*Towesie*', '*Kumsinmong*'[12]), another non-Wik feature, and there is a striking frequency of occurrence of site names beginning in /r-/ (six cases, as against one from the whole of the Wik region south of it; for example, *Rugiy, Ruchuw, Rowenem, Renanun*). There are also site names north of the Archer that reflect more than the standard five or six vowel positions of the different Wik languages, such as *Raethepen*, and *Eygegen-laem*. These, like the phonologically aberrant *Yaad* near 'Small Lake', which is unusual for being in an area long associated in historical times with a northern Wik variety not a northern Paman one, would appear to reflect the retention of names from some predecessor northern Paman language.[13]

---

[12] See map in MacKenzie (1981:213).

[13] There is mythological, ritual and oral historical evidence that a northern Paman language was, perhaps 150 years ago, the language of the estate which lies on the north side of the Kirke River estuary, but it has long been affiliated to a Wik language. *Yaad* is in the next estate north ('Small Lake'). There seems to have been a general northerly linguistic shift towards Wik language varieties going on for some time in recent prehistory between the Kirke and the Archer.

Typically this situation is true of estates that are still regarded as being formally identified with the older and more phonologically aberrant language, even if it is no longer fluently spoken. But it also applies to one or two estates where language shift has meant that the language of the owners no longer matches, in phonological structure, the language of a certain number of the names of the sites they own.

I do not think this uneven retention of ancient names of differing phonologies is merely a case of linguistic flexibility among living people, although the deep-seated multilingualism of the area has probably contributed greatly to it. One factor in retention of the voiced–voiceless contrast north of Aurukun may be that some Wik languages also have it, even though they are the southern ones (the Kugu Nganhcara subgroup). But even in the northernmost Wik languages there is non-phonemic voicing. Examples include Wik-Ngathan *eelen [e:dn]*, *nganent [ngand]*, *waj* (emphatic version of *way*), and ideophones such as *chub!* (entered water), *pubbbbb!* (Story Shark's tail beating water) and *dhrrrrrr!* (went off quickly).[14]

These occur in a language with a decidedly fortis and voiceless approach to its only phonemic stop series under most conditions. Wik-Mungkan is even more fortis and aspirated in attack when it comes to stops.[15] It is in a dialect chain with the Kugu Nganhcara varieties.[16] And it may be stretching things a bit but I am also inclined to think that the arrival of English may have had an effect on the retention of older placenames in the northern area — after all, a voiced–voiceless stop distinction, the [tɽ] sequence, fricatives and words beginning with [ɽ-], are common in English.

It is now a problem as to whether or not the older names themselves are opaque — as they always are to people who now do not know that former language — or whether they are translatable using, for example, Ken Hale's dictionary of Linngithigh.[17] And such names that might have been translatable by old people whose own languages were Wik ones, but who were polyglots with some knowledge of northern Paman varieties, back in the 1970s, may now be quite untranslatable by their grandchildren who remain fluent only in one or more Wik varieties. There may now also be loss of translatability simply on grounds of sheer erosion of specialised lexical knowledge within languages that continue to be spoken.

For a while I thought that it followed from the above that the greater the historical stability of the relation between language variant and clan estate, the greater the homogeneity of the source-language for the site names in the estate, and thus, perhaps, the greater the semantic transparency of the site names, all other things being equal.

---

[14] Note also the dorso-velar fricative in [ɣp], citation form /kap/, 'falling down'. Northern Paman languages include at least some with fricatives.

[15] This raises the interesting question as to why Wik-Mungkan is so unlike most other Aboriginal languages in this way. That is, the usual pattern is that where there is no phonemic voicing contrast, one may expect a tendency not only towards complementary distribution of voiced and voiceless allophones but also towards free variation between them under certain conditions (for example, word-initially, intervocalically). But in Wik-Mungkan /pam/ in isolation is always *[pam(a)]* never **[bam(a)]*, /wik/ in isolation is never **/wig/*, and so on. I would like to advance the idea that Wik-Mungkan's strongly voiceless and fortis stops may be a relational and areal phenomenon — that is, interaction with speakers of languages which do have the contrast may have pushed Wik-Mungkan towards avoiding its superficial appearance. Note also that where such a contrast has evolved anywhere in Cape York Peninsula, as far as I am aware it is *always* the voiced series that is innovative, deriving from intervocalic nasals, single stops following long vowels, and so on, and the voiceless series retains its basic 'original' character, unless lenited to corresponding semivowels and so on.

[16] See Smith and Johnson (2000).

[17] Hale (1997).

I am now not so sure about the stability of the language/estate relationship being likely to result in greater transparency of site names, all other things being equal. A long period of great stability would probably result in a continual reduction of many site names to the status of being opaque. Thus opacity of site names can be treated as a likely index of *in situ* time depth of the language variety of the site name, but not necessarily of that of the language currently prevailing in either the identity or the daily usage of those who claim the sites. Where the preceding language and an incoming one have the same phonological character, the difficulty of sorting out what is old versus what is merely (until recently) foreign is a very great one.

6 TRANSLATABLE MODIFIERS OF PLACENAMES

In the Wik area many placenames contain a clearly translatable reference to the feature that is being named. Prominent among these names are those translatable as 'point + X', 'creek + X', 'river + X', 'well + X' and 'swamp + X'.[18] Use of these feature generics may be optional in some cases, but where they specify a subpart or focal point of a site they have a clear and useful function. The use of preposed generics is not unique to placenames. The Wik languages are typically very consistent in classifying everything by the use of generics such as those translatable as 'meat animal', 'vegetable food', 'tree', 'person' and so on. In a few cases the site feature generic is post-posed, as in *Wayingk Thiikanen* 'Wayingk Island', actually a raised scrub on the island *Wayingk*, and *Merrek Ngamp* 'Merrek River'. The latter is a mangrove-lined waterway joining the main Kirke River channel to the large Kirke estuary, and is quite short. This device is distinct from the naming of main rivers. Modest stretches of such rivers may be named, typically from the mouth and upstream for a few kilometres, using the name of a site at the mouth and modifying it by preposing, not post-posing, a word for 'river', as in *Ngamp Thew-en* 'Lower Kirke River'.[19]

Other distinctions between places may be made using post-posed topographical modifiers, examples of which I provide below. These particular systems clearly reflect a notion of a core referent for the site name, plus a set of one or more derivative or pendant applications of the site name to nearby features. The question as to why one feature is a core referent and another is a derivative one seems mainly to be answered by the use of such sites for different camping and foraging purposes: a base camp is likely to be a core site name referent, while an associated day-shade area or 'dinner camp' may have a name derivative from that of the base

---

[18] There are also cases of 'poison ground + X' and a very large number of names which specify that a site is a totemic centre ('Story Place'), the term for the latter usually being postposed as in *Wiykath Eemoeth* ('Child Story Place', Wik-Ngathan) or *Pach Aw* ('Star Story Place', Wik-Mungkan), although a small number of totemic site names are compounds with *aw-* as the first stem, for example, *Awe-chereng* 'Greed Story Place' (*cher* = greed), but this is more likely to refer to 'plenty' in cases where, as in the instance of *Awe-chereng*, the language of the site's estate has /aw/ for 'plenty' and not for 'totemic centre'. Many totemic site names end in *-nhi(i)n* or *-nyi(i)in*, depending on language variety and degree of archaism, which is fairly evidently a reflex of the verb 'to sit'. This is not the place to enter into a full analysis of Wik placenames, however, so their other structural and semantic aspects await further study.

[19] For much of the early and mid-20th century the lower Kirke was called *Yu'engk* after a site nearby, and for a time this became the 'Yonka River' of mission records. This usage has declined, the reasons being obscure. It may be relevant that the site *Yu'engk* is now some distance from the river mouth, which could well have moved over recent decades.

camp. In some cases the core referent is a totemic centre rather than a habitation site *per se*.

The approach to this aspect of site naming rests on a logic derived from the specific geography involved. For example, along the coast north of the Kirke River there is a pericoastal dune system bearing Indo-Malaysian aquifer vine thickets which offer a superabundance of food and other resources such as wells, plus a sandy environment for clean camping, and huge shade trees, the latter doubling as rain shelters beneath which wet season huts were formerly constructed. The core site name referents along this part of the coast generally occur in this woodland environment, and a core site name (X) in this environment may be further specified as 'X on-top'. Adjacent to the east is a flood plain with grasslands and mainly ephemeral water bodies. The core site X is thus physically above both environments to its east and west, hence the modifier 'on-top'. If there are salt flats to the east of the site, the inland extension of the core site name may be 'X saltpan'. A day-shade on the beach to the west of a core site, usually a mature Casuarina tree or two, will often be called 'X beach', and a swamp to the east of the core site may be called 'X swamp'. North of the Kirke, again, there may be two 'on-top' core sites, X and Y, from which only a single beach day shade would be visited, and the latter may be selectively named 'X beach'.[20]

However, south of Kirke River, where the woodland strip is narrow and closer to the beach, and there are swamps but no flood plains to the immediate east, the core site may be called 'X (swamp)'[21] and 'X beach' is its western extension. Where the Kirke estuary reaches right to the narrow coastal woodland the core site may be specified as 'X (outside)', i.e. on the edge of an open expanse of water,[22] and its western extension may be called 'X beach'.

Just north of Knox River we see a different logic, although one still consistent with the influence of patterns of economy and physiography on nomenclature. There, the dune woodland occurs only in patches and at a considerable walking distance from the beach. In this environment the core sites are in the woodland patches and near lagoons, and the relevant beach sites to their west are called 'X road', a reflection of their status as the end points of substantial forays from base camps.

A single site name may apply to two opposed banks of a watercourse. The distinct places may be specified using cardinal directions, 'X north side', 'X south side'. The same directional distinction may also be made between sub-areas within a single large named feature such as the northern head of the mouth of the Kirke River, which is subdivided into *Thew-en Wunkenh Kungkem* and *Thew-en Wunkenh Thiipem* ('*Thew-en* north side, *Thew-en* south side'). These are both on the same side of the river, however.

A minor feature such as a small swamp or lesser watercourse which is near a focal site may be named using the core site name plus either a cardinal directional term such as 'X east side', or may be specified using a relative directional and distance-indicating term such as

---

[20] Note that all these modifiers and the head word remain uninflected.

[21] Parentheses indicate optionality.

[22] The European section of Aurukun village was for many years referred to also as 'outside', but for different reasons. This term was also once standard usage for references to the world beyond the Wik domain. Cairns, for example, was *yuun* (outside).

Wik-Ngathan *um-wetherr*, literally 'facing-midway', hence 'X *um-wetherr*', 'midway or part-way between X and the topic location'.[23]

There are cases where two adjacent places are referred to using the same core site name plus modifier(s) but where the relevant sites fall into different clan estates. Thus estate definitions are at essence place-based rather than named-site-based.

7 DEATH AND SITE NAMES

How do people refer to a place when the name is tabooed? Actually this occurs as a reasonably constant factor, for example when a placename sounds like or is the same as that of a recently deceased person, or in extreme cases is a place merely strongly associated with a person who has recently died.

A few years ago a senior Wik man died in a dinghy accident near Aurukun. His outstation was known as Stony Crossing, the preferred name, and I doubt that many local people now could give an Aboriginal name for the place. Anyhow, when I got there soon afterwards, some people at least were avoiding not only his first name but also any reference to Stony Crossing as such. Instead they referred to the outstation in Wik-Mungkan, the lingua franca, as *nhamperring*. *Nhamp* is 'name', and *nhamperriy* was the substitute term for the name of a recently deceased female. Males were *kootemat*.[24] I had not come across *nhamperring* before but -*rr* is a politeness suffix occasionally used on personal pronouns in Wik-Ngathan, (for example, *nhunterr* second person singular nominative), -*ng* is locative case in Wik-Mungkan, and my interpretation of *nhamperring* is that it is probably a bereavement cover-term for unmentionable placenames.

In a reverse kind of way, placenames are good substitutes for people, if one adds something distinctive, for example, 'head fell at [placename]', i.e. an unnamed or unnameable person who died at a certain place.

Places and people always imply each other in such a culture. If one has to avoid a site name because it resembles the name of a recently deceased person, or conjures up that person, it may be that one of the advantages of having site names that are translatable, or if opaque then substitutable, would be that one is able to refer to the site and yet avoid giving offence.

8 SAMPLES OF PATTERN

In spite of all the caveats entered into earlier in this paper, as a quick and rough exercise I have looked at site name translatability for three non-contiguous areas in the Wik region south of the Archer River, namely estate 29 of the Small Archer River, estate 1 on the mouth of Love River, and estate 6 just inland from Cape Keerweer. This is not a comprehensive sample because no southern or truly inland estates were examined, and as a proportion of all site names recorded between the Embley and Edward Rivers the ones examined here probably only amount to about 5 per cent. These figures are also qualified by the fact that I have not

---

[23] Some 34 site names in the database begin with *Um* +, but with primary stress being placed on the following lexeme these are not in such combinations single phonological words. In the case of *um-wetherr* primary word stress is on the *um-*.

[24] These usages are now archaic, being both replaced in the 20th century by *thaapich*, which functions very much like *kunmanara*, *kumunjayi* or *kwementyaye* etc. of Central Australia, and is a loan word from a northern Paman language.

had time to use the relevant published dictionaries and unpublished data to do any checking of existing translations in the site database. The translatability figures would probably rise if I did so, but in any case the figures came out like this:

ESTATE	TRANSLATABLE	OPAQUE	NOT SURE	ONE NAME TR., ONE NOT	TOTAL
29	12.0	18.0	0.0	0.0	30.0
1	37.0	57.0	7.0	0.0	101.0
6	18.0	27.0	3.0	3.0	51.0
average	22.3	34.0	3.3	1.0	60.6
%	36.8	56.1	5.4	1.6	99.9

REFERENCES

Hale, Kenneth, 1997, A Linngithigh vocabulary. In Darrell Tryon and Michael Walsh, eds, *Boundary Rider: essays in honour of Geoffrey O'Grady*, 209–246. Canberra: Pacific Linguistics.

MacKenzie, G., 1981, *Aurukun Diary*. Melbourne: The Aldersgate Press.

Smith, Ian and Steve Johnson, 2000, Kugu Nganhcara. In R.M.W. Dixon and B.J. Blake, eds, *Handbook of Australian Languages*, vol.5, 355–489. Melbourne: Oxford University Press.

Sutton, Peter, 1978, Wik: Aboriginal society, territory and language at Cape Keerweer, Cape York Peninsula, Australia. PhD thesis, University of Queensland.

— 1995, *Wik-Ngathan Dictionary*. Adelaide: Caitlin Press.

— 1997, *Wik Native Title: anthropological overview* [Word-processed 68pp.]. This has the following appendices:

Sutton, Peter (comp.), 1997, Appendix 1: Wik estates [33pp. and map].

Sutton, Peter (comp.), 1997, Appendix 2: Wik clans [68pp.].

Sutton, Peter, David Martin and John von Sturmer, 1997, Appendix 3: Supplementary Site Report [231pp.].

Sutton, Peter, David Martin and John von Sturmer, 1997, Appendix 4: Site Maps [12 GIS-generated A3 maps].

Sutton, Peter, 1997, Appendix 5: Languages [28pp.].

Martin, David and Peter Sutton, 1997, Appendix 6 [13 sample genealogies, genealogy database printouts].

Hale, Kenneth, 1997, Appendix 7: Hale Report [79pp.].

Cribb, Roger, 1997, Appendix 8: Archaeology of Cape York Peninsula and the Wik Claim Area [79pp.].

Hunter, Philip (with assistance by Rosalind Kidd and Regina Ganter), 1997, Appendix 9: Historical Document Extracts 1606 to 1970 [212pp.].

Sutton, P., D. Martin, J. von Sturmer, R. Cribb and A. Chase, 1990, *Aak: Aboriginal estates and clans between the Embley and Edward Rivers, Cape York Peninsula*. Adelaide: South Australian Museum. [Restricted desktop publication; word-processed and bound, 1000pp. x 25 copies]

Von Sturmer, J.R., 1978, The Wik Region: economy, territoriality and totemism in western Cape York Peninsula, north Queensland. PhD thesis, University of Queensland.

7   NAMES AND NAMING: SPEAKING FORMS INTO PLACE

Franca Tamisari

In 1946 Donald Thomson (1946:157) noted that 'very little has been recorded of the derivation and use of personal names among the Australian Aborigines'. Despite the significance that Australian Indigenous people in general give to the meaning and use of proper names of people and places and to the action of naming in cosmogonic events, with some exceptions this neglect continues today.[1] Thomson explains this dearth of research by the secrecy and the sacredness of proper names and toponyms which derive from their ancestral associations and by the rules of avoiding names in everyday life. However, like Keith Basso (1988:103), I am inclined to suggest that this neglect is the reflection of the prevailing preoccupation of anthropologists and linguists with the semantico-referential meanings and functions of names and language rather than with the culturally shared notions and images all names evoke, provoke and embody in the creative dialogue that people establish and continuously renew with country. As Heidegger argues for language (1971a:192–193): in order to explore the meaning and significance of Yolngu[2] names it is necessary to explain them more comprehensively than considering them as simple expressions of internal feelings and thoughts, as mere representations of reality, as vehicles by which people communicate, address each other, or in Levistraussian terms, as a means by which people classify the world and order themselves within it (see Lévi-Strauss 1966:161–190). If with Heidegger I suggest that 'language speaks' beyond expression and representation, my concern in this paper is to ethnographically explore what Yolngu names speak of, that is, what culturally shared images and notions they reveal while saying very little (Basso 1988:103). As in other regions of Indigenous Australia, Yolngu names — whether they be of individuals,

---

[1]   The following authors have given some attention to the significance of names: Hart (1930); Goodale (1980); Dussart (1988); Schebeck (1968); Stanner (1937); von Sturmer (1978); Williams (1986); and more recently Biddle (2000) and McKnight (1999).

[2]   North-east Arnhem Land extends approximately from Cape Stewart in the west to the Gulf of Carpentaria in the east and from the Wessel Islands in the north as far south as the Koolatong River north of Blue Mud Bay and includes the major settlements of Maningrida, Milingimbi, Ramingining, Gapuwiyak, Galuwin'ku and Yirrkala (see map). Yolngu (human being) is the term used by north-east Arnhem Landers to refer to themselves in relation to their western neighbours of central Arnhem Land. By extension the term Yolngu is used to mean 'Aboriginal or Black people' in contrast to Balanda, 'non-Aboriginal, White people'.

L. Hercus, F. Hodges and J. Simpson, eds, *The Land is a Map: placenames of Indigenous origin in Australia*, 87–102. Canberra: Pandanus Books in association with Pacific Linguistics, 2002.

groups, plants, animals, sacred or non-sacred objects such as cars, boats and dogs — are associated with the group's specific cosmogonic actions and the movements of Ancestral Beings who gave shape to the land, brought everything into existence at particular places, and bestowed countries and all phenomena upon specific groups of people. The general point I would like to make is that not only toponyms but all proper names are ideolocal (Casey 1996:26), that is, specific to place, as they derive from and refer to the unique cosmogonic events and actions performed by Ancestral Beings at particular places (cf. Morphy 1995:192).[3] The specificity of names is so precise that in many instances for knowledgeable men and women the proffering of a name discloses the named person's affiliation with and his/her potential rights and authority over specific places.

While the link between places, events and names has been repeatedly noted in relation to Australian Indigenous cosmologies (Stanner 1979a, 1979b; Munn 1970) and ancestral bestowal of and rights over country (Williams 1986), the interlocking of these elements, the shaping of places through ancestral events and the ways places and events are condensed in names need further detailed attention. If language speaks by itself of our being-in-the-world as Heidegger argues, I suggest that Yolngu names speak of places as events or happenings (cf. Casey 1996:27) and with places, of belonging, of self-identification with and ownership of landscape features, of rights and authority over country.

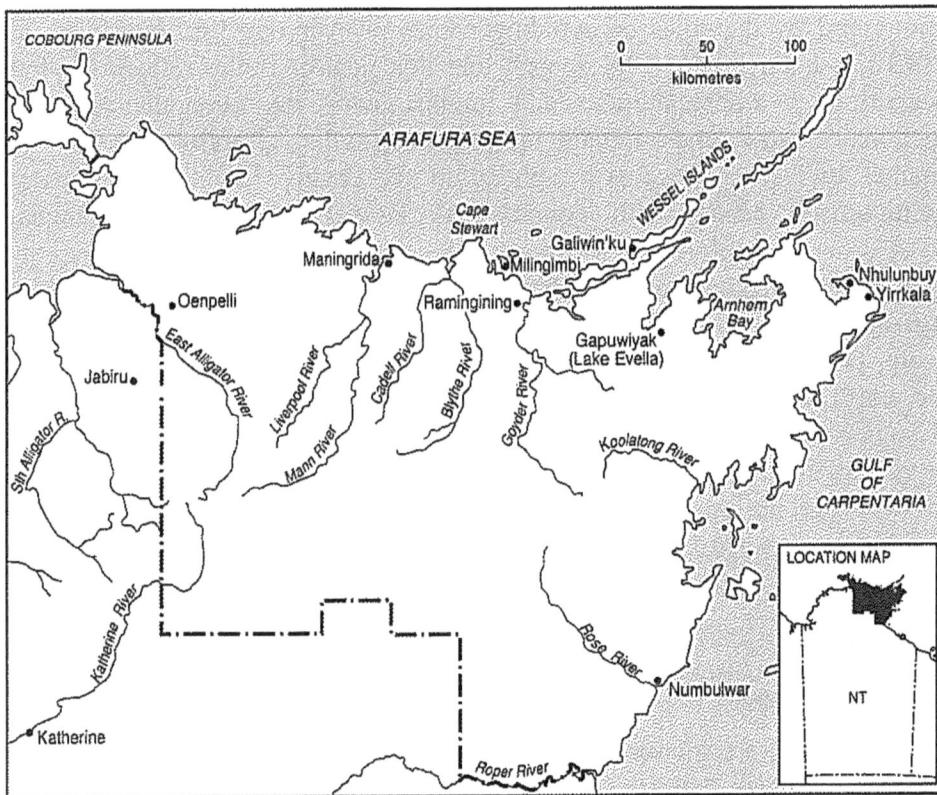

Map 1: North-east Arnhem Land

---

[3] Casey (1996:26) refers to places as ideolocal as a 'place is more an event than a thing to be assimilated to known categories. As an event, it is unique, ideolocal'.

1 'WHERE IS YOUR NAME FROM?'

The connection names establish between people and country is well illustrated by the expression 'to have a name' (*ya:ku-mirr*, literally name-having) which, in fact, underlies the link between people and their group's Ancestral Beings, stresses an association with a particular place and implies notion of belonging and owning country (cf. Williams 1986:42).[4]

Every Yolngu person is said 'to follow the father' (*ba:paw malthun*) and is born and is a member of his or her father's group. Identification with one's patrilineal country (*ngaraka* literally 'bone country') is principally expressed through 'bone names' (*likan* and *bundurr*, literally 'elbow' and 'knee') which are inherited, learnt and given during the many stages of one's gradual acquisition of knowledge from one's patrilineal relatives (both biological and classificatory). I refer to 'elbow' and 'knee names' inclusively as 'bone names' rather than 'power names' as is common in the literature, in order to maintain some of the Yolngu associations and imagery. As Yolngu talk about one's patrilineal land as 'bone country' — the country one's ancestral substance comes from before birth and returns to after death — elbow and knee names may be seen as at once conveying the sharing of bony substance with patrilineal country as well as, perhaps, comparing the main joints which give movement to the body to the dynamic *articulation* of people, Ancestral Beings and places (cf. Keen 1995:509–512). Before being conceived, a child announces his/her imminent conception to the father-to-be by appearing to him in the form, 'shadow' or 'image' (*mali'* or *wunguli'*) of an animal, phenomenon or plant which is believed to be the embodiment of an Ancestral Being (*wangarr*). This embodiment belongs to the father's group or to one of the father's patrilineally related groups. Such conception signs consist of unusual events happening at a place in the country of the father or of a patrilineally related group. One of the names given to the child often refers to such a place (*dhawal*) where the child first manifested him/herself. Just as the Yolngu term for 'to be born' translates as 'to think of (such) a place' (*dhawal-guyanga*, literally 'place of origin-think') and thus 'associates the individual with ancestral creativity' (Morphy 1995:197; see also Williams 1986:88) at a particular named place through the conception sign, so the same term is used for 'to die'. The expression 'his place of origin was silenced' (*dhawal mukthurra nhangu*, literally 'his place [was] silenced'), not only links the death of a person with a place of conception or origin but also stresses the 'interanimation' (Basso 1996:55) of place, ancestral being and person. One of the purposes of the mortuary ceremony is to send the ancestral component of a person back to his/her bone country: after a person's death until the concluding 'cleansing' ceremony (*bukulup*, literally 'face washing'), the bone country and/or the place of origin (*dhawal*) of the deceased becomes barren and his house, possessions as well as the places where he would spend most of his/her time while living, cannot be entered, frequented or used.[5]

In addition to bone names one also has several personal names (*ya:ku*), which are given at birth by several patrilineal and matrilineal relatives such as one's FF (*ma:ri-mu*), F (*ba:pa*),

---

[4] The term *wakingu* is used to indicate that a person does not belong to anyone and anywhere, thus a person without relatives who care and look after him/her. The expression *ya:kumiriw*, literally 'name-having-PRIV' is used to address and to refer to a person whose name cannot be pronounced because it is the same or it sounds like the name of a recently deceased person which cannot be proffered for a period of time from one to two years after his/her death (cf. Nash and Simpson 1981).

[5] *Dhawal* conveys several meanings at once. It refers to 'breath' and 'feeling' as well as to 'the place at which a child to be born manifest itself' through the conception sign.

FZ (*mukul ba:pa*), MM and MMB (*ma:ri*).[6] Given that several relatives may give a newborn more than one name each, a person may have from four to fifteen proper names which thus associate him or her with particular places over which the name-givers have strong or potential rights.

Generally speaking, not only toponyms but all proper names can be said to be ideolocal, that is, specific to particular places, as they derive from and refer to the unique cosmogonic events and actions performed by Ancestral Beings at particular places where they shaped the land through bodily processes. That all proper names of groups and individuals refer to specific events at particular places is well illustrated by the fact that they are said to be derived FROM (*puy*) a place (*wa:ngapuy ya:ku*, literally, 'place-FROM names') or as being IN (*ngur*) a place (*wa:ngangur ya:ku* literally 'place-IN names'). Thus, when inquiring about a person's name one asks 'what is your name from?' (*nhapuy nhe ya:ku*) or 'what is your name in or at?' (*nhangur nhe ya:ku*).[7] To these questions people answer by saying that their names are FROM natural phenomena such as the wind, clouds, and rain (*watapuy, wukunpuy, waltjanpuy*) or IN Ancestral Beings such as birds, trees, waters and spears (*warrakanngur, dharpangur, gapungur, and garangur*). While the first answer to this question can just stop at indicating the generic name of a phenomenon or a placename on a particular country such as 'from the cloud' or 'from the sea water', the clouds and the seagull are known with the group's specific names and are associated to particular ancestral bodily actions, an integral part of the ancestral events which gave shape to the country as it is today.

The following examples illustrate how names refer to specific places where particular ancestral events and often particular bodily actions took place at specific times of the year and locations.[8]

'name from (a) bird' (*warrakangpuy ya:ku*):

1) The female proper name (*) 'beak of a species of seagull' refers to the feeding of the young seagulls just after the breeding season on the shores at a particular beach belonging to the Djambarrpuyngu clan of the Dhuwa moiety.

'name from (the) clouds' (*wukunpuy ya:ku*):

2) The male proper name (*) 'long black clouds, heavy with rain on the horizon' announces the beginning of the monsoon season as observed from another shore of the Djambarrpuyngu clan.

'names from (the) rain' (*waltjanpuy ya:ku*):

---

[6] Father's father (FF); father (F); father's sister (FZ); mother's mother (MM); mother's mother's brother (MMB).

[7] What is your name? is rendered with: '*yol nhe ya:ku?*' literally 'who you name?'

[8] Please note that names are glossed with their ancestral references and not translated literally. Given the general secret/sacred nature of the different types of Yolngu names, the association they evoke, the power they summon, the authority they confer and the emotive response they might arouse when they are pronounced, throughout this paper I will omit them by inserting an asterisk (*) and by glossing them in English.

3) The female name (*) 'small rain' refers to the drizzle that follows the foggy mornings at the end of the monsoon in Wangurri country of the Yirritja moiety.

4) The male name (*) 'rain which makes one cold' brings the cold season from the east and makes everything grow on Djambarrpuyngu country.

name derived from ancestral imprinting action on the land:

5) The female name (*) 'from the head' refers to the hill shaped by Shark ancestor hitting the ground at a particular place on Djambarrpuyngu country.

name derived from ancestral externalising action (see below):

6) The male name of (*) 'low, small white triangle-shaped clouds' refers to the clouds that were spurted by the Serpent over the inlet where he rests in the deep waters of the Gupapuyngu-Birrkili group of the Yirritja moiety.

name derived from a conception sign:

7) The female name (*) 'water corm' refers to the unusually large corms through which the 'shadow' of a baby girl first appeared and thus was 'found' by her father to be at a place patrilineally related to the Wangurri group of the Yirritja moiety.[9]

Despite the fact that 'bone' and personal names and their ancestral associations with particular places and events seem to be known by most people of mature age, they are seldom used and in certain contexts avoided altogether. In addition to the strictly observed avoidance of pronouncing the proper names of certain relatives, such as the names of a man's sister (*midiku'*) and of a woman's son-in-law (*gurrung*), proper names of recently deceased people cannot be pronounced aloud in the same way that their photographs cannot be shown in most parts of Indigenous Australia. If in the first instance the speaking of a name is said to cause swelling of the speaker's lips, in the second the recently deceased person is said to be summoned among the living as a duplicate of his/her body (*mokuy*, cf. Warner 1958:446) or 'flesh soul' as I prefer to refer to it. Pronouncing the name of a recently deceased person thus disrupts the process whereby the idiosyncratic component of an individual (*mokuy*) needs to be dispersed in order to send his/her ancestral component or 'bone soul' (*birrimbirr*) to merge with his or her land of origin or 'bone-country'.[10] Whereas bone names are mainly proffered in the songs and chanted in long lists during ceremonial climaxes, personal names are seldom used as terms of address in everyday life. For everyday social interaction kinship terms,

---

9 As Ancestral Beings and associated phenomena are group-specific, their names are considered the possession of the clan. Despite the fact that most Ancestral Beings cross the countries of several groups, during their travels they change language so that the proper names derived from their actions are different.

10 Onomatopoeic words or as Yolngu say words that are 'close' (*galki*) in sound to the deceased names are also dropped from the current spoken vocabulary and substituted by alternative nouns. In the light of this it is interesting to note that the sound itself may evoke ancestral association and certainly has the power of summoning the 'flesh soul' of the individual which appears like the deceased in all aspects but it is endowed with supernatural and often malevolent powers. On a few occasions, and mainly for my own benefit, the name of the deceased was whispered to me. Similarly photographs were occasionally shown to me underneath a sheet, inside a skirt pocket or sheltered by the hand. A discussion that explores the way in which sounds, like images, can summon someone to presence cannot be pursued here.

subsection or so-called skin-names (*ma:lk*) and nicknames (*wakal ya:ku*) are preferred and widely used.[11]

While the 'kind of oneness' as Stanner (1979a:35) remarks, which includes 'notions of the body, spirit, ghost, shadow, name, spirit-site and totem', has been repeatedly noted in the Australian Indigenous lifeworld, the interanimation of these elements, the shaping of places through ancestral events and the act of naming needs to be explored in some detail. If language speaks by itself of our being-in-the-world, as Heidegger argues, I suggest that Yolngu names speak of places, and more precisely they speak of the 'corporeal connection' (Stanner 1979b:135) between ancestral bodies and country and of the sociophysical relationship of people with place which is at the basis of people's self-identification, ownership of and rights over land.

In what follows I will limit myself to explore the corporeal connection between ancestral bodies and country, how names and the act of naming are fused with processes of bodily transformation and, conversely, how these bodily transformations involve names and the act of naming. In order to explore the interpenetration of ancestral bodies, places and names, how places are shaped by names and how names embody places, it is necessary to consider Yolngu cosmogonic events, to which I now turn.

2 NAMING PLACES AND PLACING NAMES

As elsewhere in Indigenous Australia (cf. Strehlow 1947), in Yolngu cosmology, land has always existed but it was originally empty, shapeless and nameless. The landscape people inhabit was shaped into its present form by Ancestral Beings (*wangarr*) who roamed through the sky, above and below the surface of the earth, in the depths and shallows of the sea and along the rivers, thus shaping everything into existence in the geographic and climatic environment through cosmogonic movements and actions. The landscape was fashioned and oriented, transformed and moulded by the ancestral bodies through several bodily processes. These have been described as: metamorphosis, imprinting, externalisation (Munn 1970), placement and orientation.[12] Similar to the way that Munn (1970:142; 1973:132) noted the association between name, song and visible mark, which are implied in the term *yirdi* and *yini* in Warlpiri and Pitjantjatjara languages respectively, the processes of transformation, and especially imprinting, metamorphosis and externalisation, are intrinsically connected with names and the action of naming. In talking of these transformation processes and dynamics which link naming and action, language and movement, I will employ the term morphogenesis in order to avoid the misleading term of creation. As Williams points out (1986:28) quoting Stanner, in Yolngu cosmology there is no creation ex-nihilo but rather processes of generating, shaping forms into presence. At certain stages of their journeys

---

[11] It must however be noted that personal names are often used as terms of address among members of a family who reside together (cf. Stanner 1937:302). As terms of address, in this context, personal names are often shortened thus further reflecting the intimacy of these relationships. People have also 'White names' (*balanda ya:ku*) which, together with one Yolngu personal name, are mainly but not only used for bureaucratic purposes and in interacting with balanda (White) people (see Biddle 2000).

[12] To the cosmogonic processes identified by Munn (1970), Keen adds 'placement' (1978:45) or 'turning' (Keen 1994:44–45) by which the ancestors or objects used by them are transformed into a landscape feature. To this list I added 'orientation' as landscapes are not only perceived as transformations of ancestral actions but as being ordered and positioned along the path by the direction and movement of ancestral bodies (Tamisari 1998:254).

Ancestral Beings metamorphosed, imprinted and externalised whole or part of their bodies into topographic features such as a stone, a hill, the trees along a river, clouds, lightning, animals and plants. Ochre quarries are said to be ancestral faeces, a yellow clay which turns red after cooking is their blood, and clouds were formed by the water which spurted from their mouths (name no. 6; see discussion below). During their wanderings over the land, sea and sky, Ancestral Beings and the objects used by them in a certain manner, left a mark behind. Where a tree was felled to collect wild honey a long depression is to be seen today, where an ancestral bee struck its proboscis a round hole is perfectly cut into the surface of a flat rock. In Yolngu languages, these processes of shaping or imprinting are referred to as 'hitting the country' (*wa:nga buma*). Whenever an Ancestral Being hit the ground with a part of its body a mark remained to manifest its passage and action and to embody its power. Landscape features thus shaped are not only transformations of ancestral body parts but also the **embodiment of particular actions** which identify unique cosmogonic events (cf. Munn 1996:457). It is the uniqueness of these 'action features' that is condensed in the toponyms and other proper names given by the ancestors in the act of shaping the land and bestowing tracts of country upon people.[13] Thus names may be seen as localising events, manifesting actions and congealing movement. Here I use the word 'manifest' in its primary sense of 'palpable' which interestingly derives from the etymology of '*manus*' (hand) and '*festus*' (struck) as in 'of-fend' and 'de-fend'. In this sense the term manifestation is apposite in describing Yolngu cosmogony as revealing the world through form by striking the land with parts of the ancestors' bodies (see name no. 5).

I will now turn my attention to explore how this shaping of the ground involves the act of naming and, conversely, how names thus embody ancestral morphogenetic processes. I suggest that names do not only refer directly or indirectly to specific places stressing their bestowal upon particular groups, but, once again, they reveal the corporeal connection between people, places and ancestors. As I discussed elsewhere, like designs, songs and dances, names are manifestations, and as such they are visible marks or, as Yolngu would say, the footprints of the Ancestral Beings (cf. Tamisari 1998).

Each moiety has what may be referred to as a creation myth which is said to precede all other ancestral events. The Djang'kawu Sisters along the coast and the Wagilak Sisters inland are the first Ancestral Beings of the Dhuwa moiety. The Lany'tjung and Barama brothers are responsible for bringing into form Yirritja moiety countries, phenomena and people. Despite the fact that these 'first' myths may be distinguished from more clan-specific mythological stories of ancestral journeying (cf. Morphy 1990:313), they all share similar processes of morphogenesis which are characterised by the combination of naming, movement and action upon the land. Although the myth of the Djang'kawu Sisters belongs to the Liyagalawumirr clan of the Dhuwa moiety and concerns the shaping of the Dhuwa clans' landscapes, it is referred to and considered by all Yolngu people of both moieties as *the* first and most significant cosmogonic story. It was pointed out to me, more than once, that the Djang'kawu Sisters gave forms and named everything, both Dhuwa and Yirritja. To prove this claim some people went on to say that the Sisters' language, now sung in the associated songs, is a mixture of Dhuwa and Yirritja languages.[14]

---

[13] Djon Mundine (2000:100) uses this phrase to refer to the visual embodiment of ancestral events in Yolngu bark painting.

[14] On being questioned on this matter, a Yirritja man answered by recognising the role of Latjung and Barrama as the first Yirritja Ancestral Beings but also confirmed the above claims with the following somewhat puzzling statement: 'The Yirritja ancestors took the names from the two Sisters wherever they went. They

In the following brief introduction to the Djang'kawu song cycles given by Charles Manydjarri, one of the eldest owners of the Djang'kawu ritual complex, he focused on the Sisters' power of morphogenesis through naming. In his words:

> The first songs talk about where the two Sisters came from and where they are directed to. They describe how rich they are in sacred objects (*madayin*) and how they strode by using their yam sticks like walking sticks (*ngal'gam*). The first bird they saw was a black-tailed cockatoo sitting on a special tree that they called (*).[15] They gave that tree a 'elbow-name' (*likan*), it is an ancestral manifestation (*malagatj*). They named all creatures, birds and places. Where they planted (*nhirrpan*) their sticks springs of fresh water (*milminydjarrk*) sprung out of the ground and a sacred tree stood (*dha:rra*). They created everything. They created children and clans, they gave names to the land and gave the land to the people. (Charles Manydjarri: extract of conversation recorded by the author in Milingimbi, field tapes 1991)

Emphasis is given to walking and planting the yam stick in the ground; and these actions form the focus of the song text and the rhythm as well as of the movements of the public dance of the Djang'kawu Sisters.[16] Wherever they planted their walking or yam stick, a tree grew and/or a freshwater spring burst through the ground. Wherever these activities took place, the Sisters gave elbow names (*likan*) to the land and other phenomena which were then bestowed upon the human beings that the Sisters also generated in those places.[17] Along their journeying their acts were the same but their languages changed and thus the names they bestowed to places were different. Places are named as they are shaped, names are placed as they shape the landscape. Just as the Sisters imprinted the ground by making a freshwater spring flow, they also named the country where the spring is now located. Also, the walking stick planted in the ground metamorphosed itself into a tree which was given an 'elbow name'. Wherever they shaped the ground by imprinting it (*nhirrpan*) or by planting the stick which metamorphosed into a tree, the Sisters named the place. Places are shaped by 'piercing' (*nhirrpan*) the ground, and by bringing the water there (*dha:rra*) and by making the walking stick 'stand' (*dha:rra*) as a tree. The piercing of the ground contains a sexual component. In piercing the ground the actions of the two Sisters gave rise to humans who, like the trees and water springs, are thus given shape and 'stand' (*dha:rra*) to populate places.

Perhaps it is not a coincidence that proper names are 'pierced' or given (*nhirrpan*) and are said 'to stand' or belong (*dha:rra*) to people and to places. 'This is a person's name' is the translation of *Yolnguwal dhuwal ya:ku ga **dha:rra*** (literally, 'at a person this name **stands**') and 'I gave her this name' is rendered with *dhuwal nhanukal ngarra **nirrpanha*** (literally 'this name to her I pierced or planted'). Given that names are 'planted' or 'pierced' (*nhirrpan*) and

---

did not steal them, they got up onto the Sisters' skin and took them with their bodies. They did not steal them but the Sisters did not give them.'

[15] Male proper name held by a Djambarrpuyngu man, sister's daughter's son (ZDS) of the Liyagalawumirr clan, owner of the Djang'kawu song cycle.

[16] The clan songs recounting the Sisters' journeys are different in musical form from other clan songs of the same moiety as they are not performed with a drone pipe but only to the accompaniment of the sound of the clapping sticks. The result is an almost hypnotic music which echoes the regular pacing rhythm of walking. Similar to the unique 'walking' quality of the music, dancers shuffle their feet not in the characteristic 'skipping' style but in a slow forward and backward dragging movement.

[17] The same images of planting digging sticks which then metamorphose into trees are also to be found in the Yirritja moiety morphogenetic stories. Wherever any Ancestral Beings speared the ground in Yirritja countries, an ironwood tree (*maypiny*, generic) now stands.

'stand' (*dha:rra*) at places or with people, it might even be suggested that, as the shaping of the ground involves the act of naming and, conversely, names imply and embody actions of imprinting and metamorphosis, a person, an object or a phenomenon is not the bearer of a name, but its embodiment.

It is here pertinent to note that Yolngu language reflects this 'indissoluble connection' between name and person in what linguists have referred to as 'the grammar of inalienability', namely 'a permanent and inherent association between the possessor and the possessed' (Chappell & McGregor 1996:4). In relation to proper names it should thus be noted that the personal pronoun rather than the possessive is used; thus 'my name is Franca' is rendered as 'I name Franca' (*ngarra ya:ku Franca*). As Stanner (1937:301) insightfully points out, the name does not only relate to one's personality 'as the shadow or image does to the sentient body', but 'is like an intimate part of the body, with which another person does not take liberties' (Stanner 1979a:25).

The intrinsic connection between naming and the ancestral processes of morphogenesis through movement and action is further elaborated and will be more clearly explained by the meaning and practice of *guykthun*, a term I gloss with 'spurting' or 'vomiting'. This notion, to which I now turn, requires special attention as it elucidates the link between naming, names and the transformative bodily process of externalisation.

3 THE ACT OF *GUYKTHUN*

The term *guykthun* is a recurrent action through which several Ancestral Beings of both moieties are said to have given shape to the clouds.[18] The following is a simplified and abbreviated extract from a tape transcription of Lalangbuy's explanation of the story of how the Rainbow Serpent shaped particular clouds by the action of 'the spurting of words' (*guykthun*).

> I am going to sing to you what you can see there and what a Yolngu told us when we took over that snake, a water python in that inlet called (*). There at (*), the snake 'spurted words' (*guykthun*) and (then) went back home. It is here that your snake is asleep. This snake is a water python, a (*) snake. There, maybe, in his shelter he spilled sacred words towards the sea water and there stood (*dha:rra*) low white clouds (coming) from his spurting (*guykthun*). Just those clouds you might have seen in the paintings, they stand over there, they are called (*) and (*). They are formed (*mali'bakthuna*) from him, there, just from that spurting of the snake. And then the snake goes back, over there, to his home, he returns, he proceeds swaying his tail. (Lalangbuy: extract of conversation recorded by the author in Milingimbi, field tapes 1991)

---

[18] The expression *guykthun* has several meanings in different contexts. While these are important to understand the complexity of this notion, they can only be mentioned very briefly here. 1) *guykthun* can be glossed as 'cursing' in that the 'spilling' of words makes someone or something sacred or tabooed until the curse is lifted once an agreement is reached by the parties involved; 2) in preparation for circumcision the boy's own group's design is painted on his chest and water mixed with clay is sprayed on his face (*buku guykthun* literally 'face spraying'); with this act the boy becomes the recognised owner of the manifestation of the Ancestral Being(s) painted on his chest; 3) *Buku guykthun* is a synonym of *wama:rrkanhe*, an expression which is usually but not only pronounced in complimenting a singer or dancer's virtuosity in performance. In this case *guykthun* may be glossed as 'face which has been sprayed with appreciation' (see Tamisari 2000 for a detailed account of what I refer to as the 'curse of compliments'). In everyday language, the term *gakthun* means 'to vomit'.

It is by spurting sea water (implied in the word *guykthun*) and by the 'spilling of words' (also *guykthun*) that the Serpent gave shape to the clouds.[19] It is in fact by a combination of the regurgitation of water and the 'spilling of names' that the clouds are said to have been given their shape and names. In Lalangbuy's explanation quoted above, I translated *mali'bakthuna* as 'formed'. It is, however, important to specify that the verb is composed of the word *mali'* meaning 'shadow', 'image' or 'semblance' and *bakthun* which may be translated as 'to break off', 'finish off'. I suggest that this term stresses the creation process of making, shaping the likeness of clouds through a process of 'breaking off' from or externalising a body part of the ancestor.[20] This act is not only performed by naming but comes about also by the embodiment of the Serpent's morphogenetic actions through externalisation. The clouds thus shaped can be said to be a manifestation of the Serpent through words and actions. These clouds are sacred and, in paintings, they are usually represented together with the Serpent in the form of triangles. It follows that the names of these clouds embody this manifestation and with it the morphogenetic power of the Ancestral Being. The name that constitutes these clouds through this particular action is thus considered a 'big name' (*yindi ya:ku*). The 'spurting of water' is not only a metaphor for uttering words but the means by which the manifestation of the Serpent's body through spurting is externalised and becomes forms.

What I would like to stress here is that the act of naming is made powerful and performative through externalisation and suggests a complex series of transformations in which body, language and place constitute each other. The ancestral externalisation of power in action and movement, therefore, gives words their morphogenetic power which is expressed in the term *guykthun*. Similarly, the term is also used in other contexts in which words acquire specific potency and have particular effects. In the several contexts in which the term *guykthun* is used, for instance as in cursing (see fn 18), the words pronounced by particular actors are not only performative as defined by Austin (1962), but they are better described as a poiesis, a disclosing, a bringing into presence.[21]

4 THE KINETIC QUALITIES OF NAMES

Having considered how the act of naming and ancestral bodily actions give shape to places, I would like to briefly consider the ways in which names condense these actions in describing their topographic and kinetic characteristics. For this purpose I will refer to four names of a place at which ancestral fresh water first appeared to begin its journey across the country of a Yirritja group. The names of the place where the water bubbled up from inside the ground refer to specific aspects of this event, they are:

---

[19] Other Ancestral Beings of both moieties are said to have created natural phenomena such as clouds, and/or places through 'the spurting of water and words'. One such example is a Sea Snake of a Dhuwa group.

[20] See Morton's (1987:108) discussion of a similar Arrernte term. Another Yolngu term, (*), refers to the action of 'biting' (literally 'lock in mouth') as well as to naming. Another expression is (*), literally, 'swim-bladder-speak'. By throwing a part of its body onto the beach a fish ancestor of one clan of the Yirritja moiety named and shaped landscape features such as trees in the form of the organ that had been externalised. The terms for these actions are omitted as they are now part of names of places thus shaped.

[21] Austin defines performatives as 'doing something as opposed to just saying something' (1962:133), however stressing that 'it is not to perform an act in some specially physical way, other than in so far as it involves, when verbal, the making of movement of vocal organs' (1962:134).

1) (*): this name renders as 'to swell up to shake and tremble' and refers not only to the force and impetus with which the ancestral water came up but also to the effect produced on the surface of the ground;

2) (*): evokes the sound produced by the water rushing to the surface of the opening;

3) (*): this name can be translated as 'fresh water pandanus-LOC' and provides an image of the surrounding landscape where the ancestral action took place;

4) (*): this name refers to an ancestral fish which, with the water, emerged from the opening in the ground.

These four names evoke rather than describe particular details of an action. Name number one evokes the kinetic energy of the water pressing from underneath the ground; number two is an onomatopoeic name which suggestively reproduces the sound of gushing water; name number three describes the adjacent environment to the action; and name number four stresses the direction of the action from inside to outside and introduces another Ancestral Being which, transported by the water, will cross the length of this group's country shaping and naming the country. Here I would like to focus on name number four, a name that evokes movement rather than sound. It is in reference to this aspect that I speak of 'kinesiopoeic' names and nouns: a neologism I propose for those names and nouns that evoke the movement associated with the thing or action designated. In the context of ancestral travelling over long distances, these kinesiopoeic names highlight the significance of movement that connects places along the route of a particular journey. The sites at which Ancestral Beings have emerged and/or stopped to shape the land are connected by paths marked by the ancestral body in movement. I suggest that these kinesiopoeic nouns (which may also be names of people and places) convey the incessant movement of what may be referred to as the 'site-path flow' (Munn 1973:137) of ancestral journeying. These nouns capture and reproduce either a linear motion that traces a path between places or evokes a localised bodily movement at a particular place. Yolngu song texts are replete with these nouns. Here it is important to note that these nouns are often but not always associated with and performed as dance movements.

Some examples of names and nouns that evoke linear and localised movement are listed below:

linear movement

(*): also a proper male name, this noun evokes the water displaced by a fish ancestor at a particular site. As the fish advances towards the land, the water surrounding him forms a wave which precedes him to the shore. This noun is repeatedly sung while a line of dancers advances frontally interlocking their spears at thigh-height.

(*): evokes the swimming motion of the turtle in the water. This noun is repeated in the song text and refers to both the linear trajectory of the turtle through the water as well as the wake or track she leaves behind in the water and on the land at specific places. The turtle's characteristic movement is also imprinted on the ground by the advancing steps of the performer of this dance (see Tamisari 2000).

localised movement

(*): this noun refers to the strength and might of the sea waves at a particular place in a storm following the heavy monsoon rains. More precisely this term compares the upright, curved shape of the wave to the sudden and violent movement that arches the body in the act of retching.

(*): also a female proper name, this noun evokes the bobbing movement of driftwood floating on the little waves of the calm sea water after a storm at particular places. *Dawu pudat*, one of the several expressions frequently repeated in the driftwood song units, combines sound with movement in a suggestive way. *Dawu* refers to the sound of water slapping the floating driftwood, while *pudat* evokes its 'floating dance along the calm water' as the song text says. In the corresponding dance, performers hold a stick at either extremity and lift it up and down parallel to the ground.

CONCLUSION

I have considered elsewhere the image of the footprint and especially its ontological significance in relation to the accumulation and transferral of knowledge and its negotiation in the context of individual and group identity (Tamisari 1998). As with landscape features, Yolngu people talk of stories, kinship, songs, painting and dances as 'footprints' (*djalkiri*) of the ancestors. *Djalkiri* or *luku* (literally 'foot' and by extension 'footprint' and 'step') is the term often used to refer to different aspects of Yolngu Law (*rom*) which originated from ancestral journeys and actions that shaped the land. *Djalkiri*, the footprint, thus refers to all visible marks left by the Ancestral Beings such as landscape features (*wa:nga*), kinship relationships among groups derived from their positioning along an ancestral trajectory (*gurrutu*), to the stories that recount the Ancestral Beings' journeys and actions (*dha:wu*, CF. Marika-Mununguritj 1991). These visible marks or 'footprints' are continuously manifested and retraced in paintings (*dhulang*), songs (*manikay*) and dances (*bunggul*). In this context I argue that the image of the footprint has ontological significance in that it is not simply the result, objectification or inscription of ancestral and human experiences and events in and on the land. In phenomenological terms the 'footprint', I assert, is a 'living body' and a 'knowing body', an embodied consciousness and perception of and in the world, simultaneously a fragment and an agent of place, a product and actor of social relations, a subject and object of action and experience. *Djalkiri*, the 'footprints' of the ancestors, the manifestation of ancestral creativity, not only fuse the body, place and event in an indissoluble whole, but mark connections between places, establish relationships between people, visualise movement, unravel narratives, and embody names. In this paper, I have further shown how this fusion of body, place and event is condensed in names, and conversely names are manifestations of ancestral morphogenetic actions. As 'footprints' not only placenames but proper names in general reveal once again the corporeal connection between people, places and ancestors. As a 'footprint' is static but implies movement, is localised but can be considered only in relation to other footprints which together form a journey, names localise and at the same time connect. While names embody an event into place and identify a place with a particular group or individuals, they often imply that these connections are temporary stops along an ever-moving trajectory which distributes unique ownership of country within a shared knowledge and authority over wider regions.

From this perspective, I suggest that Yolngu names may thus be considered, as Heidegger argues, as 'signs of things' (1971b:97), however not in its 'debased meaning — lines on a surface' (1971b:121). The naming of the landscape which brings a localised feature or phenomenon into existence through the processes of metamorphosis, imprinting and externalisation, as originating from the ancestral body, should be approached 'de-signs' in which the word sign should be understood in its etymological meaning of cutting a trace from the Latin *secare* as in saw, sector, segment and section. A de-sign says, shows, discloses, lets appear and determines things as things in the world (1971b:121ff). If names say the world into existence rather than representing it, bring it to presence rather than communicating it, it is this saying that lets one see and understand. Like poetry, which Heidegger describes as revealing and fixing the play between words and things, language and being, Yolngu names participate in the process of emplacing things into presence which, through bodily transformations, become an integral part of the world. I suggest that the significance and power of names thus reside in their corpo-reality or bodiliness: indeed their ability of 'de-signing' and thus showing and making appear, or in Yolngu terms 'piercing' and 'spurting' a visible mark which fixes acting Bodies, their history and experiences into the landscape features and localises phenomena that constitute places.[22] Thus by combining the act of naming with bodily transformations and fusing names with things as indissoluble manifestations of ancestral morphogenesis, Yolngu cosmogonic processes may be more appropriately referred to as a morphopoiesis, that is speaking forms into place, the making of place through names.

ACKNOWLEDGMENTS

I wish to thank all Yolngu people of Milingimbi community and especially my adoptive family for teaching me. Although I omitted proper names and avoided other details, I hope my discussion has respected their trust and the cultural sensitivities they are required to live under or observe.

---

[22] The corporeal interpenetration of places, people and ancestral events which is exemplified in the processes of 'piercing' and 'spurting' the world into being is not limited to names. The act of painting is referred to as *miny'tji yarpuma*, literally, 'paint/colour jabbing or spearing' (cf. Morphy 1989:24). The significance of this terminology has been recently noted in the sand designs made by women in the Balgo area (Watson 1997:110).

REFERENCES

Austin, John L., 1962, *How to Do Things with Words*. Oxford: Oxford University Press.

Basso, Keith H., 1988, 'Speaking with names': language and landscape among the Western Apache. *Cultural Anthropology* 3(2):99–130.

— 1996, Wisdom sits in places: Notes on a Western Apache landscape. In Steven Feld and Keith Basso, eds, *Senses of Place*, 53–90. Santa Fe: School of American Research Press.

Berndt, Ronald, M., 1952, *Djanggawul: an Aboriginal religious cult of north-eastern Arnhem Land*. London: Routledge & Kegan Paul.

Biddle, Jennifer, 2000, Writing without ink: methodology, literacy and cultural difference. In Alison Lee and Cate Poynton, eds, *Culture and Text: discourse and methodology in social research and cultural studies*, 170–187. St Leonards, NSW: Allen & Unwin.

Casey, Edward, S., 1996, How to get from space to place in a fairly short stretch of time: phenomenological prolegomena. In Steven Feld and Keith Basso, eds, *Senses of Place*, 13–52. Santa Fe: School of American Research Press.

Chappell, Hilary and William McGregor, 1996, *The Grammar of Inalienability: a typological perspective on body part terms and the part-whole relation*. Berlin: Mouton de Gruyter.

Dussart, Francoise, 1988, Notes on Warlpiri women's personal names. *Journal de la Société des Oceanistes* 96(1):53–60.

Goodale, Jane, 1980, *Tiwi Wives: A study of the women of Melville Island, North Australia*. Seattle: American Ethnological Society; University of Washington Press.

Hart, C.W.M., 1930, Personal names among the Tiwi. *Oceania* 1(3):280–290.

Heidegger, Martin, 1971a, *Poetry, Language, Thought*. Trans. Albert Hofstander. New York: Harper & Row.

— 1971b, *On the Way to Language*. Trans. Peter D. Hertz. New York: Harper & Row.

Keen, Ian, 1978, One ceremony one song: an economy of religious knowledge among the Yolngu of north-east Arnhem Land. PhD dissertation, Canberra: The Australian National University.

— 1991, Images of reproduction in the Yolngu Madayin ceremony. *The Australian Journal of Anthropology*, Special Issue 1:192–207.

— 1994, *Knowledge and Secrecy in an Aboriginal Religion*. Oxford: Clarendon Press.

— 1995, Metaphor and the meta-language: 'groups' in Northeast Arnhem Land. *American Ethnologist* 22(3):502–527.

Lévi-Strauss, Claude, 1966, *The Savage Mind*. London: Weidenfeld and Nicolson.

McKnight, David, 1999, *People, Countries, and the Rainbow Serpent: systems of classification among the Lardil of Mornington Island*. New York: Oxford University Press.

Marika-Mununguritj, R. 1991. How Can Balanda (White Australians) Learn About the Aboriginal World. *Batchelor Journal of Aboriginal Education* July:17–23.

Morphy, Howard, 1989, From dull to brilliant: the aesthetics of spiritual power among the Yolngu. *Man* 24(1):21–40.

— 1990, Myth, totemism and the creation of clans. *Oceania* 60(4):312–328.

— 1995, Landscape and the reproduction of the ancestral past. In Eric Hirsch and Michael O'Hanlon, eds, *The Anthropology of Landscape: perspectives on place and space*, 184–209. Oxford: Clarendon Press.

Morton, John A., 1987, Singing subjects and sacred objects: more on Munn's transformation of subjects into objects in Central Australian myth. *Oceania* 58(2):100–118.

Mundine, Djon, 2000, *The Native Born: objects and representations from Ramingining, Arnhem Land.* Sydney: Museum of Contemporary Art in association with Bula'bula Arts, Ramingining.

Munn, Nancy, 1970, The transformation of subjects into objects in Walbiri and Pitjantjantjara myth. In Ronald M. Berndt, ed., *Australian Aboriginal Anthropology*, 178–207. Perth: University of Western Australia Press.

— 1973, *Walbiri Iconography: graphic representation and cultural symbolism in a Central Australian society.* Ithaca and London: Cornell University Press.

— 1996, Excluded spaces: the figure in the Australian Aboriginal landscape. *Critical Inquiry* 22(Spring):446–465.

Nash, David and Jane Simpson, 1981, 'No-name' in Central Australia. In Carrie S. Masek, Roberta A. Hendrick and Mary F. Miller May, eds, *Papers from the Parasession of Language and Behavior*, 1–2:165–177. Chicago: Chicago Linguistic Society.

Schebeck, Bernard, 1968, Dialect and social groupings in North East Arnhem Land. Unpublished manuscript. Canberra: Australian Institute of Aboriginal and Torres Strait Islander Studies.

Stanner, William E.H., 1937, Aboriginal modes of address and reference in the north-west of the Northern Territory. *Oceania* 3(7):300–315.

— 1979a, The dreaming. In William E.H. Stanner, *White Man Got No Dreaming. Essays 1938–1973*, 23–40. Canberra: Australian National University Press (first published in 1953).

— 1979b, Religion, totemism and symbolism. In William E.H. Stanner, *White Man Got No Dreaming. Essays 1938–1973*, 106–143. Canberra: Australian National University Press (first published in 1962).

Strehlow, Theodore G.H., 1947, *Aranda Traditions.* Melbourne: Melbourne University Press.

Tamisari, Franca, 1998, Body, vision and movement: In the footprints of the ancestors. *Oceania* 68(4):249–270.

— 2000, The meaning of the steps is in between: dancing and the curse of compliments. In Rosita Henry, Fiona Magowan and David Murray, eds, *The Politics of Dance.* Special Edition 12, *The Australian Journal of Anthropology* 11(3):274–286.

Thomson, Donald, 1946, Names and naming in the Wik Monkan tribe. *Journal of the Royal Anthropological Institute* 76(2):157–167.

von Sturmer, John, 1978, The Wik region: economy, territoriality and totemism in western Cape York Peninsula, North Queensland. PhD dissertation, University of Queensland.

Warner, Loyd, 1958, *A Black Civilisation: a social study of an Australian tribe.* New York: Harpers and Brothers.

Watson, Christine, 1997, Re-embodying sand drawings and re-evaluing the status of the camp: the practice of iconography of women's public sand drawings in Balgo, Western Australia. *The Australian Journal of Anthropology* 8(1):104–124.

Williams, Nancy, 1986, *The Yolngu and Their Land: a system of land tenure and the fight for its recognition.* Canberra: Australian Institute of Aboriginal Studies.

8 'I'M GOING TO WHERE-HER-BRISKET-IS': PLACENAMES IN THE ROPER

Brett Baker

1 INTRODUCTION[1]

In the United Kingdom and elsewhere in Europe, we find many placenames that can be characterised as a compound of a *generic* term for a topographic feature or habitation, together with a *specific* or *modifying* term characterising that place with reference to a person, a characteristic, historical or mythological event, or some other topographic or habitation term; some examples are presented in (1).[2]

(1) a.   Salt Creek, Roper River

     b.   East Hills

     c.   Chilton, Dutton, Petersham

     d.   Sherwood

[1] I would like to acknowledge the assistance of Jen Munro in researching the material for this paper, and the Northern Land Council, under whose auspices much of the research contained in this paper was prepared. I thank also Peter Johnson, School of Geosciences, Sydney University, for Map 1. Various people have contributed to the ideas presented here, notably Mark Harvey, David Nash and Michael Walsh. None but myself should be held accountable for any errors of fact or interpretation. Fieldwork on Ngalakgan was supported by AIATSIS grants L95/4932 and 93/4657 and the University of Sydney. Thanks especially to people who patiently discussed their own placenames with me: *Nyulpbu* (Doreen Duncan), *Golokgurndu* (Roy James), *Gerrepbere* (Splinter James), Sandy August, Barney Ilaga, the Joshua sisters, and Stephen Roberts.

[2] Abbreviations: 1, 12, 2, 3: 1st, 1st incl., 2nd, 3rd person; NC: noun class (I-IV); O: object; pl: plural; S: subject; sg: singular; sp: species; ABL: ablative; ALL: allative; AUX: auxiliary, finite verb stem; COM: comitative; DAT: dative; DEF: definite; DU: dual; ERG: ergative; F/FE: feminine noun class/gender; F/FUT: future; NP: Non-Past prefix; LAT: lative; LOC: locative; M/MA: masculine noun class/gender; N/NE: neuter noun class/gender; O: object; OBL: oblique; PC: past continuous; PNEG/PRNEG/FNEG: past/present/future negative suffixes; POSS: possessive; POT: potential; PP: past punctual, present perfective; PR: present (continuous); RECIP: reciprocal; RED: reduplication; REL: relative/subordinator; RR: reflexive/reciprocal; VE: vegetable noun class; Boundary symbols: '+' separates verbs from their tense inflections, '=' separates coverb from auxiliary; '−' is the general morphological boundary.

L. Hercus, F. Hodges and J. Simpson, eds, *The Land is a Map: placenames of Indigenous origin in Australia*, 103–129. Canberra: Pandanus Books in association with Pacific Linguistics, 2002.

Many such names — *Salt Creek,* for instance — are transparent in meaning to a speaker of English. Others — such as *Chilton, Dutton* — are entirely opaque, though their original meaning can sometimes be determined through etymology. Still others — *Sherwood, Petersham* — have *parts* that are meaningful (*-wood, Peters-*), but which are not entirely transparent (assuming that 'ham' is obsolete for most speakers of English). In all of these, however, we observe that the general structure — specific+generic — remains the same, regardless of whether the name is now analysable or not. This indicates a continuity of strategies for placename formation in the English language.

My aim in this paper is, firstly, to discuss the range of construction types that are typical of placenames in one area of Indigenous Australia: the Roper River, which flows along the southern border of Arnhem Land in the Northern Territory. These constructions include species names (§2.1), locative-marked nouns (§2.2) and clauses (§3), which record the presence of the Dreaming. One of the interesting aspects of placename analysis in the Roper is the large proportion of names that are not entirely analysable in modern languages. In some cases, names are partially transparent and partially obscure. In others, we can recognise morphemes or structures, but these are put together in ways not found in modern languages, suggesting dialect or language differences. My second aim then is to derive inferences, based on these partially analysable names, about the likely distribution of language groups in the past, compared to their present-day distributions. I discuss some placenames, which may constitute our only record of an extinct language of the area, Yukgul, in §3.2. In §5, I discuss the extent to which placenames can provide evidence about long-term residence of a linguistic group in an area. Finally, in §6, I examine the linguistic use of toponyms: their reference, and their interaction with morphology. First, I give a brief overview of the languages of the area, in §1.1.

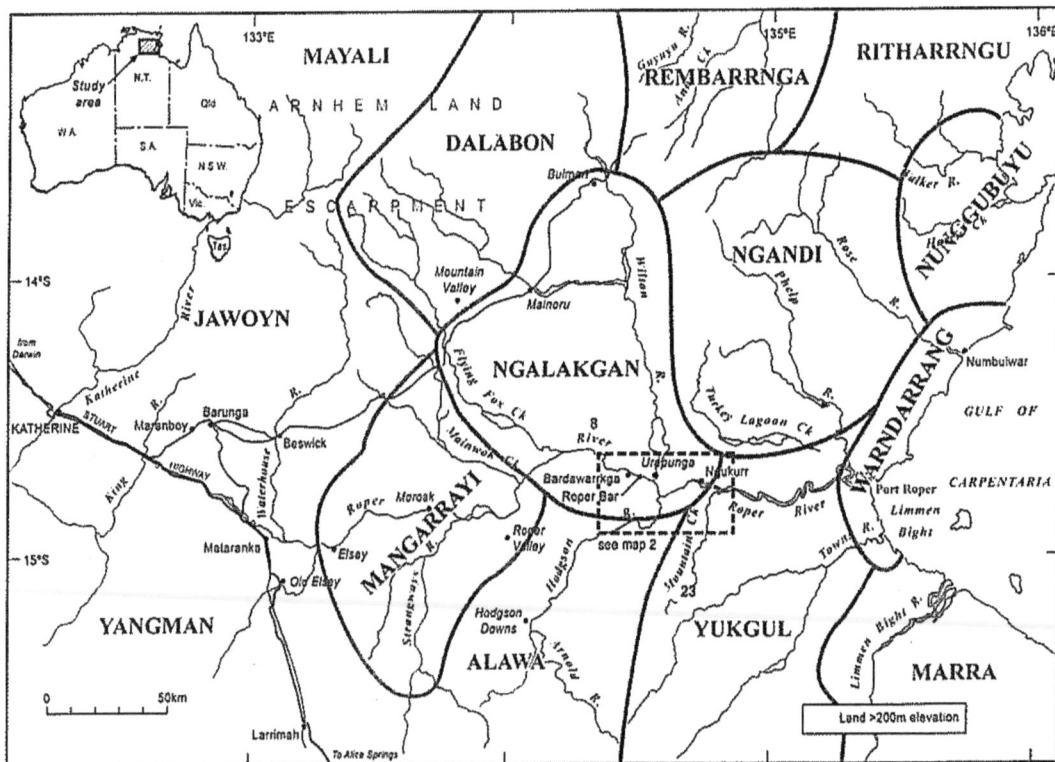

Map 1: Land–language associations in the Roper

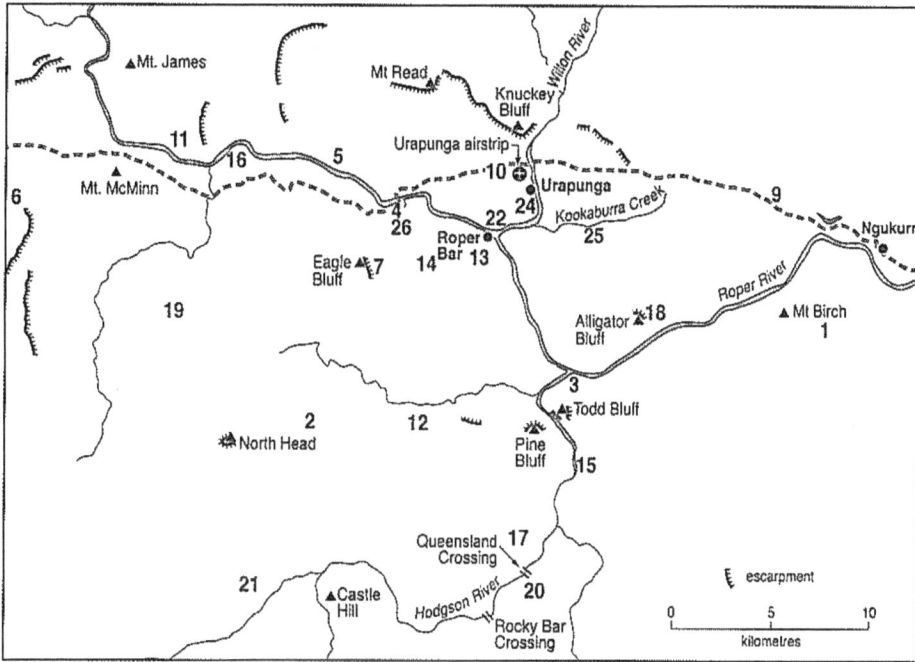

Map 2: Placenames around Roper Bar

Table 1: Key to Placenames in Maps 1 and 2[3]

1	Awarabankawinjin	14	Nawarnbarnkulyi
2	Balalayarrurru	15	Ngalardarra binkulinma
3	Baltjjardatbutjjinygah	16	Ngurruboy
4	Bunditjgah	17	Riwanji
5	Burrhburrminygah	18	Walanji
6	Golotdoh	19	Wandarrganiny
7	Gurdanggapbul	20	Yandah-jandah
8	Jalboy	21	Wankarnangintji
9	Jarrburdetjbutjjinygah	22	Wararrirr
10	Jilmiyunginy	23	Wudbudbalanji(n)ji
11	Larriboy	24	Yinbirryunginy
12	Nabarlmantji	25	Yurende
13	Nabordopburlani	26	Yurlhbunji

---

[3] A practical orthography for Roper languages has been developed by Diwurruwurru-Jaru (the Katherine Language Centre) and Batchelor College and is employed throughout the paper: syllable-initial *b, d, rd, j, g* are short and mostly voiced stops: bilabial, apico-alveolar, apico-postalveolar, lamino-alveopalatal and velar respectively. Syllable-finally, these stops are represented with the corresponding voiceless symbols: *p, t, rt, tj, k*. The digraphs *pb, td, rtd, tjj, kg* represent geminate (long, tense, and voiceless) versions of the same stops. The symbol *h* represents a glottal stop following sonorants, and lamino-dental articulation following the symbols *t, d* in Ngandi, Nunggubuyu. Nasals corresponding to the stops are *m, n, rn, ny, ng*, laterals *l, rl*, tap *rr* and labio-velar, retroflex and palatal approximants *w, r, y*. Digraphs: retroflexion is indicated once only in clusters: *rnd, rtd*, represent [ɳɖ], [ʈʈ]; *nk* represents an alveolar nasal + velar stop [nk], contrast *ngg* for [ŋg], and similarly *nj* and *ntj* represent homorganic [ɲɟ] and heterorganic [nɟ] respectively. Various conventions have been followed with respect to heterorganic nasal+stop clusters in the area, the convention followed here is not ideal but is consistent.

1.1 The linguistic context[4]

There are six language groups associated with the Roper River drainage basin. The list in (2) gives a rough location for each (and see Map 1).[5] The order proceeds from the Roper source to the mouth.[6]

(2)   **Mangarrayi**, centred on the community of Jilkmirnkan (Duck Creek, Jembere) near Elsey Station, and up to Maiwok Creek;

Ngalakgan, around the south-flowing tributaries of the middle Roper: Flying Fox Creek and the Wilton River, centred on the community of Urapunga and the outstation of Bardawarrkga;

Alawa, centred on the community of Minyerri (Hodgson Downs), and around the north-flowing tributaries of the Roper: the Hodgson and Arnold Rivers;

Ngandi, traditionally spoken in an area north of the Roper around the upper portions of the Phelp and Rose Rivers, and around Turkey Lagoon Creek just north of Ngukurr;

Warndarrang, spoken around the Roper mouth and along a narrow strip of coastline north and south;

Marra, traditionally spoken south of the Roper on the coast around the Limmen Bight and Towns Rivers.

Warndarrang (Heath 1980) became extinct in the 1970s. Clans that formerly spoke Warndarrang are now affiliated with Ngandi or Marra language groups (Bern 1974), and hence these languages can now be said to be affiliated more closely with the Roper than they were formerly. In addition, a language called 'Yukgul' (i.e. [júkkʊl]) by Ngalakgan and Alawa speakers, or [júgʊl] by Marra speakers, is said to have occupied the area around Ngukurr (on the Roper, about 50 kilometres from the Gulf of Carpentaria), and the country to the south. This language is unrecorded, but is said (by Marra speakers) to have been 'like' Marra.

The question of what happened to Yukgul speakers is not easy to answer. There are several extant clans which are associated with estates within the territory said to have been formerly Yukgul. The members of these clans in every case have switched linguistic affiliation to one of the neighbouring groups — Marra, Alawa, Ngalakgan, or Yanyuwa. What this means is

---

[4]   My conventions for citing placenames are as follows:

— when a placename (Indigenous or introduced) is discussed as a name it is given in italics;

— when a placename (Indigenous or introduced) current in English or other European languages is mentioned as a place it is given in plain type;

— words of Indigenous languages are given in italics;

— glosses for meanings of placenames are given in single inverted commas;

— when the phonetic or phonological status of words of Indigenous languages is under discussion, these are given in plain IPA font.

[5]   This map represents an estimate of land–language associations immediately before European contact. Territoriality in the area is the subject of ongoing research, therefore the map should not be taken to be definitive in any sense. Numbers on Map 1 and Map 2 correspond to the placenames in Table (1), which represents a selection of unanalysable placenames discussed in the paper.

[6]   Language names, abbreviations and sources for language data are as follows: Alwa: Alawa (Sharpe 1972); GN: Gunwinyguan (Alpher, Evans and Harvey to appear); Dlbn: Dalabon (Merlan 1993, Evans 1995); Jwyn: Jawoyn (Merlan 1989b); Mgry: Mangarrayi (Merlan 1989c); Mrra: Marra (Heath 1980); Myli: Mayali (Evans 2000); Ngkn: Ngalakgan (Merlan 1983, Baker 1999); Ngdi: Ngandi (Heath 1978b); Rmba: Rembarrnga (McKay 1975); Wndg: Warndarrang (Heath 1980).

that these people are now said to be 'Marra' or 'Ngalakgan' (whether or not a person actually speaks a language is irrelevant to this kind of identification: Merlan 1981). The way in which this switch came about was perhaps through temporary succession to the estate of a Yukgul clan by a neighbouring clan of the same semi-moiety (Bern & Layton 1984:74).[7] By 'temporary', I mean that the knowledgeable Elders of the neighbouring clan agree to assist with the spiritual duties expected of clan members weakened for whatever reason.[8] Some placenames which possibly include Yukgul morphology are examined in §3.2.

Ngalakgan and Ngandi are related to each other, and to a large group of languages spoken in central Arnhem Land given the name 'Gunwinyguan'. The basis for the claim that these languages are related rests on their shared verb inflectional morphology (see Alpher, Evans & Harvey to appear, Heath 1978a). More immediately, Ngalakgan is very closely related to Rembarrnga, spoken to the north from Bulman to the middle Blyth River. Ngandi is more closely related to Nunggubuyu, spoken to the north of Warndarrang around Numbulwar (Heath 1978a, Baker 1996). Mangarrayi is argued by Alpher, Evans and Harvey (to appear) to be a Gunwinyguan language also, albeit a distantly related one.[9]

Much of the discussion in this paper originated in the process of preparing reports on two land claims lodged by the Northern Land Council (NLC) under the *Native Title Act*: St Vidgeon's Pastoral Lease (PL), and Urapunga Township. St Vidgeon's PL covers a large (600,000 sq. km.) area of land whose northern border is the Roper River. The township of Urapunga is a small parcel of land, immediately to the south of the Roper Bar crossing, which was gazetted in 1887. (This particular area is dotted with some important sacred sites.)

All of the languages of the Roper region are becoming moribund. Ngalakgan has just three first-language speakers, and has not been acquired by children since the 1930s or 1940s. The state of Ngandi is even more dire. Alawa and Marra are not much better off. Therefore, the linguistic uses of placenames spoken of here should be understood to refer to the limited context in which these languages are now spoken.

2 THE MORPHOLOGY OF PLACENAMES

There are no special morphological features that identify Ngalakgan placenames as such. However, in Ngalakgan, Alawa and Marra placenames behave differently from other nouns with respect to

---

[7] One of my main language consultants, one of the last proficient speakers of Ngalakgan, has said on various occasions that his father's father was Yukgul, but married to a Ngalakgan woman. My consultant belongs to a clan owning an area of land north-west of Roper Bar which has apparently been associated with Ngalakgan speakers for a considerable length of time. In this case then, a Yukgul man's sons were adopted into a Ngalakgan clan of the appropriate ritual classification (i.e. semi-moiety). And indeed, now this man is an accepted authority on Ngalakgan language and ritual matters.

[8] Almost certainly the whole area immediately around the Roper River and along the Queensland cattle route (which passed through St Vidgeon's) suffered early onslaught from disease and shooting parties (Merlan 1978).

However, Heath (1978a) claims that Alawa, Marra and Warndarrang form a distinct genetic group, based on pronominal paradigms, and noun class/case morphology; and Sharpe (nd) and Merlan (1989a) also include Mangarrayi in this group. It should be noted that verb morphology does not constitute part of their evidence for this claim and, indeed, very little can be attested in the way of specific similarities in the verbal morphology of these languages (beyond general typological similarities and a few widely distributed roots).

locative and allative case marking.[10] Two endings found on a few placenames in the Ngalakgan area are otherwise not found elsewhere in the morphology. In §4, I argue that these derive from old derivational suffixes still found in neighbouring related languages such as Jawoyn. Placenames have the form of a variety of word classes in Ngalakgan: bare noun, affixed noun, inflected verb. Examples of the various types are given below in each section.

2.1 Bare nouns as placenames

Placenames that have the form of a bare noun are typically the names of species (trees/plants, birds, fish, or animals). Species names, in this area (Arnhem Land), are commonly shared among many languages of a *Sprachbund* (Heath 1978a), so it is not possible to ascribe them distinctively to any language without research into their distribution within family-level linguistic groups; examples (in Ngalakgan orthography) of placenames like this are presented in (3):

(3) a. *Dubal*   'Leichhardt tree' (*Nauclea coadunata*) Ngkn, Alwa, Mrra

b. *Jendewerretj*   'willy wagtail' Ngkn, Alwa

c. *Jalburrgitj*   'brown honeyeater' Ngkn, Alwa

d. *Ganjarri*   'bonefish' (herring sp.) Ngkn, Ngdi

e. *Mirnitjja*   'shade tree' (*Cathormion umbellatum*) Ngkn, Alwa, Mrra

f. *Jininggirrijininggirri*   'willy wagtail' Alwa, Ngkn

g. *Garrinji*   'jabiru' Alwa, Mrra

h. *Mayngu*   'red ochre, red stone' (Marra; Ngkn *mayngoh*)

i. *Motjo*   cf. Ngkn *mutjju* 'coolibah tree' (*E. microtheca*); Alwa, Mrra *mudju*

The placename *Mirnitjja* is an alternative for the longer name *Mirnitjja-ngoji-kgah* examined below. A common term in the Roper for the 'shade tree' found around billabongs and flood plains is *mirnitjja*. The word for coolibah tree in Ngalakgan is *mutjju* [muccu]. This word is subtly different to the Alawa and Marra term for the tree, *mudju* [mutcu], which has a cluster of /t/ plus /c/ sounds, whereas the Ngalakgan word has a long lamino-palatoalveolar stop /cc/, written *tjj*. The placename *Motjo* on the Roper is different from both of these, it is pronounced [mɔtco] by Ngalakgan speakers, i.e. with the same /t/ plus /c/ cluster as Alawa (written differently in Ngalakgan). This name appears to be a phonological adaptation into Ngalakgan of the term for 'coolibah' in Alawa and Marra, rather than a straight translation of the Alawa term to the Ngalakgan one (which appears to be an old borrowing).[11] We speculate that the placename preserves an older form of the word, just as the

---

[10] For example, in Marra, Heath (1981:92) notes that placenames and demonstratives take two special suffixes: noncentripetal *-nyindi* and a rare suffix *-nyinkarr* ('from...to'), and that cardinal direction adverbs and placenames take a special allomorph of the ablative suffix.

[11] Alawa and Marra lack the vowel /o/ as a phoneme. Ngalakgan words with /o/ correspond to Alawa or Marra words with /a/ or /u/, e.g. Ngkn *morrortdinh* Mrra *ngurrirdin* 'bush banana' (*Leichhardtia australis*). The correspondence m:ŋ is sporadically found among languages in the area.

English placename *Each* preserves the dative singular of the word for 'oak' (æc) in Old English (Gelling 1984:218), which became obsolete in Modern English.

There remain a number of placenames that appear to contain lexical stems but which are not entirely analysable. One name (4a) was said to contain an Alawa word for 'stone knife' *limbirr* (this word does not occur in Sharpe's dictionary of Alawa), and the place is associated with a stone dreaming, but the other part has no identifiable meaning. Example (4b) appears to have undergone initial lenition, lenition of /g/ word-initially is a well-attested phenomenon in Alawa (Sharpe 1972). The placename is associated with a stone in the water at that place.

(4) a. *Wundalimbirr* Alwa *limbirr* 'stone'

     b. *Wabarnda* cf. Alwa *gabarnda* 'white stone' (Ngkn *gapbarndah*).

Other placenames have more irregular forms, some examples are presented in (5).

(5) a. *Jimaju* Translated as *gimaju* 'milky way' in Alawa, Marra.

     b. *Andawu/Arnawunban* ('Lancewood yard' in Hodgson Downs) cf. Alwa *arnawun* 'lancewood', and *banban* 'flat woomera', a weapon commonly made from the wood.

At least in the case of the Ngalakgan placenames in (3), those names that are tree species terms (e.g. *Motjo* 'coolibah') refer to actual trees of that species which still stand at that place, or did until recently. I examine the reference of these tree names further in §6. As noted, these named, specific trees are the synchronic manifestations of the dreaming ancestors. In other cases, such as *Ganjarri* ('bonefish'), the name refers to the dreaming ancestor itself associated with that place. More common than bare species names are nouns inflected for locative case. These are the subject of the next section.

2.2 Inflected nouns as placenames

Aside from pure lexical items, many names along the Plains Kangaroo track are clearly Ngalakgan in origin, and cannot be derived from any other nearby language. These are names that carry distinctively Ngalakgan morphological material such as affixes. While in some cases the affixes may be shared with neighbouring languages such as Dalabon or Rembarrnga, the combination of lexical items and affixes in most cases can only be Ngalakgan.

A number of names (shown in 6) carry the Ngalakgan feminine dative pronominal enclitic *-ngoji*, either alone or in combination with a following locative suffix *-kgah ~ -gah*.

(6) a. *Berre-ngoji-kgah*
         brisket-hers-LOC
         'At her brisket' or 'Where her brisket is'

     b. *Bolkgotj-ngoji-kgah*
         backbone-hers-LOC
         'Where her backbone is'

     c. *Mirnitjja-ngoji-kgah*
         shade.tree-hers-LOC
         'Where her shade trees are'

d. *Gu-Jambay-ngoji-kgah*
    IV-flat.rock-hers-LOC
    'Where her flat rock is'

e. *Manga-ngoji*
    throat-hers
    'Her throat'

f. *Jawarnda-nowi*[12]
    whiskers-his
    '[Where] his whiskers are'

g. *Giyarrk-ngoji-kgah*
    tooth-hers-LOC
    'Where her tooth is'

All these names are transparently derived from modern Ngalakgan nouns and nominal morphology. The feminine possessive enclitic *-ngoji* 'her(s)' is particular to Ngalakgan and does not occur in any of the neighbouring languages with this meaning: Rembarrnga, Mangarrayi, Alawa, Jawoyn, or Dalabon.[13] Likewise the locative suffix *-kgah ~ -gah* 'at' is Ngalakgan, though not distinctively so: it also occurs in neighbouring Dalabon. The noun class prefix for neuter class *gu-* in (6d) also occurs in neighbouring Ngandi. Similarly, the placename *Dubal* in (3a) typically occurs with a vegetable noun class prefix: *mu-Dubal*. I return to the characteristics of nominal morphology on placenames in §6.

The names record the actions of dreaming ancestors in placing various body parts at the named locations. Aspects of the landscape (certain trees, billabongs, rock formations) represent these manifestations of the ancestors. In this case, the feminine possessive suffix is a reflection of the gender of the dreaming ancestor *garndalpburru*, the term for 'female plains kangaroo' (*M. antelopinus* 'antelopine wallaroo'). Although this ancestor was said to have travelled in a mob of male (*jardugal*) and female kangaroos, only placenames with the feminine suffix are associated with these ancestors.[14]

The locative suffix in these placenames is to be interpreted as deriving a headless relative existential clause, following Harvey (1999:174): e.g. 'where her brisket is'. Relative clauses are commonly headless in Ngalakgan, and nominals can function either as (existential)

---

[12] The name recorded by Morphy and Morphy (1981) is *Jawarnda-ngoji-kgah*, with a feminine dative clitic. This is a place on the track of the Quiet Snake (olive python) ancestor *gurrijartbonggo*. Heath (1981:358) records a version of the taipan myth (in Marra) in which the dreaming pulls out his own whiskers.

[13] Rembarrnga /-*ngadə* / and Ngandi /-ʔŋṭṭayi/ are related to the Ngalakgan form by regular sound correspondences.

[14] If the placenames recorded the actions of male ancestor kangaroos we would expect forms such as **mirnitjja-nowi-kgah*, with masculine dative enclitic *-nowi*, rather than feminine *-ngoji*. Similarly, a mixed group of kangaroo ancestors would take either the masculine suffix (which is the unmarked gender) or possibly the plural dative enclitic *-borre*, although plural marking of non-human arguments is subject to well-known restrictions in Australian languages (see e.g. Merlan 1983). I note that in neighbouring related Ngandi, the cognate feminine Dative suffix *-hngutdhayi* is used for all non-human referents, as well as human females (Heath 1978b:57). This may have formally been the case in Ngalakgan also. The reasons for the exclusive use of the feminine suffix in these placenames deserves further investigation.

predicates or arguments.[15] The locative suffix (and other nominal suffixes) can occur on verbs in Ngalakgan in just this sense.

Similar patterns are found in neighbouring areas. Most of the placenames to be discussed below come from the St Vidgeon's pastoral lease or contiguous areas of the Roper, in country that is marginal to the areas considered to be the Alawa and Marra 'heartlands'. Consequently, many of these names contain what looks like Alawa or Marra morphology, but cannot be straightforwardly derived from the modern languages. Most of the St Vidgeon's area is said to be owned by clans that were formerly Yukgul-speaking, a language that is now extinct. Since Yukgul is claimed (by Marra, Alawa speakers) to have been 'like Marra', we expect to find Marra-like morphology in these names.

Many placenames of the St Vidgeon's area immediately contiguous to Hodgson Downs (which is traditionally affiliated with Alawa) contain reflexes of the Alawa locative suffix. Of the 290 names in the Site Register for the St Vidgeon's claim, 250 were able to be checked and confirmed with the claimants. Of those 250, 79 (or 32 per cent) ended in one of these sequences. A few of these placenames are examined below, with suggestions as to their derivation from Alawa.

The locative suffix in Alawa takes a number of allomorphs, depending on the final consonant of the stem (Sharpe 1972), and the placenames conserve this relationship. I repeat Sharpe's summary here (with orthography adjusted to that of this paper, and ignoring irrelevant aspects):

Table 2: Alawa locative suffix allomorphy (after Sharpe 1972).

STEM	STEM+SUFFIX
-i	*-irri*
-V	*-irr*
-Y	*-Ynji*
-C	*-Cji*
-Vrr	*-VndV*
-VL	*-VLdV*

[Key: Y: semivowels {y, w}; C: oral, nasal stops; L: {l, rl, r}, V: any vowel, V2 in a suffix is the same as V1]

The first group contains names that are straightforwardly derived from Alawa, that is, their meaning is clear to a speaker of the language:

(7) a. *Leguldu* 'Waterhole on creek that runs into Cox River'. Alwa *legul-du* 'deep depression in ground'-LOC

b. *Linylinji* cf. Alwa *linyliny* 'special type of sloping stone found on ridges'-LOC, Ngkn 'flat stone found on ridges'

---

[15] For example, when requesting further identification of an animal from another speaker a common strategy is to use a headless relative clause:

mu-darda   gu-mu-ngunu-ngu+n-gun?
III-sugarbag   NP-III-RED-eat+PR-REL
'[the one] that eats native honey?'

c. *Ngarrgalirr* 'A hill to the west of *Walgundu* road'; cf. Alwa *ngarrgala* 'high bank, cliff'-LOC (*ngarrgalirr* is the regular outcome)

d. *Erldi* 'bottom; down'

Such names as those in (6) and (7) are not necessarily good evidence for long-term residency, as Harvey (1999) notes. This is because the names have the same linguistic status to speakers of Ngalakgan and Alawa as their English translations do: they are still recognisable as words or phrases. When recording the Plains Kangaroo dreaming story the narrator explained the meanings of the placenames by saying such things as "*Berre-ngoji-kgah, imin libim briskit*" (the Kriol means '(s)he left his/her brisket'), in both Ngalakgan and Kriol. Such names may be direct translations — a calque — of the name in a previous language. For example, the Gaelic name *Cearamh Meadhonach* has an English version: *Middlequarter*, a direct calque from the earlier Gaelic (Nicolaisen 1976:54).

Better evidence for long-term residency, as Harvey (1999) again notes, comes from names that contain language-specific morphological patterns, but which are not entirely susceptible to analysis by speakers, like the English placename *Chilton* for instance. The ending *-ton* is commonly associated with English placenames, but the derivation of *chil-* is not obvious to the average person.

In the area around Roper Bar, for instance, there are several names that contain either the feminine possessive suffix *-ngoji* or the locative suffix *-kgah*, attached to stems that are not recognisable; two are shown in (8).[16]

(8) a. *Nana-ngoji* ?-hers

b. *Bunditj-gah* ?-LOC

The Morphys (1981) record a placename *Nanangoji* which seems to contain the same feminine suffix *-ngoji*, but the stem *nana* to which it is attached is unknown.[17] Similarly, the name *Bunditjgah* appears to contain the locative suffix *-kgah* ~ *-gah*. The name refers to a place in the Roper River where a firestick was left in the water in the Dreamtime. The water here churns at low tide and is said to 'boil' from the action of the dreaming firestick. But the stem *bunditj* is not used in Ngalakgan for 'firestick' or 'fire', indeed it has no meaning. Notably, the word cannot be found in any of the surrounding languages (Alawa, Mangarrayi, Marra) either. Therefore, names such as *Bunditjgah* and *Nanangoji* indicate that speakers of Ngalakgan have been using these names for long enough so that their original linguistic meaning has been lost. Now, they are placenames like *Chilton*, *Petersham* which are partly recognisable (because of the elements *-ton*, *-ham* ≈ *-gah*, *-ngoji*) but whose meaning cannot now be reconstructed by speakers.[18]

Like the partly opaque names in Ngalakgan, many names in the St Vidgeon's area contain what look like Alawa nominal morphology, but which are not regular or transparent in the modern language.

---

[16] A selection of unanalysable placenames is provided in Map 2. Again, this map should be treated with caution. Sources: Northern Land Council, Morphy and Morphy (1981), own fieldwork.

[17] This name is unlikely to derive from a single morpheme, since the vowel /o/ in Ngalakgan has a highly restricted distribution. With two exceptions in the lexicon (one onomatopoeic term and one recent borrowing), /o/ never occurs in medial syllables unless it also occurs in edge-most syllables in a word (see Baker 1999:72).

[18] §6.1 provides evidence that names like *Bunditjgah* contain elements that are still meaningful to speakers.

(9)	a.	*Walgalirr*	cf. Alwa: *gelerr* 'skull'
	b.	*Walngirr*	?Alwa: *ngirr* 'red'
	c.	*Wulbulirr*	cf. Mrra *bulabula* 'shoulder blade' + Alwa *-irr* LOC
	d.	*Yumanji*	Al. *yumarr* 'good'; locative form should be *yumanda*

I omit a full discussion of these names. The only placename in this set that was translated by informants was (9d): *Yumanji*. This name was translated as the 'good place', presumably because of the Alawa word *yumarr* 'good'. However, the placename does not have the predicted form for *yumarr*+LOC, that would be *yumanda*. Instead, it has the locative allomorph associated with stems ending in a stop or nasal, e.g. **yumany*, **yuman*, or a semivowel [j]: **yumay*. The name is therefore irregular. It is unclear what speakers' translation of the name signifies, whether it should be regarded as folk etymology, or whether it represents a continuation of the meaning of an archaic and/or irregular form.[19]

The name for the Wilton/Roper junction is *Wararri* or *Wararrirr* (the former is that used by Ngalakgan speakers, the latter by Alawa speakers).[20] The junction is rocky and creates eddies in the water. The name was not translated by speakers, but there are some tantalising possibilities. It resembles both Alawa *wari-rri* (hole-LOC) 'at the hole' and the Alawa non-finite 'coverb' *warirr* 'churn up, disturb water'. The vocalism difference in either case is unexplained. One further possibility is that the name derives from a reduplicated form of the related Marra noun *warirr* 'agitated or bubbling water' on the Marra pattern: *war-'arirr* 'many eddies', with perhaps dissimilation of the consonants *r* [ɹ] and *rr* [r].

3 INFLECTED VERBS AND CLAUSES AS PLACENAMES

Many names of the Roper contain verbal morphology. As we have seen with the nominal forms, some of these are transparent to speakers, but many are not. There are also a number of names that are suggestive of earlier language forms in the area.

3.1 Verbs as placenames in the Ngalakgan area

The Ngalakgan area immediately surrounding the Roper Bar and nearby communities on the river has several placenames which can be identified with verbs or clauses. In Ngalakgan, Alawa, and Marra, an inflected verb can take the place of a clause. The verb is prefixed with pronominal elements that indicate the grammatical person, noun class and number of the subject and object or indirect object. The verb is suffixed for tense/aspect/mood. In Ngalakgan, as in many languages of this region, verbs often consist of two parts: one of a small class of directly inflecting root verbs — the 'finite verb' — which bear the tense

---

[19] Given the tendency for Alawa to underdetermine syllable final [r] and [ɹ] (see the next footnote), and the areal pattern of alternations among approximants [ɹ] and [j], the hypothesis that *yumay* represents an older form of the word cannot be discounted.

[20] This name seems to have a variable form, which makes it difficult to determine the quality of the consonants involved. Sharpe (1972:15, 18) notes the difficulty in distinguishing [r] from [ɹ], [l] from [r] and [ḷ] from [ɹ] in Alawa, particularly in syllable-final position. I have had the same experience in working with Alawa speakers. The ending *-irr* which is very common in placenames of this area (being one allomorph of the Alawa locative suffix) varies between tap and (alveolar) approximant articulation.

inflection (like auxiliary verbs such as 'do', 'will', 'can' in English), and one of an open class of non-finite 'coverbs', which carry the weight of the semantic meaning. Many of the verbal placenames in this region have this kind of form. Some examples are presented in (10).

(10)   a.   *Wandarrganiny* (Morphy & Morphy 1981) cf. *ganiny* 'take'+PC

       b.   *Yinbirryunginy* cf. *yony*, *yongoniny* 'lie'+PP, PC

       c.   *Jilmiyunginy* or *yenginy* 'put'+PC

       d.   *Burrhburrminygah*
            burrhburr-mi+ny-gah
            'clap.hands'-do+PP-LOC
            'where [they] sang corroboree' (lit. 'where they clapped hands')

       e.   *Jarrburdetjbutjjinygah*
            jarr-burru-detj+bu+tjj+iny-gah
            ?leg-3PL-?cross+[hit+]RR+PC-LOC
            'where [they] crossed their legs'

       f.   *Julyurrhminygah*
            julyurrh-mi+ny-gah
            ??-do+PC-LOC
            'where they *x*-ed?'

       g.   *Berrhberrminygah*
            berrhberr-mi+ny-gah
            ?step-do+PC-LOC
            'where [they: wallabies] stepped'

       h.   *Baltjjardatbutjjinygah*
            baltjjardat-*bu+tjj+iny-gah*
            ??-[hit+]PC-LOC
            'where [they]*x*-ed'

As with the nominal examples in §2, we again find both completely analysable forms and those which are more or less obscure in part. One name that is identical to a modern Ngalakgan verb is (10d) *Burrhburrminygah*. When I recorded this name from an informant as part of his recitation of the Plains Kangaroo dreaming track he translated it with both the gesture of tapping clap sticks together, and, in Kriol, as *thei bin kilim fingga* ('they hit their hands together', i.e. clapped).

Other placenames are obscure while showing what appears to be Ngalakgan verbal morphology. Example (10f) *Julyurrhminygah* ends in the same sequence *-minygah* as *Burrhburrminygah*, but the meaning of *julyurrh* is not known to speakers.[21] Similarly (10a) *Wandarrganiny* (recorded by the Morphys) ends in the Ngalakgan form of the verb 'to take' in Past Continuous tense. This verb is also used as a causative. The meaning of *wandarr* is again unknown. The verb inflectional endings *-niny*, *-nginy* are highly distinctive of Gunwinyguan languages (Alpher, Evans, Harvey to appear). Alawa and Marra have no inflecting verbs that end in this kind of suffix.

---

[21]   The Ngalakgan auxiliary root *-mi+* is the commonest inflected verb in the language; over half the recorded verbs take this root as an auxiliary.

The name *Jarrburdetjbutjjinygah* was translated with the meaning given here by informants.[22] The form of this name is a good indication of its age. The word for 'thigh', 'upper leg' in modern Ngalakgan is *jarrpbitj*, not *jarr*. However, *jarr*, or some variant of it, is the word for 'thigh' common to other Gunwinyguan languages, such as Jawoyn to the north-west.[23] The name may therefore preserve an older form of this word. Names such as those in (10), then, indicate a considerable length of time for Ngalakgan speakers in the area between the Hodgson River and the western edge of Eagle Bluff. That is, the traditional 'border' between Ngalakgan and Alawa appears to be confirmed by the evidence of placename etymology.

3.2 Verbs as placenames in the Alawa, Marra and Yukgul area

In this section I examine transparent examples of phrasal placenames incorporating verb morphology, in the southern Roper. I also speculate on the etymology of some names that are not currently translatable by speakers, but which bear some tantalising similarities to the form of Marra and Alawa placenames, suggesting names derived from earlier forms in these languages, or else from another related language.

A large number of placenames in the eastern and south-eastern area of the St Vidgeon's claim area contain an ending *wawurlu* or *wawulu*, which in Marra means 'he/she/it sits', and refer to an ongoing physical manifestation of the ancestors.[24] (As in other Australian languages, Marra lacks a copula and uses stance verbs to indicate existence, therefore, the names mean something like 'he/she/it is located'.) That their presence is continuing is overtly indicated by the present tense of the verb. (Note that this is in contrast to Ngalakgan-derived names, which, if they include verbal morphology, are always in past tense.) In some cases, the word or stem to which *wawurlu* is affixed can be found in Marra, in other cases not. This again indicates some depth of age for these names. The examples in (11) are those that are, or probably are, combinations of a noun together with *wawurlu* 'sits' of which it is the subject. The translations given here are those supplied by informants.

(11)	a.	*Yirriga wawurlu*	?*yirriga*, but cf. Wndg *yirrwa-* 'sister'
	b.	*Yulbaranyi wawurlu*	?*yulbaranyi*
	c.	*Ginggarra wawurlu*	(on Limmen river) poss. *ginggirra* 'wild rice'
	d.	*Jagurl wawurlu*	*jagurl* 'testicles'
	e.	*Mirnijar wawurlu*	*mirnijar* 'salt'
	f.	*Wardanggurr wawurlu*	?*wardanggurr*
	g.	*Wujula wawurlu*	Mrra *wujula* 'ordinary woomera'

---

[22] I have been unable to confirm the location to which this name refers. Its position on Map 2 is my guess based on the evidence of dreaming track recounts. However, it appears in two dreaming tracks, and one of these would place this location around 'North Head', south of *Wandarrganiny* on Map 2.

[23] In this respect, Ngalakgan agrees with Ngandi and Nunggubuyu, which have /ṯarppic/, /ḷarpic/ respectively. The correspondence Ngkn /c/ : Ngdi /ṯ/ : Ngby /ḷ/ word-initially is regular. The Rmba form *darrama-* may also reflect *jarr* (correspondence between PGN *j* and Rmba, Myli, Dlbn *d* is regular: Harvey (to appear)).

[24] In these names, *wawurlu* always takes a stress on the first syllable, regardless of the length of the preceding word or stem, and hence retains some independence.

h.	*Jangguyala wawurlu*	Mrra *jangguyarla* 'lilyseed damper'
i.	*Maranggarlba wawurlu*	Mrra, Ngkn *maranggarlba* 'green tree snake'
j.	*Guduwaru wawurlu*	cf. Alwa *gudaru* 'stone'
k.	*Narrwar wawurlu*	?*narrwar*

Example (11j) preserves an older form of the word *gudaru* 'stone, rock, hill' which occurs in Alawa, but is obsolete in Marra (Marra informants knew the older form, however).

There is also one example of the stance verb *wajurlu* 'stands', given in (12). The name refers to a billabong just north of the Old St Vidgeon's stationhouse, in the heart of what was formerly Yukgul territory. Informants claimed that the word *jalbalbay* referred to the female genitals.[25] Note that the word *jalbalbay* has the form of a Mangarrayi-Alawa-style reduplication of **jalbay*. This kind of reduplication is unknown in Marra, and rare in Alawa. Therefore, the existence of this highly distinctive type of reduplication together with an inflected Marra verb suggests that the language from which this name derives also had this type of reduplicative pattern.

(12) *Jalbalbay* *wa-ju+rlu* 'genitals stand'
      'genitals' NP-stand+PR

Other placenames in the Marra-speaking area take the form of transitive clauses, where a noun realises the object, and the subject is expressed only in the pronominal agreement marking on the verb. Some of these examples are presented in (13).

(13) a. *Warrirdila wayi-wuyi* '(s)he abandoned a hook boomerang'
          warrirdila *wayi=wu-yi*
          hook.boomerang abandon=3sg-AUX+PP

      b. *Warurrgu yib-ganyi* '(s)he hid a nulla nulla'
          warurrgu *yib=g-anyi*
          nullanulla hide=3sg-AUX+PP

      c. *Mindiwaba gub-ga-wu+rlu* (Mrra)/*Mindiwaba dad-gawurlu* (Wndg)
          baler.shell=??-NP-sit+PR

Example (13c) is worth discussing (see also Heath 1981:317). This place (in the sea off the coast from the Cox River Land Trust) was translated by claimants as 'where he turns belly down' in Marra. It is said to have both a Marra and a Warndarrang version. In fact, neither verb *gub* nor *dad* is recorded in either language with this meaning, and *wawurlu ~ gawurlu* is not a recorded auxiliary form in Warndarrang.[26] The form *mindiwaba* is 'baler shell' in both languages.

---

[25] This word cannot be found in Heath's (1981) dictionary of Marra. The word recorded for 'vulva' is similar: *jarlbarr*. And compare the **y:rr* correspondence noted above between *Yumanji*, implying a stem **yumay*, and modern Alawa *yumarr* 'good'.

[26] Alternation between glide-initial and stop-initial forms of prefixes and auxiliaries is regular in Warndarrang and Marra. Glide-initial forms are found after continuants (liquids and glides) and stop-initial forms are found after nasal and oral stops.

(14)   *Wurr-ngumbarnarra*   *ngarl-ngarl=warri-yi+rlana*
        3DU-brothers.in.law.DYAD   RED-speak=3DU-AUX.PP+RR
        'two brothers-in-law spoke to each other'

Example (14) is another placename from the Cox River area; it has the translation given in Marra (cf. Heath 1981:316). It can also be referred to by the shorter forms *Wurrngumbarnarra* and the anomalous *Wurrngungbarnarra*. In addition, claimants also gave the form *wurr-ngungbarnarra ngarl-ngarl=warlijinji*, claiming this was a Warndarrang version. This does not appear to fit with the description of Warndarrang we have in Heath (1980) however. The nominal class/case prefix for dual in Warndarrang is not *wurr-*, but *yirri-*. The dyadic kin stem *ngumbarnarra ~ ngungbarnarra* does not occur in Heath's grammar, though it may be unrecorded. The non-finite coverb *ngarl* meaning 'speak' does occur in Warndarrang, and takes the auxiliary *-ja-* 'to tell'. The form we would expect in this case would be *ngarl=wad-ja-yi*. Heath (1980:83) records *-ji-* as an archaic form of the Warndarrang reciprocal.

The name therefore cannot be derived from the descriptions we have of these languages. If the name derives from a Yukgul source, then Yukgul appears to have had a reciprocal suffix inflection more like Warndarrang (or Nunggubuyu) than Marra.[27]

(15)   *Wudbud=bala-nji(n)ji*   'they were leaning'
        RED-lean=3pl-stand(DUR)+PP

This placename (15) seems to preserve what looks like a similar ending to (14) *ngarl-ngarl=warlijinji*. This place is south of Old St Vidgeon's, and again is probably in the Yukgul/Alawa/Marra border country. It is said to be the place where the mermaids were chased by mosquitoes. The portion *wudbud* is what we would predict for the reduplication of a coverb root *wud* in Warndarrang or Marra. In Alawa, the verb *wud=neni* means '(he) rests', in Marra the apparently related form is *wud=jinji* meaning 'to lean against'. The form *wudbud=bala-ji+nji* is Marra for 'they were leaning (or: they rested)'. The placename itself is subtly different, having *n* (i.e. [ɲ]) between the prefix *bala-* and the auxiliary *-ji(n)ji*. There is no explanation for this [ɲ], it may be an archaic feature (other auxiliaries, such as *-mburlmarli* 'do this', begin in clusters). The name was also recorded as *Wudwudbalanjiji* (Heath (1981:53) notes a nasal cluster dissimilation rule operating in certain inflected verbs in Marra).

(16)   *Àwarabankawínjin~winygin/Nàwarabanggarínygin*
        ʔawaran bang=ga-gu-?
        lightning split=NP-3SG-
        '?[where] lightning split [it]'

This name was recorded in several variants. It refers to a place where Lightning dreaming struck the rock at a ridge south of the Roper, the place is visible from Ngukurr store: again in traditional Yukgul country. The word for lightning in Alawa is *awaran*. The name as a whole looks like the kind of name we have seen so far: Noun+Coverb+Prefix(es)+Auxiliary. It is possible that the name includes a form related to the Marra verb stem *bang* 'split s/thing': so the name would mean something like 'lightning split it (i.e. rock)'. The *ga-* is possibly the Marra Non-Past prefix (which precedes third person *wu- ~ gu-*), but the rest of the auxiliary is

---

[27] Analysis is further complicated by our limited knowledge of Warndarrang, which became extinct in the 1970s.

not derivable from Marra. Indeed, assuming the auxiliary (if that is what it is) is in the Present tense, as we have found for the other Marra verbal forms examined so far, there are no Marra or Warndarrang inflected verbal auxiliaries that end in a nasal. Therefore, if the name is from Yukgul, and it represents a structure ending in an inflected auxiliary, then the inflectional system of Yukgul was quite distinct from both Marra and Warndarrang.

4 DISTINCTIVE ENDINGS FOR PLACENAMES

Two recurrent endings are found on placenames in the Ngalakgan area which cannot be derived from any current morphemes in the language. These prove interesting for the internal reconstruction of the Gunwinyguan family. One of these endings is *-boy*, found on a number of placenames in the area of the Roper River, as well as places further to the north-west, such as *Maranboy* (south-east of Katherine on the Central Arnhem highway).

(17) a. *Larriboy* (Morphy & Morphy 1981)

b. *Ngurruboy* Roper *ngurru* 'lesser salmon catfish'

c. *Martboy* ? Ngkn *mardu* 'coolamon'

d. *Waluboy* ? Roper *bala ~ wala* 'side; bank'

e. *Jalboy* ? cf. Jwyn *jarr* 'thigh', Ngkn *jarrpbitj*; or Ngkn *jala* 'mouth'

Both *Larriboy* and *Ngurruboy* are just west of Bardawarrkga, near a Ngalakgan-speaking clan centre on the Roper River. *Jalboy* is currently applied to one of the tributaries of the Roper, on the western edge of Ngalakgan-speaking territory. *Waluboy* is in the vicinity of Roper Bar. The location of *Martboy* is obscure.

Only a couple of these names have a plausible analysis. Most are opaque to Ngalakgan speakers, with the exception of *Ngurruboy*, which was said by claimants to refer to where the Catfish ancestor (*Ngurru*) crossed the Roper. The element *-boy*, which forms the ending of these words, is mysterious: in none of the nearby languages do we find a potential source for this element. For instance, *-boy* does not look like the locative suffixes of any of the local languages.

One possibility deserves some attention here. That is that *-boy* in placenames preserves some kind of lative 'across, around, along' or Associative/Comitative 'having, with' element. The nearest morphological analogy comes from Jawoyn, which has an adverbial prefix or bound stem *boy-* with a number of senses, one of which is 'transitivity and motion across, especially perlative and transfer via another party' (Merlan 1989b): *nga-boy-bim-bunay* 'I painted it right across'; *wal-boy-bi-jungay* 'I told you sg. via/across (someone)'.[28] Ngalakgan and Rembarrnga have (per)lative suffixes *-wi*, *-wə* respectively. Ngandi has *-pbitj ~ -bitj*, and Nunggubuyu *-baj ~ -waj*.

---

[28] A correspondence between a suffix in one language and a prefix in another language is not unheard of in Gunwinyguan: cf. the Ngkn, Rmba nominal instrumental/comitative suffix *-yih*. In Mayali, this suffix is specialised to Instrumental and Proprietive functions in nouns in some dialects and is absent in others (Evans 2000:211), but is found as a productive Comitative prefix on verbs. Presumably, nouns inflected with this suffix were incorporated into the Mayali verb, and the suffix was retained as a comitative marker. This prefix is not used in Ngkn and Rmba, rather the comitative prefix *bartda-* (which co-occurs with *-yih* on nouns: *bartda-birn-yih* 'with/having money') is used as the verbal comitative. Presumably something similar could have occurred in Jawoyn at an early stage.

Internal analysis in these languages indicates an original form of the lative suffix for these languages like *-bay* or *-boy*, which was retained as *-boy* in Jawoyn, but underwent various changes in the other southern Gunwinyguan languages. The form *-boy* in these placenames then presumably represents an archaic form which became the modern forms *-baj*, *-bitj*, and *-wi*; synchronically the original form is preserved intact only in Jawoyn and Warray (Harvey pers. comm.).[29]

Heath (1984:207) characterises the Nunggubuyu Pergressive *-baj ~ -waj* as defining 'a zone or field in which or through which some entity is located or in motion'. An element having this kind of interpretation seems like a highly plausible candidate for the derivation of placenames. The name *Ngurruboy* would then mean '[place] where Catfish (*Ngurru*) passed through or is located', again with a headless relative interpretation. It is interesting that the Jawoyn prefix preserves what may be the original form of the suffix in Rembarrnga, Ngalakgan and Ngandi.[30]

The placename *Jalboy* has been extended to the modern (European) name of one of the western tributaries of the Roper: 'Jalboi Creek'. The initial element *jal* is possibly a form of the same stem *jarr* 'thigh' that we find in other names such as *Jarrburdetjbutjjinygah*, as well as (18a) below.[31]

(18) a. *Jarrmunu* cf. Jwyn *jarr* 'thigh'

  b. *Weyamunu* cf. Ngkn *weya* 'shade'

The element *-munu* is, like *-boy*, unknown in Ngalakgan. In Jawoyn there is a proprietive ('having') suffix *-muna*. Proprietive suffixes, like latives/pergressives, are plausible endings for placenames, and presumably have a similar force to the Dative pronominals used on Ngalakgan names such as *Berre-ngoji-kgah* 'her brisket'. Hence, (18a) *Jarrmunu* would mean 'having a thigh', and (18b) *Weyamunu* 'having shade', though the vocalism difference between Jawoyn *-muna* and the ending *-munu* is unexplained. In modern Ngalakgan, the proprietive is expressed analytically, with the Comitative prefix *bartda-* and the Instrumental/Ergative suffix *-yih*. One or both of these elements are found over a wide area including Ngandi, Rembarrnga, Mayali and Dalabon, perhaps suggesting a recent diffusion of this morphology. It is noteworthy that the suffix *-muna* is rare in Jawoyn according to Merlan (1989b).

---

[29] There are other correspondences Ngdi, Ngby final /c/: Ngkn, Rmba /y/ or /ʔ/ which make this story more plausible.

Ngalakgan	Ngandi	Nunggubuyu	
ngey	*ngitj-*	*mitj-*	'name'
-hwi	*-hwitj*	*-watj*	pronominal emphatic
-(k)gah ALL/LOC	*-gitj* ALL	*-watj ~ -gatj*	PERL
-bugih	*-bugih*	*-wugitj*	clitic 'just, only'

[30] This seems to be a common characteristic of prefixal stems in Gunwinyguan, since some prefixal or initial bound stems preserve archaic morphemes no longer used independently. Besides the Myli comitative prefix *yi(h)-* mentioned above, the Ngdi form *ngitj-* 'name' occurs only as a bound stem in verbs and compound nouns. In Ngalakgan the cognate *ngey* can occur both independently and in compounds.

[31] I have no explanation for why the liquid coda should be *l* in *Jalboy* but *rr* everywhere else. The other possibility is that it is related to the common Gunwinyguan term for 'mouth': *jala* in Ngalakgan.

5 OVERLAPPING RANGES

Around half to perhaps two-thirds of the placenames in the Ngalakgan-speaking territory have no derivation in any language, a figure comparable with those reported elsewhere (e.g. Sutton, this volume).[32] A number of these names have what look like Alawa or Marra or Yukgul derivations, including the tell-tale Alawa locative suffix allomorphy, as well as phonotactic patterns not found in Ngalakgan. The question these names raise is, does Ngalakgan represent a recent arrival in this area? This is a pertinent question given the extensive range of Gunwinyguan languages throughout Arnhem Land, and the large number of Ngalakgan placenames that are transparent. What we find, however, is that although many names in the Ngalakgan area appear to have Alawic or Marran features, *none* of them are synchronically derivable from either of these languages with any degree of certainty.

5.1 Phonological and phonotactic patterns

It has been noted by Harvey (1999:162) that phonological patterns are of little use in determining placename–language associations, since the majority of Australian languages have very similar phonological systems. There are just a few features that distinguish the various languages of the claim; these are set out in (19).

(19) a. **Vowel inventories**. Ngalakgan and neighbouring Mangarrayi and Ngandi have a five-vowel system /a, e, i, o, u/. Alawa has a four-vowel system: /a, e, i, u/, with restrictions on the distribution of /e/. Marra and Warndarrang have a three-vowel system: /a, i, u/.

   b. **Consonant inventories**. Ngalakgan, Ngandi and Mangarrayi contrast five supralaryngeal places of articulation in stops and nasals, and two in laterals. These languages also have a phonemic glottal stop. Alawa, Marra and Warndarrang contrast stops and nasals at the same five places, except that Alawa in addition contrasts a palatal lateral (Sharpe 1972:18 notes this is rare). Ngalakgan and Ngandi contrast simple (short) from geminate (long) stops. Mangarrayi, Alawa, Marra and Warndarrang lack such a contrast. Sharpe (1972) proposes that Alawa formerly had a contrast between simple and pre-nasalised stops however. (Initial nasal+stop clusters are also found in Marra and Warndarrang, but it is not clear that these are pre-nasalised stops.)

   c. **Phonotactics**. Ngalakgan disallows clusters of liquid plus alveolar stop within roots, and such clusters are rare inter-morphemically also. Alawa and Marra allow such clusters both root-internally and inter-morphemically. All languages allow a wide range of word- and morpheme-final consonants and clusters.

All three kinds of patterns could potentially distinguish the origin of placenames in this area. The problem is again that speakers translate between phonological systems with ease. One example is the placename *Narakgarani* (no derivation), a site on the bank of the Roper River

---

[32] For example, on the land claim map of Morphy and Morphy (1981) there are 51 names in an area approximately 540 km^2 (i.e. roughly the area shown in Map 2). Of these, just 11 have a regular translation (from Ngalakgan) of which we can be confident. None of the other names has any regular translation in any other language.

downstream from the Roper–Wilton junction. This name has a long stop for Ngalakgan speakers (symbolised with the digraph 'kg'): [nàɹakkaɹáni]. But for speakers of Marra, as well as for Ngalakgan people who are not speakers of Ngalakgan, the name is pronounced [nàɹagaɹáni], where the Ngalakgan geminate stop is equivalent to a simple stop in Marra.[33]

One form which could prove useful is the name *Rono*. This is a place several kilometres north-east of Urapunga community, probably on the earlier border with Yukgul. The name has two /o/ vowels which we have noted are not found as phonemes in Alawa, Warndarrang or Marra. Hence, the form of the name would tend to support its origin in a language such as Ngalakgan. However, this name has no derivation in Ngalakgan.

The fact that the name has two instances of /o/ does not necessarily provide good evidence for its origin however. Sharpe (1972:19) notes that in Alawa both /a/ and /u/ are realised with low back allophones ([ɔ] and [o] respectively) in certain environments. Sharpe's statement is as follows:

(20) /a/   [ɔ]   occurs infrequently fluctuating with [a] following velar consonants,
                    and between retroflex consonants (including [ɹ]).

      /u/   [o]   only occurs contiguous to a liquid, alveolar semivowel [ɹ], or palatal,
                    or in one syllable closed syllable words.

It is possible, then, that [ɹono] is the phonetic realisation of an earlier form */ruwunu/ or */runu/. Other placenames from the Ngalakgan area are realised with variable vocalic quality depending on the speaker. Examples are given in (21).

(21) a.   Golotdoh [gɔ́lɔtto?]   (Spring behind Mount McMinn)

      b.   Gorongah [gɔ́ɹɔŋa?]   (Spring north of Urapunga)

Marra speakers pronounce these two placenames as [gʊludu], [gʊɹuŋa], respectively. Speakers appear to translate foreign phonological contrasts into their own inventory. The existence of a phonological contrast in a placename is no guarantee that the placename derives from a language that maintains that contrast in its lexicon. All three of these names with the vowel /o/ could have derived from names like the Marra forms given, since in each case the phonological environment in the name is one that is conducive to realising /u/ as [o].

As Harvey (1999:165) finds, distinctive phonotactic patterns are potentially more useful in determining placename derivation than are phonological contrasts. An example of this from the Roper is the placename *Yirriwurlwurldi*. This place is at the mouth of the Roper, where the water is disturbed and forms whirlpools. The phonotactics of this placename — with an [ɭɖ] cluster — constitute negative evidence that it is unlikely to derive from a language such as Ngalakgan or Ngandi, both of which disallow clusters of liquid plus alveolar stop intramorphemically. The form of this name also suggests that *-di* is a suffix, since reduplication in Warndarrang and Marra targets a disyllabic portion of the base, unless the base is monosyllabic.[34] Neither Ngalakgan nor Ngandi has a suffix of this form. Therefore, an

---

[33] Alawa speakers tend to have long and voiceless stops, at least in some positions in some words, very similar to geminates in Ngalakgan. This may be a reflection of the stop contrast which Sharpe claims existed formerly in the language.

[34] As we have seen in Table (2), *-dV* is a characteristic locative ending for nouns in Alawa, where the stem ends in *-rl* or *-l*. However, in Alawa this allomorph of the locative copies the final vowel of the stem: **Yirriwurlwurldu* is the expected form.

origin in a language like Warndarrang or Marra, which allows such clusters, is indicated in this case.

There are several names in what is now regarded as the territory of Ngalakgan-speaking clans which have phonotactic patterns foreign to Ngalakgan. One placename recorded by the Morphys (1981) as *Balalayarrurru*, just to the south-west, has a phonotactic pattern that is extremely rare in Ngalakgan but quite common in Alawa. Ngalakgan disprefers sequences of liquids, especially sequences of the tap [ɾ] or glide [ɹ]. Sequences such as **yarrurru* do not occur in Ngalakgan.[35] Such sequences are well represented in Alawa, however, which has forms like *gurrurreru* 'dingo; wild dog'.

5.2 Foreign morphology

The placenames in (22) have the general appearance of being derived from Alawa, Marra or a related language. All of them are on the south-eastern borders of the territory generally accepted as Ngalakgan.

(22)
a.	*Wirlbirlbarri*		Knuckey Bluff near Urapunga
b.	*Yurlhbunji*		Old Roper Bar police station area
c.	*Walanji(wurr)*		'Alligator Bluff'
d.	*Yurende*		A swamp east of the Wilton–Roper junction
e.	*Wararrirr*		Wilton–Roper junction
f.	*Nabòrdopburláni*		A creek east of the Roper Bar airstrip
g.	*Náwarnbarnkùlyi ~ Wàrnbarnkúlyi*		Roper Bar airstrip
h.	*Ngalardarra* archer.fish	*bìn=gu-línma* ??=3sg-CAUS+PP *~ Warrba binggurlimba*	Tidal limit of Hodgson River
i.	*Nàbarlmántji*		'Pine Bluff'
j.	*Wànkarnangínji*		Spot on Hodgson River west of Castle Hill
k.	*Riwanji*		Swamp near Queensland Crossing

One of these placenames — *Walanji* (22c) — has three variants: *Walji*, *Walanji* and *Walanjiwurr*; the last with an ending *-wurr*. This ending would appear to be a suffix of some kind, since we have observed optional suffixes in examples like *Mirnitjja-ngoji-(kgah)* 'where her shade trees are'. However, none of the extant languages of this area preserves a suffix of exactly this form. Marra has an allative/locative suffix *-yu(rr) ~ -nyu(rr)*. The Alawa allative case is similar, but like the locative discussed earlier it takes a range of allomorphs, depending on the final syllable of the stem. One allomorph ends in *-(n)dVwurr*, that for stems ending in a

---

[35] In a database of 1,268 Ngalakgan roots, only 10 such roots occur with an identical sequence of liquids. In every case, roots were of the form $CV_1LV_1LV_1$, where every vowel was the same, e.g. *mululuk* 'conkerberry ('dog's balls')'.

liquid or retroflex glide. The allomorph for stems ending in a palatal nasal (which would be the predicted form if the stem is *walany) is -njirru.

My consultant commented on placename (22a) thus: 'thet Wankarnangintji thet kro bin ibap' ('Wankarnangintji is where that crow heaped up', or possibly 'heaved up').[36] This name appears to be a combination of one Marra term for 'crow' wanggarnangin, with Alawa locative suffix allomorph -ji. (Possibly my rendition of the name is a mistranscription of the first nasal-stop cluster). Wankarnangintji, Yurlhbunji and Walanji all end in the sequence [ɲɟi], which as we have seen is a locative allomorph in Alawa for words ending in -y, or a nasal or oral stop. None of these names can be derived from Alawa however. Yurende is possibly yuruwerr+LOC 'bowerbird place' (yuruwende would be the regular outcome, and jurerr is an alternative term for bowerbird which also occurs in Ngalakgan); as far as I know there are no bowerbird associations with this place.

Therefore, in the territory now associated with the Ngalakgan language we find names, such as Bolkgotj-ngoji-kgah, which are transparent in the language, as well as names which are probably derived from some other language similar to Marra or Alawa. Taken together, these facts would tend to indicate a recent origin in this area for Ngalakgan. This idea is even more attractive given that Ngalakgan is so closely related to Rembarrnga, which is spoken a long way to the north. Nevertheless, names such as Jarrburdetjbutjjinygah and Wandarrganiny indicate a considerable length of time for Ngalakgan speakers between the Wilton and Hodgson rivers, and the western edge of Eagle Bluff, south of the Roper River, since they contain morphology that is distinctively Ngalakgan, yet they cannot be derived regularly from the language. In addition, Jarrburdetjbutjjinygah is the name used by Alawa and Marra speakers also. Therefore it cannot be a recent calque. It is also within this area that we find names, such as Balalayarrurru and Riwanji, which phonotactic patterns and morphology foreign to Ngalakgan. That is, within the same local area, we find names seemingly from a range of different linguistic strata. Indeed, many names (perhaps around a half to two-thirds) in the Ngalakgan area have no clear derivation in any language. It is notable that, of the names in the Ngalakgan area that look like Alawa, none is in fact transparent synchronically.

Such patterns are common elsewhere in the world. In Scotland for example, we find Northumbrian English names such as Eldbotle and Morebattle (<OE eld 'old' and mere 'lake' plus bōðl 'a dwelling') (Nicolaisen 1976:77). We also find many names derived from Scottish Gaelic such as Baldornoch (Gael. baile 'township' dornach 'pebbly') (Nicolaisen 1976:139). In addition, we find names such as Shiel, which cannot be easily derived from Germanic or Celtic sources (Nicolaisen 1976:189). The British situation is not strictly comparable however, since small group and linguistic affinities to particular tracts of land appear to be more important and more enduring in Australia. One reflection of this difference is the fact that dreaming myths and placenames sometimes make direct reference to language names, such as the placename Marra ngarl-ngarl-n-amban '[where] he started talking Marra'. As Rumsey (1993:200ff) has noted, such myths and placenames set up a cosmological association between land and language with which it is difficult to find analogies in the European context.[37]

---

[36] I am not sure what 'heave up'/ 'heap up' refers to here.

[37] Of course, there is commonly in the UK a deeply-felt connection between localities and dialect variants. So the difference is one of degree perhaps.

Having discussed the morphological patterns and derivation of a range of placenames in the Roper area, in the final section I address the issue of how these names are used by speakers.

6 THE LINGUISTIC USE OF PLACENAMES

In this section I discuss the interaction of placenames with the morphology of the languages, as well as their extension in the actual world.

6.1 Grammatical status

The status of placenames in the language — as transparent or opaque forms — is reflected in the way they are used by speakers. The placename *Mirnitjja-ngoji-kgah,* for instance, can also be referred to as *Mirnitjja-ngoji* 'her shade tree', or simply *Mirnitjja.* Similarly, the place *Dubal* 'Leichhardt tree' can also be referred to (with noun class prefixes) as *mu-Dubal,* or *mun-gu-Dubal,* where *mu-* and *mun-gu-* have a function like that of the English articles 'a', 'the' respectively. If I want to say 'I'll meet you at *Dubal',* then I say '*mu-Dubal-kgah*' 'at the Leichhardt tree'. Similarly the placename *gu-Mirnitjja-ngoji-kgah* takes the neuter noun class prefix *gu-* appropriate to the term *mirnitjja.*

Names like *Bunditjgah* are quite different. Recall that this name is untranslatable, though it appears to contain the Ngalakgan locative suffix *-kgah ~ -gah.* This name can take articles also: *gu-Bunditjgah.* Here, the name takes a neuter class prefix. Neuter class is the noun class associated with topography in general. This name cannot lose the locative suffix *-gah* however, unlike *Mirnitjja-ngoji-kgah* above: *Bunditj* is not an equivalent for *Bunditjgah.* The name *Bunditjgah* is interpreted by speakers in all three of locative, allative and nominative case roles, without any change in the name itself.

The suffix *-kgah ~ -gah* has both locative and allative functions, but Ngalakgan also possesses a specific allative suffix *-kgagah ~ -gagah* (possibly a frozen reduplication of *-kgah ~ -gah*). This case suffix, which is infrequently used, can occur on placenames; Merlan (1983:46) has the example *Warnbarnkulyi-kgagah* (cf. 22g above).

The occurrence of the allative is apparently limited to those placenames which do not already contain the locative suffix as a typical or possible part of the name itself. Informants specifically rejected the form **Bunditjga-gah* [bùndickága?] 'at/to *Bunditjgah*' in either of the locative or allative senses.[38] Informants also rejected this suffix on **Mirnitjja-ngoji-kgagah* 'to her *mirnitjja* trees'. At first glance this seems odd, since the name is entirely transparent, and the sequence *mirnitjja-ngoji-kgagah* is otherwise perfectly acceptable in Ngalakgan.

The unacceptability of the allative form fits with Harvey's (1999) analysis that names with locative suffixes are to be interpreted as headless relative clauses. The locative suffix can derive such headless relative interpretations, but the allative cannot. Although the name could be interpreted as *mirnitjja-ngoji-kgah-gah,* that is, as a LOC/ALL form of a headless relative, such multiple case 'stacking' is ruled out in Ngalakgan. The unacceptability of **Bunditjgagah*

---

[38] Stop-initial suffixes, such as *-kgah ~ -gah,* in Ngalakgan alternate between geminate and simple stop realisations, depending on the prosodic and segmental structure of the preceding stem. These alternations can be local (as here) or non-local; see Baker (1999, §4.3) for discussion. The process is also found in Ngandi and Rembarrnga.

in addition indicates that this name also retains a relative clause interpretation for speakers, even though it is not clear in this case what the clause is.

6.2 Prefixes on names

All of the languages of this area (Mangarrayi, Ngalakgan, Ngandi, Alawa, Marra and Warndarrang) have a series of prefixes for person and/or gender (and, in the case of Mangarrayi, Alawa and Marra, case) which seem to appear on a number of placenames in the two claim areas examined here. Fifty-two, or just over 20 per cent, of attested placenames in the St Vidgeon's area begin in the sequence *na-*, *nga-*, or *nya-*, all three of which are prefix forms in Marra and/or Alawa. Again, while the placenames often have the right morphology and prosody (stress on the stem) to be prefixed words, often the stem cannot be identified with a meaning. I examine a few examples below; discussion follows the example in each case.

(23)   *Nya-Márranguru*

This is the name of a place on the south side of the Roper River, south-east of *Wardangaja* lagoon. Once again, the place lies within what was probably Yukgul-speaking country. The name is transparently Marra *nya-marranguru* 'head'. The prefix *nya-* is the oblique form of the neuter prefix in Marra, which normally requires a suffix *-yu(rr) ~ -nyu(rr)* in the locative. Heath (1981:92) notes that placenames, exceptionally, are realised with the oblique prefix in a locative sense, without the locative suffix.

The stress in *nya-Marranguru*, falling on the first syllable of the noun stem, is the characteristic stress pattern of nouns with polysyllabic stems in all the languages of the Roper (see Baker 1999). Its retention in the placename indicates that the word is not opaque to speakers: stress is normally initial in unprefixed stems in Marra.

(24)   a.   *Nya-Mayigarl*   N.OBL-pandanus.nut

        b.   *Na-Mala-yurr*   M.OBL-cloud-LOC

Example (24a) is an island in the Towns River about 4.5 kilometres downstream from the road crossing. The example (24b) signifies a big swamp. The word is transparently 'at the cloud' (Mrra), or else 'at the navel; where the navel is' but in the latter case with anomalous class assignment. Note that for this name the locative suffix is part of the citation form of the placename. Heath (1981:92) further notes that besides names such as *Wiliyurru*, with optional prefix, there are others, such as *Nawarrwarr* (Mount Moore), where the prefix is frozen and cannot be omitted.[39] Stress on this name reinforces this analysis: [náwarwar]. However, I previously mentioned (at example 22) the name *Náwarnbarnkùlyi ~ Wàrnbarnkúlyi*, which is unanalysable, but which nevertheless also has both prefixed and unprefixed forms (and note that the prefixed form has initial stress). At least for Ngalakgan speakers, these forms are in free variation and do not seem to derive distinct case functions. Indeed, this name can take Ngalakgan prefixes and case-suffixes: *gun-gu-Warnbarnkulyi-kgah* 'at W'. In this case, the Marra prefix is always omitted.

The grammatical structures in which placenames occur confirm what we have found in terms of etymology: placenames have a range of morphological transparency in the languages concerned, from fully analysable to completely opaque. Even where a name is not fully

---

[39]   See Heath (1981:91–93) for further discussion of the use of morphology with placenames in Marra.

analysable (*Bunditjgah, Nawarnbarnkulyi*), speakers seem to recognise (in some cases) the existence of morphological structure, and treat the names accordingly.

6.3 Referential range

For the placenames of which I have some experience — those around Roper Bar and the area surrounding Bardawarrkga and Urapunga communities — there are primarily two types of names: those that name large-scale regions or topographic features (such as rivers, hills and clan territories), and those that name a specific (usually small) area within such a topographic area, such as a rock waterhole, a deep pool in the river, a stand of tree species, a billabong or the like. Generally, such names have an identifiable range or limit. The name *Yawurlwarda*, which is applied to several billabongs around the Roper, refers to the billabong and the immediately surrounding bank. The reference of the name is bounded by the extent of the topographic feature to which it refers. Similarly *Marrangarh*, the name for the Wilton River, is bounded by the river and its banks.

The places *Yawurlwarda* and *Marrangarh*, being bodies of water, can only have this kind of bounded reference. But terms that refer to objects — trees, rocks, hills, caves — tend to have both a specific localised reference to that feature, and also a more general reference to the surrounding area. Such is the case with the placenames that are tree species names: *Mu-Dubal* (Leichhardt tree), *Mirnitjja* (*Cathormion umbellatum*), and *Motjo* (≈ Coolibah *mutjju*). In each case, the name refers to the specific tree, or stand of trees, that rest at that specific spot. This kind of name seems to encompass as many trees of that species which are found at that place, so in the case of *Mu-Dubal* it is currently one, *Motjo*, also one (dead one), and for *Mirnitjja* it is a grove of *Cathormion* trees. The name also refers to the area around these trees: both *Mu-Dubal* and *Motjo* stand on the bank of the river. At the Roper Bar boat landing called *Jinji*, there used to stand two sacred paperbark trees (*mu-bulpbul*). These have died and fallen down, but others are growing in their place and are referred to in the same way as manifestations of the Plains Kangaroo. This confirms that it is the link between particular species and particular sites which is important to the Ngalakgan, not the actual tree which happens to be the (temporary) embodiment of the ancestor during its lifetime.

7 ROPER PLACENAMES AS UNIQUE IDENTIFIERS

I claimed in the introduction that the major formal difference between the placenames in the Roper and placenames in Britain is the scarcity of topographic terms in the former. In Britain, Nicolaisen (1976) and Gelling (1984) characterise the most common placenames as a compound of a generic and a specific, qualifying or modifying term (Germanic names have the order specific+generic, and Celtic names generic+specific). The generic is a topographic or settlement term such as *mountain, hill, pool, creek, town, ridge* or their equivalents in other languages such as Gaelic. Transparent examples from the UK are such names as *Black-pool*, *Christ-church, Sliabh-sneacht* ('mountain [of] snow'), *Penfro* ('head/end-land'). This kind of naming has been transferred with British settlers in Australia: *Peat's Ridge, Roper River*, etc.

There are no names like this in the Roper. I know of no names in the Ngalakgan area that include a topographic term, whether analysable or not. These terms certainly exist in Ngalakgan — *birn* 'rock, hill', *bo* 'river', *langga* 'billabong', *gabogaboh* 'plain, open place', *ruwurr* 'ridge' — and are frequently used by speakers to refer to the landscape. They are never found in placenames in combination with modifying terms however.

In the St Vidgeon's area, there are a handful of Alawa placenames that are topographic terms, these are presented in (7) above; e.g. *Ngarrgalirr* 'high ridge, bank'+LOC, and *Leguldu*, meaning 'deep depression'+LOC. Again, there are no names combining generic and modifier terms. Furthermore, the majority of Alawa placenames end in one of the locative allomorphs, implying the same kind of headless relative clause interpretation we find in Ngalakgan.[40] In that case, *Ngarrgalirr* is not simply 'high bank', but 'where [he/she left] a high bank' or possibly 'where [he/she did something] at the high bank'.

As well as the lack of placenames formed through compounding of topographic terms, in actual use placenames do not co-occur with topographic terms in a phrasal construction either. Speakers never (to my knowledge) say such things as *gun-gu-birn-Walanji* (IV-DEF-rock-[toponym]) 'Walanji Hill' or *gun-gu-langga-Yawurlwarda* 'Yawurlwarda Billabong'. They *do* use such constructions in English and Kriol: *Mabiligalu Plain, Walanji Hill, Walgundu Billabong, Wurrinjal Cave*.[41] The lack of generic+modifier constructions in placenames does not derive from any morphological restrictions in Ngalakgan: generic+modifier compounds are productive and very common, e.g. *langga-ganyah* 'small billabong', *weh-balkginy* 'salty water; beer', *gurndu-mah* 'good country'.

I suggest that the lack of topographic terms, and in particular compounds involving topographic terms, in placenames in the Roper reflects a difference in the basis of naming, and more deeply, of conceptions of place. In the Roper, places that are named constitute the manifest embodiment of the ancestral creator beings. What non-Indigenous people to the area call the 'Hodgson River', the locals think of simultaneously as the track of the Mermaids *gilyirringgilyirri*. The ridge of ground above Roper Bar is not simply a local geological inclination, but is also the point where the Plains Kangaroo mob left themselves as *mutjju*, a coolibah tree which grows there.

It is the fact that topographic features are meaningful in the Roper cosmology which makes sense of the fact that they tend to have names reflecting cosmological events and manifestations. Placenames in the Roper have many of the characteristics of personal names, indeed, most personal names are derived from placenames within the clan territory. In this respect, it makes sense that placenames do not co-occur with generics, since we do not find constructions such as *nu-gu-bigurr-Golokgurndu* (I-DEF-human-[personal name] '*Golokgurndu* man') either. That is, the typical Roper placenames cannot be combined with a generic because they are unique identifiers, like personal names. The use of a generic would be redundant, given that the name uniquely identifies its referent.[42]

---

[40] I have been unable to confirm whether Alawa and Marra speakers allow such constructions.

[41] One (Alawa) speaker used such constructions frequently, other speakers never used them.

[42] There are (rare) cases in which two distinct places share the same name. An instance of this in the Roper is the name *Yawurlwarda*, which is applied to a billabong near Roper Bar, as well as one near Ngukurr, dozens of kilometres away. In a text recounting the Plains Kangaroo myth, the narrator distinguished them with an adjoined clause:

gun-gohje-bugih	*gu-langga*	*gu-gurndu-ngey-bu+n*	*'Yawurlwarda'*
IV-there-just	IV-lagoon	NP-country-name-[hit+]PP	[toponym]

That drimin bin kolim that bilabong 'Yalwarda', lil bilabong.
'That billabong is called "Yawurlwarda".'

langga-ganyah,	*nomo*	*gun-gohje*	*gu-yenipbi*	*gun-gu-ngolkgo,*
lagoon-small,	NEG	IV-there	IV-whatsit?	IV-DEF-big

CONCLUSION

The discussion has shown that Roper names take a wide variety of forms: bare noun, noun inflected for case/possession/class, inflected verb; many names have an interpretation equivalent to a clause in English. One interesting factor is the large proportion of names (throughout the Roper) that have either no translation, or are not entirely analysable in any of the languages now spoken in the area. I have suggested that in some cases these names point to the presence of earlier languages, such as Yukgul and Warndarrang. In many other cases, these names bear no resemblance to any of the commonly found morphological patterns for toponyms in the area. There remain a significant number of names that are only partially analysable, pointing to a long continuity of occupation for the language groups now found in the Roper.

REFERENCES

Alpher, Barry, Nicholas Evans and Mark Harvey, to appear, Proto-Gunwinyguan verbal suffixes. In Nicholas Evans, ed., *Studies in Comparative Non-Pama-Nyungan*. Canberra: Pacific Linguistics.

Baker, Brett, 1996, Verb classes and subgrouping in Gunwinyguan. MS, University of Sydney.

— 1999, Word structure in Ngalakgan. PhD thesis, University of Sydney.

Bern, John, 1974, Blackfella business/whitefella law. PhD thesis, Macquarie University.

Bern, John and R. Layton, 1984, The local descent group and the division of labour in the Cox River land claim. In L.R. Hiatt, ed., *Aboriginal Landowners*, 67–83. Oceania Monograph No. 27.

Evans, Nicholas, 1995, Dangbon materials. MS, University of Melbourne.

— 2000, Bininj Gun-wok: a pan-dialectal grammar of Mayali, Kunwinjku and Kune. MS, University of Melbourne.

Gelling, Margaret, 1984, *Place-names in the Landscape*. London: J.M. Dent & Sons.

Harvey, Mark, 1999, Place names and land-language associations in the western Top End. *Australian Journal of Linguistics* 19(2):161–195.

— to appear, Gunwinyguan phonology. In N. Evans, ed., *Comparative Non-Pama-Nyungan*. Canberra: Pacific Linguistics.

Heath, Jeffrey, 1978a, *Lexical Diffusion in Arnhem Land*. Canberra: AIAS [AIATSIS].

— 1978b, *Ngandi Grammar, Texts, and Dictionary*. Canberra: AIAS [AIATSIS].

— 1980, *Basic Materials in Warndarang: grammar, texts, and dictionary*. Canberra: AIAS.

---

onli   gu-langga-ganyah,   gohje-bugih
only   IV-lagoon-small,   there-just
Onli that bilabong lilwan, nomo that bigwan - onli lilwan.
'It's a little billabong, not that big one, only that little one.'

   This is one instance where we might expect to find a construction such as *Yawurlwarda-ganyah* 'small Y', or *Yawurlwarda-ngolkgo* 'big Y', but no such examples are attested.

— 1981, *Basic Materials in Mara: grammar, texts, and dictionary*. Canberra: Pacific Linguistics.

— 1984, *A Functional Grammar of Nunggubuyu*. Canberra: AIAS [AIATSIS].

McKay, Graham, 1975, Rembarrnga: a language of central Arnhem Land. PhD thesis, Australian National University.

Merlan, Francesca, 1978, Making people quiet in the pastoral North. *Aboriginal History* 2(11):70–106.

— 1981, Land, language and social identity in Aboriginal Australia. *Mankind* 13(2):133–148.

— 1983, *Ngalakan Grammar, Texts and Vocabulary*. Canberra: Pacific Linguistics.

— 1989a, The genetic position of Mangarrayi: evidence from nominal prefixation. Paper presented at the Australian Linguistic Society Conference. [To appear in Evans, ed., *Studies in Comparative Non-Pama-Nyungan*.]

— 1989b, Jawoyn grammar, texts and vocabulary. MS, Australian National University.

— 1989c, *Mangarrayi* (Croom Helm descriptive grammars). London: Routledge. [First published in 1982, Amsterdam: North-Holland.]

— 1993, Ngalkbon materials. MS, Australian National University.

Morphy, Howard and Frances Morphy, 1981, *Yutpundji-Djindiwirritj Land Claim*. Darwin: Northern Land Council.

Nicolaisen, W.F.H., 1976, *Scottish Place-names: their study and significance*. London: B.T. Batsford.

Rumsey, Alan, 1993, Language and territoriality in Aboriginal Australia. In Michael Walsh and Colin Yallop, eds, *Language and Culture in Aboriginal Australia*, 191–206. Canberra: Aboriginal Studies Press.

Sharpe, Margaret, 1972, *Alawa Phonology and Grammar*. Canberra: AIAS [AIATSIS].

— n.d., The evolution of Alawa: internal and external evidence. MS, University of New England.

9 THE ARCHAISM AND LINGUISTIC CONNECTIONS OF SOME YIR-YORONT TRACT-NAMES

Barry Alpher

1 PEOPLE, COUNTRIES AND LANGUAGES[1]

Yir-Yoront (pronounced *Yirr-Yorront*)[2] speakers' country lies around the mouths of the Mitchell River in western Cape York Peninsula, Queensland. Their land tenure is organised in terms of named tracts, many of considerably less than a kilometre in breadth.[3] Sets of tracts, usually but not always contiguous within the territory of the Yir-Yoront language, make up the estate of a patrilineal clan (Sharp 1934a and b; in more recent and exact terminology, a group in which membership is assigned through serial patrifiliation; see Sutton 1998:24). Some of these clans are represented among speakers of other languages and in territory other than that of Yir-Yoront (Sharp 1958), but Yir-Yoront appears in all recorded sources and, from information obtained currently, to have been spoken within a contiguous territory within which only one very small enclave is recognised as the territory of speakers of another language.

[1] Acknowledgements: I thank Jane Simpson for editorial advice on content, and I thank the Australian Research Council for its support (Grant 'Analysing Australian Aboriginal Languages', administered through the University of Melbourne Department of Linguistics and Applied Linguistics) of the fieldwork (1997–2000) during which I recorded the Yirrk-Mel placenames I report on here.

[2] Yir-Yoront has stops *p, th, t, rt, ch, k,* and *q* (glottal stop), nasals *m, nh, n, rn, ny, ng*; liquids *l, lh, rr*; glides *w, r, y*; vowels *i, e, a, o, u,* and *v* (shwa). Compounds with a hyphen written in the middle have an unstressed first part and a primary-stressed second. Compounds with primary stress on the first part and secondary on the second are written with a plus sign in the middle (in the text; on the map they are written with both parts capitalised and no space in between: *KawnYumlh*). I have also written '+' to separate inflection material from stems where the analysis is relevant. In this paper I am using the same symbols (less glottal stop) and conventions to represent examples from neighbouring languages, despite variations in the sound systems. Simplex words in Yirrk-Mel always stress the second full vowel (not shwa), but stress is unpredictable in Kokapér as written here and is indicated with an acute accent. Kokapér initial (and after a nasal in a monosyllable) /t/ is frequently realised as [trr], an apical voiceless stop with voiced trill release; by convention I have written *Trvpénvmvn* and *Tarrch-Manéngk*, although the latter in particular is often heard with [trr].

[3] See in this connection Sinnamon and Taylor (1978), who show the tract-like rather than point-like nature of most (not all) sites in this area.

L. Hercus, F. Hodges and J. Simpson, eds, *The Land is a Map: placenames of Indigenous origin in Australia*, 131–139. Canberra: Pandanus Books in association with Pacific Linguistics, 2002.
© Barry Alpher

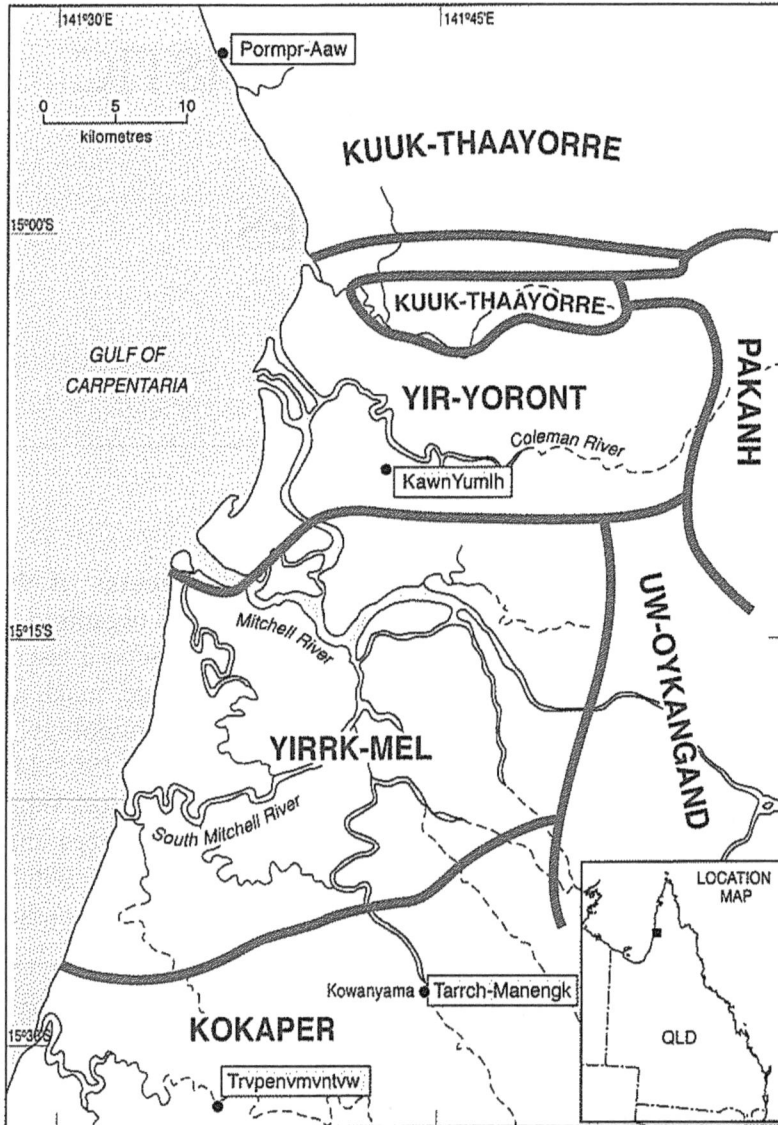

Map 1: Major places and language areas mentioned in the text. Language boundaries are those reported for the 1930s (Sharp 1937) and are schematic only. Sample Indigenous names (those shown on other publicly available maps, and that of the Kowanyama location) are enclosed in rectangles; their locations as shown are approximate only.

South of Yir-Yoront are the territories of its sister-dialect (or sister-language) Yirrk-Mel (Yirrk-Thangalkl, in Yir-Yoront called Yirr-Thangell [thángedl]) and of the closely related Kokapér;[4] north of Yir-Yoront is the territory of the closely related Kuuk-Thaayorre; and to its north-east that of the less closely related Pakanh. Much of the south-eastern part of Yir-Yoront territory was Yirrk-Mel-speaking in the not-too-distant past, and Map 1 reflects the situation as of 1934.[5] In this study I draw on corpora of properly transcribed placenames

---

[4]   This language name is also spelt Koko-Bera, Koko-Pera and Kok-Kaper. I use here the transcription currently favoured by Paul Black, who is preparing a description in depth.

[5]   Land-tenure maps of parts of Yir-Yoront and Yirrk-Mel territory can be found in Sharp (1937) and Sinnamon and Taylor (1978).

numbering roughly 100 for Yir-Yoront, roughly 50 for Yirrk-Mel, and rather fewer for Kuuk-Thaayorre and Kokapér.

2 THE INTERNATIONAL NATURE AND COGNATION OF SOME OF THE PLACENAMES

Each of these languages has names for tracts in the territories of the others, but the names are not simple duplicates. Often these names are direct translations of the foreign names:[6] Kuuk-Thaayorre *Pormpvr-Aaw* (the contemporary settlement Pormpuraaw, formerly Edward River Mission) and Yir-Yoront *Ngolt-Thaw* name the same place and both mean 'door'. The Kuuk-Thaayorre form consists of *pormpvr* 'house' and *aaw* 'mouth'; the Yir-Yoront form consists of *ngolt* 'house' and *thaw* 'mouth'. In some cases, the elements of a placename in two or more languages not only translate each other but appear to be cognate: they continue a single form, affected only by sound-change, from the language ancestral to the modern languages. In the case of the name *Pormpvr-Aaw/Ngolt-Thaw*, the second element is cognate in Kuuk-Thaayorre and Yir-Yoront, continuing a form **tjaawa* from Proto-Paman.[7] Other examples of partially or wholly cognate tract-names follow:

(1) The Kokapér site *Trvpénvmvntvw* (Trubanaman, the original site of Mitchell River Mission in 1904), also named by Yir-Yoront *Lipalh*; both mean 'termite-mound'; each contains a suffix (the suffixes are non-cognate); the stems preceding the suffixes, *tvpen* in Kokapér and *lipn* in Yir-Yoront, are cognate and continue Proto Paman **tipan*.

(2) The Kokapér site *Tarrch-Mvnéngk*, also named by Yirrk-Mel *Tarrch-Mono* and Yir-Yoront *Larry-Munu* (also *Larry-Monql*);[8] all mean '*Morinda citrifolia* fruit [stuck] in the throat'; the first element of each designates the fruit and continues ancestral (local) **taarrtja*, and the second means 'in the throat' and continues ancestral **manu-ngku*.

(3) The Yirrk-Mel site *Thetethrr-Thak*, also named by Yir-Yoront *Therteyrr-Tha*; the first element means 'sandpaper fig'; both elements correspond regularly.

---

[6] I avoid the term 'loan-translation' here because at least some of the time these pairs of names are references to commonly held mythic material not necessarily having priority in the language proper to the named place.

[7] The forms that I am calling 'ancestral' are drawn from the set reconstructed by Hale (1964, 1976a, b, c) and augmented by myself, for the protolanguage called 'Proto Paman' postulated as ancestral to the contemporary languages of the Peninsula region. Regularly corresponding cognate sets, for which a reconstruction (a putative ancestral form) is possible but which cannot be said to have belonged to Proto Paman or any identifiable subgroup, I designate 'local' (Alpher & Nash 1999).

[8] The principal site adjacent to the settlement Kowanyama. The latter name is an English rendition of Yir-Yoront *kawn yamar* 'many waters', evidently used since the 1916 move of Mitchell River Mission from Trvpénvmvntvw, where the water sources had become saline.

3 THE GRAMMATICAL NATURE AND DEGREE OF SEMANTIC TRANSPARENCY OF THE NAMES

A great many Yir-Yoront tract-names are transparent in meaning (as for example those in (1) and (2)); some, however, like *Puyvl* (Puyul), *Mirtayrr* and *Kunqvmvl* resist interpretation. As a very rough guess, simplex placenames with no other designation than the place itself constitute no more than 20 per cent of the total. For those whose meaning is interpretable, the meaning is usually elicitable in a straightforward manner from native speakers and often depicts an event in a myth connected with the tract. For some, however, only the first element can be given a ready gloss while the second cannot; examples are *Therteyrr-Tha* (*therteyrr* 'sandpaper fig'; example (3)) and *Kawvn+Yumvlh* (*kawvn* 'water').

All tract-names are proper nouns. Most of them contain more than one part, or constituent. Of these constituents, the first always belongs to the noun category. The second constituent can be a verb with tense-ending, or a noun. Tract-names with a noun as the second constituent take the form of noun-plus-noun or noun-plus-participle (a participle being a noun derived from a verb) compounds. Single-word tract-names, however, are not uncommon: *Purr*, *Therlh*, *Puyvl*, *Mirtayrr*, *Kunqvmvl*.

Tract-names whose second constituents are nouns often contain an overt LOCATIVE case-suffix (such are the *+vmvntvw* and *+lh* of (1)). With common nouns, like *yalq* 'road, track', the locative suffix indicates 'at' or 'on' or sometimes 'to' the referent of the stem: *yalq+a+lh* 'on the road'. The unmarked, or citation form (the ABSOLUTIVE), *yalq*, can function as the subject of an intransitive verb or the object of a transitive verb and is the form normally given in response to a question of the type 'How do you say "road"?' With tract-names, on the other hand, the locative suffix appears to be frozen in place and to have lost the ability to contrast semantically with its absence. So it is, for example, with *Wangrr-Yalqalh* 'ghost-track', a noun-plus-noun compound with *wangarr* 'ghost', *yalq* 'track, road', and the locative suffix *+a+lh*; the corresponding form with no locative suffix, *wangarr-yalq*, is understood as a common noun, 'ghost's track', and not as a tract-name.[9]

It is likewise with tract-names whose last element is a participle: *Larr-Low+Pannan* 'Crying-Place', with *larr* 'place', *low* 'crying' (a noun), *pann* 'crying' (a participle; *low+pann* 'crying' is a participial phrase), and the locative suffix *+a+n*. Single-word tract-names continuing a locative form of a noun include items like *Pirrmvrr*, naming a place where trees of the species *yo-pirmarr* (coffee bush, *Breynia oblongifolia*) are abundant. Since elicitation of *pirmvrr* as the locative or ergative of the tree name *pirmarr* can be difficult to impossible, the placename *Pirmvrr* counts as an archaism.[10] In other names, like *Puyvl*, there is an element (here *l*) that could be interpreted as a locative suffix, but no independent attestation of the putative stem (here **puy*; i.e. *puy* is not recorded as an ordinary word whose meaning can

---

[9] Proto Paman, from which Yir-Yoront has descended, distinguished an Ergative case, with both subject-of-transitive-verb and instrumental functions, from a locative case. The suffixes that marked the former all had the vowel **u*, as in **lu*, **ngku*, **nytju*, and others; those that marked the latter had **a*, as in **la*, **ngka*, **nytja*, etc. Yir-Yoront has lost the final vowels of the earlier Proto Paman, and hence a form like *+lh* continues both **nytju* and **nytja*. I am calling this form 'locative' in this presentation; elsewhere (e.g. Alpher 1991) I call it 'Ergative'. The *+a* preceding the ending *+lh* in *yalq+a+lh* is the THEMATIC VOWEL; see Alpher (1991:11 and 1999).

[10] Nouns that inflect according to the pattern (Absolutive, Ergative) *kanharr*, *kanhvrr* 'crocodile', *ngunyan*, *ngunyvn* 'waves' regularly continue ancestral trisyllables, but this pattern is losing ground to simple suffixation.

connect in any way with the tract *Puyvl*). And there are tract-names, typically doubling as the common-noun names of animal species, which contain no suffix: *Ngar-Manl* (Rutland Plains Station homestead; 'catfish' as a common noun), *Yoq-Minh-Mor* ('jabiru tree' as a common noun: *yoq* 'tree', *minh-mor* 'jabiru').

A tract-name, whether its final element is a tensed verb or a noun (with or without locative ending), is used as follows: the citation-form means 'at' or 'to' the place: *Puyvl* 'at Puyul'; 'to' or 'towards' the place is the citation-form plus the Allative case-ending +*uyuw*: *Puyvl+uyuw* 'towards Puyul'; 'from' the place is the citation-form plus the Ablative postposition *ngorvm*: *Puyvl ngorvm* 'from Puyul'. A sample of usage in context is the following:

(4)   *I   pal   thalvnh,   Puyvl+uyuw.*
      there hither   returned   (place)+ALLATIVE
      '[They] were coming back this way, towards Puyul.'

(5)   *Ngethn oylt, artm   athan ngethn oylt   nhinvnh,   Puyvl.*
      we   there mother my   we   there stayed   (place)
      'We, my mother and I, were staying there, at Puyul.'

There is evidence of a historical drift towards the locative marking of the citation-forms of tract-names in recent times: as time passes, more and more tract-names contain a recognisable locative suffix. Lauriston Sharp (pers. comm.) recorded one site as *Minh-Tholhth Warrch* 'Ritually Potent Bird' (*minh-tholhth* '(little) bird', *warrch* 'bad, ritually potent'). I recorded the name of the same site in 1966 as *Minh-Tholhth-Wirrchir*, with *wirrch+i+r* the locative of 'bad'. The form with locative suffix evidently once had contrastive value as 'at' the place, as opposed to the citation-form, used for example (6 [1933 usage]) as the subject of a verbless predicate. The possibility of a case contrast of this kind (a meaningful difference between *Minh-Tholhth-Warrch* and *Minh-Tholhth-Wirrchir*) is evidently now lost.[11]

(6)   1933: *Minh-Tholhth-Warrch,   larr   ngerr-yap.*
      1966: *Minh-Tholhth-Wirrchir, larr   ngerr-yap.*
            (place)   place good
            'Minh-Tholhth-Warrch is a good place.'

Yir-Yoront forms new lexical items primarily by compounding. The most frequent type of nominal compound is that of an unstressed noun followed by a stressed one (the hyphen as a notational device indicates this stress pattern). With names of plant and animal species, the unstressed element is a generic one and the stressed one a specific: *minh-lalpm* 'wallaby', *minh-mirtin* 'possum', *minh-purrq* 'bandicoot', etc., where *minh* is the generic for most furred animals, as well as reptiles and birds. Similarly, for fish the generic is *ngart*, or *ngar-* in context: *ngar-kurr* 'barramundi', *ngar-marq* 'diamond-scaled mullet', etc. With life-form terms, the generic is normally inseparable from the specific.[12]

---

[11] Tract-names do occur as subjects of transitive verbs, in such sentences as 'Place X makes me sick', 'Place Y makes me homesick (for some other place)'. When used thus, the name occurs in its citation-form.

[12] In compound nouns of this type (the phrasal type, with the second element bearing the main stress), a case-ending can appear on the first element rather than on the second (or endings can appear on both

The generic term for 'place' at its most general is *larr*. It has an ancient heritage in the Pama–Nyungan language family, with cognates like Warlpiri *rdaku* and Pintupi *taku* 'trench', and Diyari *rdaku* 'sandhill', and possibly Kala Lagaw Ya (Western Torres Straits Islands) *laaga* 'island', as well as numerous cognates meaning 'place' in Cape York languages; in Paman languages it continues a form reconstructible as *taakurr. *Larr* in Yir-Yoront is very polysemous (it has many distinct but related senses): 1. place, site, tract, estate; 2. spot, campsite; 3. terrain, country, land, zone, domain, distance; 4. ground, earth, soil, dirt; 5. ground as opposed to air or water; 6. substance, stuff; 7. time; 8. day, daylight; 9. period of day or history; 10. occasion, event; 11. story; 12. custom, style, fashion; 13. season; 14. weather; 15. condition of an entity; 16. world, cosmos. *Larr* is, however, used as a generic in rather few compound tract-names — an example being *Larr-Low+Pannan* 'Crying Place'.

A more popular generic for tract-names is *pin*, also highly polysemous: 1. ear; 2. generic for ear-shaped things; 3. ear as seat of intelligence and attention; 4. leaf of a plant; 5. hair; 6. vulva; 7. site, place; 8. home-place, country. The metaphor relating the last two of these senses to the first is probably that of concentric circles, as suggested by the cartilage of the ear and as represented pictorially in the art of Aborigines of the Centre.[13] As a generic *pin* is usable freely with most tract-names: *Pin-Ngar-Manl* (Rutland Plains Station homestead), *Pin-Marrcha*, etc. Unlike the generics of life-form terms, however, *pin* in this usage is freely omissible: *Ngar-Manl, Marrcha*.

4 GRAMMATICAL ARCHAISMS

The grammatically archaic tract-names to which the title of this paper makes reference are among those whose last element is a verb with a tense ending.[14] Examples containing no archaisms are *Thum+Tharrarr* ('heats up'), with *thum* 'fire' and *tharrarr* 'rises, does', and *Par+Thayl* ('burns the head') with *par* 'head' and *thayl* 'burns'. The tense-endings in these forms are the Nonpast +*rr* and the Nonpast +*l* respectively. They are the regular contemporary Nonpast endings for verbs of certain categories, the RR and L conjugations, respectively.

The marking of Nonpast tense for three of the other conjugations of contemporary Yir-Yoront, the NH, the LH, and the Deponent-L, is zero, i.e. the absence of a suffix: respectively *thur* 'laughs', *parrng* 'blows', *than* 'stands'. There is good comparative evidence,

---

elements): *kalq* 'spear', Ergative *kalqalh*; *kal-kow* 'barb of spear', Ergative *kalvlh-kow*; *kawn* 'water', Ergative *kowolh*; *kawn-wil* 'salt water', Ergative *kowlh-wila* (or *kawn-wila(lh)*); *minh* 'animal', Ergative *minhal*; *minh-thaml* 'animal foot', Ergative *minhvl-thamarr*. No other form of splitting of compounds occurs (without resulting in the loss of the item's lexical identity). For compounds that name an animal or plant species (note that a compound like *kawn-wil* 'salt water' consists, like biological names, of a generic and a specific), splitting by case-endings has not been recorded: *minh-lalpm* 'wallaby', Ergative *minh-lalpalh*; *minh-purrq* 'bandicoot', Ergative *minh-purrqa*.

[13] Yir-Yoront *pin* continues Proto Paman *pina, with attestations in Cape York meaning 'ear' or 'leaf'. Outside of the Paman subfamily, Yuwaalaaray has *bina* 'ear'. But the connection of ear to place recurs with the Proto Pama-Maric etymon *yampa, continued as (for example) Umpila *yampa* 'ear', Yaraikana *yampa* 'leaf', and Bidjara *yamba* 'camp, humpy, place, time'.

[14] The noun-like use of clauses with finite verbs is more general in the language, as in *yoq-themrr+payl* 'cockatoo-eats-it tree' (red ash); see Alpher (1991:72) for a brief discussion in context.

however, that verbs of these categories once marked their Nonpast with a suffix, +*m*. This evidence comes from Yirrk-Mel, in forms like *thanvm* 'stands', *nhinvm* 'sits'.

Tract-names whose last element is a verb of one of these conjugations also mark that verb's tense, if it is Nonpast, with +*m*. Such are *Pulpal+Thanvm* 'cotton-tree stands', *Tholprr+Thanvm* 'stands [up to the chin] [in water]', and *Minh-Kerrqel+Thanvm* 'hawk stands'. It is evident that the tract-names retain a form of the verb that has been eliminated from the rest of the language. The elimination of +*m* as a productive Nonpast ending was a matter of grammatical change (dealt with at length in Alpher 1999) and not sound-change. This is to say that it was not a matter of a shift in pronunciation: at no time did forms like *thanvm* become unpronounceable.[15]

Contemporary verbs of the L and RR conjugations mark their Past tense with a zero suffix: *ye* 'cut', *tha* 'rose, went up, did'. But the evidence is strong that these once marked the Past with a suffix, +*r*; this suffix originated as **ntV*[16] and continues as +*t* in Yirrk-Mel *yakát* 'cut', and as +*nt* in Kokapér *yakánt* 'cut'. Its loss in Yir-Yoront was, like that of +*m*, a matter of grammatical change and not sound-change (Alpher 1999). And tract-names with a final Past-tense verb of the relevant conjugation retain the +*r*: *Then+Ya+r* 'cut the penis' and possibly also *Ngaml+Tha+r* 'star rose'.

Some of the relevant tracts, like *Tholprr+Thanvm* and *Then+Yar*, are in the territory of people who formerly spoke Yirrk-Mel and now speak Yir-Yoront. It could be argued that these site-names simply continue the Yirrk-Mel form unchanged and do not constitute archaic Yir-Yoront. Consistent with this position is the fact that the vowel of *yar* 'cut' in the tract-name has failed to undergo a specifically Yir-Yoront change to the *e* of the ordinary Yir-Yoront verb *ye* 'cut'. In favour of the archaism interpretation, however, are (i) that some of the names in question are from territory that has always been Yir-Yoront in language; (ii) that tract-names are almost always translated from language to language and almost never demonstrably borrowed, and all Yir-Yoront tract-names from the territory in question are distinctively Yir-Yoront in phonological form and not Yirrk-Mel (as, for example, *Kowlh-Yalqalh* whose *lh* and *q* identify it as Yir-Yoront, where Yirrk-Mel would have the corresponding *th* and *k*, respectively); and (iii) that the vocalism in a form like *yar* can perfectly well be regarded as a further archaism.

5 DO COGNATE PLACENAME SETS REVEAL ANYTHING ABOUT PRIOR OCCUPATION OF THE PLACES NAMED?

Evidence that land occupation continues from ancient times has enormous relevance in contemporary Native Title deliberations, and it is tempting to see cognate placename sets as providing such evidence. In this regard, however, data like these need to be approached with extreme caution. Compound words, for example, do not necessarily continue *as compounds*

---

[15] Other tract-names in *m* exist to whose second element no contemporary ordinary verb corresponds exactly. Examples are *Lorrql+anvm*, *Pernt+anvm*, *Yum+anvm*. It is clear for various reasons that these names have two constituents. One reason is that sequences of vowels like *o...a* in successive syllables, such as in *Lorrql+anvm*, do not occur in single-word common nouns. Another reason is that polysyllabic words in final *m* are rare among common nouns. It is possible that the +*anvm* parts of these forms continue **than+m* 'stands' in changed form, or continues some other verb now lost. There is at present no compelling evidence for this putative history, however.

[16] The 'V' indicates a vowel of unknown quality whose probable presence is inferable from Yir-Yoront data (Alpher 1999).

from a single ancestral compound. Clearly, the *parts* continue as cognates, but it is possible and even quite likely that the placenames are compounded independently in the two languages according to similar grammatical rules, to form names that translate one another. So evidence of this kind falls short of forcing a conclusion that (for example) the speakers of a language ancestral to Yir-Yoront, Yirrk-Mel and Kokapér frequented the places cited in (1–3), respectively, calling them *Tipanhtha*, *Taarrcha-Manungka*, *Thitithirr-Thaaki*. The same caution holds for names freely formed under other grammatical constructions.

The case for continuation of a single placename from the protolanguage of the people who lived there is more compelling with sets of single-word placenames that are both (i) cognate but sound different — the more different, the more convincing — and (ii) without evident designation other than the place itself. Such might be Yirrk-Mel *Mitathrr*, Yir-Yoront *Mirtayrr* and Yirrk-Mel *Kunkvmal*, Yir-Yoront *Kunqvmvl*. As these tracts are in Yirrk-Mel country, however, and it is possible for Yir-Yoront speakers to adopt the name and shape it to their sound system and to apparent regularity of correspondence ('correspondence mimicry'; see Alpher and Nash 1999), even these examples fall short of being conclusive: it is conceivable that these places came to be recognised as Places (to be named) after the languages had diverged.

It is nonetheless possible that sets of cognate placenames will turn up that fairly unequivocally indicate prior occupation by speakers of the protolanguage before its differentiation. Because such a situation is plausible and because near-examples are known at present (see above), I think intensive research is likely to be successful. Such findings would be of enormous weight in considerations where continuity of residence and tenure is at issue.

REFERENCES

Alpher, Barry, 1991, *Yir-Yoront Lexicon: sketch and dictionary of an Australian language.* Berlin: Mouton de Gruyter.

— 1999, The origin of ablaut as a grammatical process in Yir-Yoront. MS.

Alpher, Barry and David Nash, 1999, Lexical replacement and cognate equilibrium in Australia. *Australian Journal of Linguistics* 19.1:5–56.

Hale, Kenneth, 1964, Classification of Northern Paman languages, Cape York Peninsula, Australia: a research report. *Oceanic Linguistics* 3:248–265.

— 1976a, Phonological developments in particular Northern Paman languages. In Sutton, ed., 1976:7–40.

— 1976b, Phonological developments in a Northern Paman language: Uradhi. In Sutton, ed., 1976:41–49.

— 1976c, Wik reflections of Middle Paman phonology. In Sutton, ed., 1976:50–60.

Sharp, R. Lauriston, 1934a, The social organization of the Yir-Yoront tribe, Cape York Peninsula. *Oceania* 4:404–431.

— 1934b, Ritual life and economics of the Yir-Yoront of Cape York Peninsula. *Oceania* 5:481–501.

— 1937, The social anthropology of a totemic system in North Queensland, Australia. PhD dissertation, Harvard University.

— 1958, People without politics. In V.F. Ray, ed., *Systems of Political Control and Bureaucracy in Human Societies*, 1–8. Seattle: American Ethnological Society.

Sinnamon, Vivian and John Taylor, 1978, Summary of fieldwork, Kowanyama. *Australian Institute of Aboriginal Studies Newsletter*, New Series, 10:34–73.

Sutton, Peter, ed., 1976, *Languages of Cape York (RRS 6)*. Canberra: Australian Institute of Aboriginal Studies.

— 1998, *Native Title and the Descent of Rights*. Perth: National Native Title Tribunal.

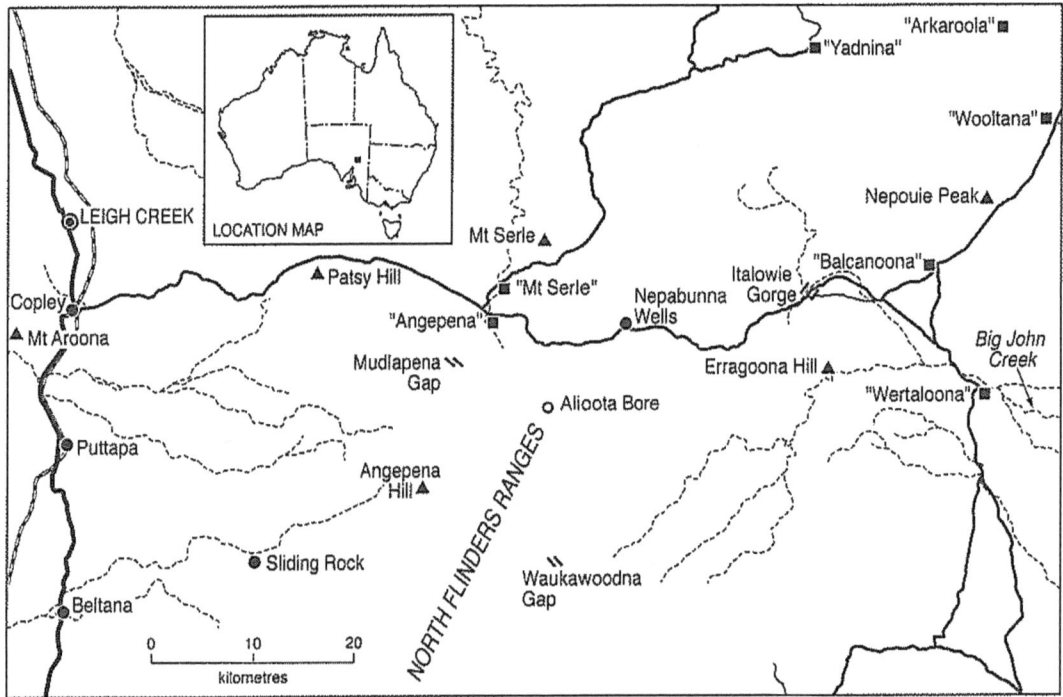

Map 1: The Northern Flinders Ranges

10   SOME REMARKS ON PLACENAMES IN THE FLINDERS

Bernhard Schebeck

1 'DESCRIPTIVE NAMES': A CURSORY ANALYSIS OF A SAMPLE[1]

The Northern Flinders Ranges are the home of the Adnya-mathanha people, 'the people of the rocky hills', whose language is related to the Kaurna language of the Adelaide Plains. The cursory analysis presented here is based on a sample of 483 placenames in the Flinders Ranges.

This number is subject to the understanding that it is not always clear whether two entries are really two names or rather two 'spellings' of the same name. But for the present brief survey I simply ignore such questions. I shall also disregard the question of 'one versus two words', for instance spellings such as **Idhiawi** versus **Idhi Awi**. This might appear to be a minor decision, but I shall briefly argue below that the decision of what is a genuine name is perhaps not so clear-cut.

Of these 483 names I have assigned 139, almost one-third of the total, to the 'descriptive class'. In order of frequency I have found the following:

1.1 Names containing awi 'water'

41 end in **-awi** 'water'. These names represent almost one-third of the 'descriptive class':
5 end in **awi urtu** 'waterhole'. So, this is quite a small subclass.
3 (or perhaps 4) begin with **awi** 'water'. So, this too is a small subclass.

---

[1]   This paper is a sequel to the research of Hercus and Potezny (1999). In mid-1999 I wrote a letter to Luise Hercus, in which I made some comments on this article, whereupon I was encouraged to work those remarks into a paper. This is the result. I wish to thank Luise Hercus also for providing me with a draft copy of her contribution to the present volume.

Due to external circumstances, my own material is not fully accessible to me and, therefore, the bulk of the examples is drawn from what is known through the publications of other people, mainly John McEntee and Dorothy Tunbridge. My own recordings are still only partially accessible to me at the time of writing. In other words, if an explanation given to me is quoted here, this is done only from memory.

L. Hercus, F. Hodges and J. Simpson, eds, *The Land is a Map: placenames of Indigenous origin in Australia*, 140–153. Canberra: Pandanus Books for Pacific Linguistics, 2002.

1.2 Names containing vari 'creek'

(including one name ending in **-parinha**):

22 end in **vari** or **vari-nha** 'creek'. That is, over 15 per cent of the 'descriptive class'.

1.3 Names containing vambata 'hill'

20 end in **vambata** 'hill'. That is, over 14 per cent of the 'descriptive class'.

This means that between them these three subclasses comprise about two-thirds of the descriptive class entries. The numbers in the remaining subclasses are quite small and for several of them there is just a single entry. Only the following two subclasses surpass 5 per cent of the total.

1.4 Names containing inbiri 'gap'

(**irnbiri** 'gorge, gully', in McEntee and McKenzie 1992):

8 end in **inbiri/irnbiri** 'gap; gully, gorge'.

1.5 Names containing urtu 'hole'

7 end in **urtu** 'hole'. If the 5 names listed above as ending in **awi urtu** 'water-hole' are included, then the number is 12, i.e. still less than 10 per cent.

For the sake of completeness, I list the remaining subclasses with the numbers of entries given in brackets:

ending in

ithapi/ithipi	'hole, hollow, cave' (4)
warldu	'saddle' (4)
yurru	'range, ridge' (4)
adnya	'stone, rock' (3)
vitana/vithana	'plain, flat' (3)
madapa	'valley' (2)
mati/marti	'hillside, slope' (2)
wami	'bend in creek' (2)
yukari	'rock wall' (2)

The rest are found in only one name each, specifically:

ending in

itala/itarla	'hole; crack, gap'
manyi	'slate; regosol, coarse sand'
murrka	'stony slope, steep scree'
n(h)iirri / n(h)iarri	'side, edge (of cliff)'

vinki(r)ti / virngarti-nha	'junction (of a road), fork (in a creek)'
warru	'red clay'
yaurru / yaurra	'peak, pointed hill'

It is clear that the **-awi** names fit into this wider pattern of compound names. There are still other similar compound names, which do not quite fall into this category.

First of all, there are names containing a term denoting a natural species — be it flora (mainly **wida** 'river red gum', but also **arlku** 'black oak') or fauna (mainly **urdlu** 'red kangaroo'). Often the species name is the first element, but not always, for instance:

Nharni wida-nha	'Mathewson's Spring and Creek'
Urlkambarra-nha-madlha	'mallee patch behind **Wayarl-aralya**', both mentioned by McEntee and McKenzie (1992:32).[2]

One may or may not include here names containing **Akurru/Akurra** 'the mythical Rainbow Serpent'.[3] In this case the decision of whether it is a 'name' or not is particularly difficult, because of the notion of an **akurru awi** 'serpent water(hole)', presumably one created, but not necessarily inhabited, by the serpent.

In this context I take up again the question of what is 'really a name'. Often we find a name, and also that name with the addition of a natural feature word such as 'creek', 'hill', etc. Tunbridge (1988:161) has the name **Adlyundunha** and also **Adlyundunha Awi** 'Alioota Bore' as well as **Adlyundunha Vari** 'Alioota Creek', and many others; sometimes it is unclear whether the occurrences of a name with and without a given natural feature word refer to the same place or not. Thus we find **Irrakanha,** which McEntee (1992:22) gives as 'Erragoona Hill', but Tunbridge (1988:162) as 'The John Waterhole', for which she also has **Irrakanha Awi** 'The John Waterhole'; McEntee however lists **Irrakanha Vari** 'Big John Creek at Wertaloona'. Therefore a certain amount of 'fusion' may be required for a compound to qualify as a 'name': it may be more than sheer accident that of all the compound names those ending in **-awi** are particularly frequent in English (i.e. names in '-owie'). A more complete list of names would be required for further study of this and similar questions.

2 A FEW FURTHER COMMENTS ON THIS FIRST ANALYSIS

As already hinted in the last paragraph, this analysis poses a certain number of questions, of which the following three are singled out for further comment:

---

[2]   From here on this work is referred to briefly as 'McEntee'.

[3]   Note that Tunbridge consistently writes **akurra**, where Andrew Coulthard with whom I worked always recorded **akurru**. C.P. Mountford's spellings 'acaru' and 'arcaru' and the English placename Arkaroola, quoted by both McEntee and Tunbridge, hint at a pronunciation 'akarru'. This is but one example of minor variations in my own recordings and those made by Tunbridge and, perhaps to a lesser degree, by J. McEntee. However, these variations do not appear to be important enough to deserve the label 'dialectal', and I prefer to speak of 'family traditions'.

a)   What is a placename?

b)   What is a descriptive placename?

c)   What is the meaning of a placename?

2.1 What is a placename?

This question is of course by no means peculiar to the region under consideration. Some brief comments on various categories of placenames are given here.[4]

2.1.1 *'General terms'*

A general term, such as for instance the word for 'river', may function as a placename, as in the case of the Diamantina/Warburton being called simply **Karla**, i.e. 'The Creek' by the Simpson Desert Wangkangurru people.[5] No similar example seems to have been reported from the Flinders Ranges — unless we take the very name **Adnyamathanha** as an indication that **Adnya** 'stone, hill' may have functioned as a sort of name for 'The Ranges', or, alternatively, it may hint at a case where 'a general term is used because there IS no specific term' (Hercus, in the present publication), a decision which I find difficult to make.

2.1.2 *'Intermediate' names*

There are no clear reports in the Flinders Ranges of names of the type Hercus calls 'Intermediate', in which a feature characterising a given place indicates the locality to people who are already 'in the know'. However, a term such as for instance **Arlku vari** 'Black Oak Creek' (Tunbridge 1988) may be, or originally have been of such an 'intermediate' type. The above-mentioned expression **akurru awi**, which also appears as the 'name' **Akurra Awi** in Tunbridge (1988:161), is doubtlessly a serious candidate for this type, as there are said to be many **akurru awis**. Unfortunately we do not know enough about the use of this expression in daily speech; but some of the recorded texts do indeed hint at such 'intermediate' usage.

---

[4]   There do not appear to be any placenames in the Flinders based on possible misunderstandings of the type listed by Hercus as 'silly names'. see however Hercus and Potezny (1999:175), about the reported misunderstanding concerning the 'name' Wundowie. Although this paper is specifically about placenames, I may point out here that the name 'Nimbalda', found in Smith (1879:87) for the language of the Adnyamathanha people, is probably of this type: **nimba** 'such' plus **arlda**, an archaic word for 'language'. In this context we may mention the expression **Yura ngawarla**. This, I suspect under the guidance of European linguists, has in recent years become quite fashionable as the name for the language of the Adnyamathanha people, whereas the older term **Adnyaarlda** is practically forgotten today. The new term is akin to this type — although, as I have argued elsewhere, it is based on a certain carelessness rather than outright misunderstanding.

[5]   As an example from English, compare 'The Gap' in Alice Springs.

Figure 1: Edeowie Gorge. Idhiawi Inbiri. *Photo: D. Kelly*

2.1.3 *'Derivative' names*

What may be called a 'derivative' type results from the addition of a general term such as for instance 'creek', 'waterhole', 'gap', 'saddle', or whatever, to a placename.[6] Thus, besides the

---

[6] The orthography of such names calls for comment: why write for example **Wakarla-udnanha Inbiri** 'Waukawoodna Gap', but **Murdlu Udnanha** 'a big hill', 'Murdlu's droppings', as is done in Tunbridge (1988:164), rather than **Wakarla Udnanha Inbiri** or **Murdlu-udnanha**? My guess is that this somewhat haphazard treatment is reflected in the corresponding transcribed 'English' names, e.g. 'Waukawoodna' versus cases where there is no explicit reference as in 'big hill', or cases where there is a partial translation, as in 'Balcanoona Creek' for **Wildya Vari**. The treatment appears to be capricious and no clear reason is manifest for variations even within the same publication as, for instance, **Urta Awi** 'Weetowie Waters, and spring above' and **Urtaawi** 'a spring' in Tunbridge (1988:166); not to mention variations between authors, comparable to other types of variations in orthography, e.g. **Itarl-awi** 'Italowie' (McEntee 1992:19)) as against **Ithala Awi** 'Italowie Well' (Tunbridge, 1988:162). To come to some clearer conclusion on this, one

name **Idhiawi(nha)** (Tunbridge 1988:162 writes **Idhi Awi**, 'Edeowie') we also find the name **Idhiawi Inbiri** which may be interpreted either as the expression 'the gap at Edeowie' or, alternatively, as the name 'Edeowie Gorge'. In Tunbridge (1988) several series of such 'names' appear as for instance **Adlyundunha**, **Adlyundunha Awi**, and **Adlyundunha Vari**. Clearly, **Adlyundunha Awi** 'Alioota Bore' differs in formation from **Idhiawi** 'Edeowie' (lit. 'Finch water'), just as **Adlyundunha Vari** differs in formation from **Wildya Vari** 'creek branch flowing into Balcanoona Creek', (lit. 'Night/darkness Creek'): there are no independent placenames *****Idhi(nha)** or *****Wildya(nha)** — not that such simple formations are unknown in the area. But how do we decide whether a given 'name' belongs to this 'derivative' type or not? For instance, the link between **Valdhanha** 'Beltana' and **Valdha Marti** 'name of rock face in the Mount Serle area' (Tunbridge 1988:166) is purely linguistic (cf. **valdha** 'rug; clothes'), there is no geographical nor, as far as can be ascertained, any mythological link. Therefore, we may decide that **Valdha Marti** (lit. 'Rugs Slope'), is a 'genuine placename', rather than one of the 'derivative' type; this decision is considerably strengthened by the fact that there is a mythological explanation for that name (see §2.2 below). I do not have any ready answers to such questions.

2.1.4 *Secondary names*

What may be called 'secondary' names are not 'official' names, but rather allusions, they resemble the 'intermediate names', discussed above. However, in this case, the allusion is usually to a story. For instance, Mount Serle has the 'official' name **Atuwarapanha**, but it is said to 'be' **Wildu** the eagle, because of the associated story. Therefore, if the word **Wildu** 'Eagle' may be extended to be used as a reference to the place, as if it were a placename, we would have such a 'secondary' usage. A name such as **Kilawila** may belong to this category: in the myth of Mount Termination (**Kakalpunha**, in Kuyani country) **Kilawila** is the name of the killer of the cannibal **Papurdityirdityi**. The cannibal is said to 'be' the larger of the two main hills of Mount Termination, while **Kilawila** is said to 'be' the smaller hill.

2.2 What is a 'descriptive' placename?

The 'descriptive class' discussed above is nothing more than a selection of names that contain a word for a natural feature of the landscape but not, for instance, those names which contain words for a natural species. A name such as **Ithala Awi/Itarl-awi** 'Italowie' (Tunbridge[7]/McEntee) 'lit. hole-water', i.e. water running through a hole or gap in the rocks, may reasonably be accepted as delivering a sort of general and vague description of the place. A name such as **Idhi Awi** 'Edeowie', 'Finch water' — presumably because finches came

---

could invoke stress patterns, but this might not deliver clear-cut decisions. On the other hand, we could say that 'derivative' names should be written as two words, and 'genuine names' as a single word. Thus, we would have **Valdhamarti** rather than **Valdha Marti**. However, if it turned out that a 'genuine name' decision can be justified, would everybody be happy, for instance, with the orthography **Idhiawiirnbiri**? The main drawback of such a suggestion has to do with the fact that the orthographic decision would depend on the semantic interpretation which may be uncertain.

[7] Tunbridge here refers to Tunbridge 1988.

there to drink — may lack descriptive power, because finches may well drink from other waterholes in the area, and because this feature may not always be evident. Similarly **Adata Madapa** 'Nepabunna Valley', explained by Tunbridge (1988:161) as 'Frost Valley', will reveal its alleged distinctive feature only at certain times, without necessarily appearing unique. On the other hand, the meaning of the above-mentioned name **Valdha Marti** 'Rugs Slope' remains totally obscure if one is unaware of the mythological reference. Finally, what is presumably a description might prove quite ephemeral, as the landscape changes: I am thinking of a name such as **Arladu-vuyuvuyunha**, explained to me by A. Coulthard as meaning 'tea-trees all twisted/torn into pieces', which would rather hint at the final stages of existence of the defining feature of the description.

The use of the term 'descriptive' in the examples discussed is therefore appropriate only in the sense that the name contains the word of the natural feature, such as 'Hill', 'Gap', 'Creek', 'Waterhole/Pond', etc. which it names. And, as we have already seen, it is this very characteristic — namely a sort of 'self-reference', to somewhat abuse an important concept — which raises the general question of 'is it a name at all?'

2.3 What is the meaning of a name?

First of all, it will be noticed that in Adnyamathanha, the natural feature word, when present, usually appears at the end of the name.[8] It is often preceded by some other word which appears to qualify or perhaps better, specify the following term — no matter what the allusion in that 'qualification' — as for instance in the names quoted in fn. 9: **Warturli Widanha** (lit. 'ringneck parrot gum tree'), as against **Vira Widanha** (lit. 'moon gum tree'). Such interpretations can be definitive only in those cases where the first word can be linked to the known vocabulary of the language. Even if such a link is established, the 'meaning' is not always obvious: while a literal meaning may be established, a clear reference is not always available. In many cases, sheer guesswork will establish a 'meaning'. Thus, for instance, in the case of **Adata Madapa** 'Frost Valley' we may establish a 'meaning' by guessing that winters are particularly severe in that valley, or at least that the valley has that reputation, a little like 'Clare Valley' in the area west of Adelaide. Similarly, we may guess that the name **Idhi Awi** 'Finch Water' hints at the idea that this is a favourite spot for finches; and so on. However, in the absence of clear information on this, all these remain just guesses.

The case of **Valdha Marti** 'Rugs Slope' is different: either we run out of guesses altogether, or we will resort to highly implausible or 'contrived' ones; we can grasp the reference only by discovering the story, in which an Ancestor pegs out wallaby rugs **valdha** for drying on that slope **marti**. The example of **Adlyundyanha** 'Mundy Waters' in fn 10 is probably similar although the presence of myrtle trees may have contributed to the naming.

This distinction between literal meaning and reference shows that there are 'layers of meaning'[9] to be considered. Furthermore, there is also the problem of how reliable a given interpretation by a 'native speaker' is, especially in the case of more complex formations. Thus the hill or range called **Wayalayala/Wayarlayarla** was explained to me by Andrew

---

[8] As already pointed out by Hercus and Potezny (1999:177), there are some exceptions to this.

[9] Moreover, in the case of mythological references at least, we may never be sure of how many layers there are or were! Thus, for example, I know of at least one 'deeper' reference of the repeatedly quoted name **Valdha Marti**.

Coulthard as a derivation of the verb **wayali-** 'to turn (around)', hence as meaning something like 'twisted in all directions': i.e. it was ultimately interpreted as a simple description of the shape of this geological formation. By contrast, McEntee (1992:113) explains it literally as a contraction of a reduplicated form of **wayu-warla** 'shelter-break', but gives no further information as to what this refers to. This example clearly shows that an explanation by a native speaker cannot necessarily be relied upon as 'the correct interpretation'. Apart from cases of popular etymology, where the literal meaning has simply been forgotten, there may also be sheer ad hocery, especially when 'informants' are pressed for answers under the assumption that there must be a simple answer and that it must be known to them. Although the threat of re-analysis by popular etymology or sense-making can never be fully excluded, we do certainly stand on much firmer ground where the meaning is embedded in a traditional story, as in the example of **Valdha Marti** mentioned above. This fact shows just how deficient our knowledge of many placenames really is.

Therefore, in the absence of deeper knowledge, we simply have to fall back on literal meanings — vulnerable to misinterpretation as they may be. Such literal analyses will often be able to come up with only a partial answer. Thus the name **Adlyundyanha** 'Mundy Waters' can be partially translated by the word **adlyu** 'myrtle tree',[10] while the remainder of the name cannot be linked to any known lexeme. A similar remark applies to the name **Adlyundunha**, given in Tunbridge (1988:161) and many others. Whereas the difficulty of translation strengthens the status as a 'real name', the lack of analysis may call into question the correctness of the recording. Thus the name for 'Mount Serle' has been recorded as **Atuwarapanha** by myself and by J. McEntee, whereas Tunbridge (1988:161) has **Arta-wararlpanha/Artawararlpanha**. Although we may say that two recordings weigh more than one, the absence of any further analysis of the name adds to our uncertainty.[11]

3 A FEW FURTHER SUGGESTIONS

I propose to discuss a few types of names containing some special morphemes or, in the last case, a special construction.

---

[10] **Adlyundyanha** 'Mundy Waters' (recorded from Andrew Coulthard) does not refer to myrtle trees, **adlyu**, nor, as one might think to 'a place with myrtle trees'. It refers to a place 'where an Ancestor's myrtle spear went down'. Similarly **Warturli Widanha** 'Mount Lyndhurst' (McEntee 1992:108) presumably does not refer to a gum tree, but to a place where it stands or once stood; nevertheless such self-references are not necessarily absent in this type of name. The entry in Tunbridge (1988:70), **Vira Widanha** 'gum tree in Wilpena Creek', appears to attest to this. Thus, our attempts to discover clear rules are frustrated yet again.

[11] Notice also the name 'Batuarapunna' in Taplin (1879:87), which is identified there with Mount Freeling. If it is correct to link this name with the one under discussion, then the apparent dropping (rather than 'weakening' to v-) of the initial B- remains unexplained, whereas the reference to Mount Freeling would appear to be due to 'mis-identification' rather than to a historic change of identification. Such a change appears unlikely, because of the relatively short period of time between the report in Taplin and those of others, such as for instance N. Tindale (who writes in his diary 'Artu|warapana'), and also because of the considerable importance attached to Mount Serle by Adnyamathanha people.

   I have never been tempted to link the first part of that name to the word **artu** 'woman'. I do not know whether or not Tunbridge had in mind a link with **arta** 'grass tree'. Without speculating any further about the (literal) meaning of this name, it must be pointed out that the transcriptions alone imply the hypothesis that the second term begins with the syllable **wa-**, rather than with **a-** as would be possible in the case of the spelling 'Batuarapunna' reported in Taplin.

3.1 Names in -pi-nha

First of all, back to the names ending in **-awi(-nha)** which, after all, triggered the comments made in the present paper. Hercus and Potezny (1999) point out that there are two forms for 'water', namely **-awi** and **-api**, and that the second variant form is apparently totally absent in Adnyamathanha. However, it may be suggested that the few placenames ending in **-pi-nha** are formed with **-api**. The three names that appear in publications on Adnyamathanha are:

Mudlhapinha	Mudlapina Gap/Mudlapena Spring	(McEntee 1992:89), (Tunbridge1988:163)
Tharrapinha	Terrapinna Waterhole	(McEntee 1992:48)
Thudupinha	waterhole west of **Var-ardlanha**	(McEntee) 1992:49)
	Arkaroola: hill near Freeling Heights	(Tunbridge 1988:167)

Although there is no agreement on the identification of the places (namely 'Gap' as against 'Spring' in the first case, and 'waterhole' as against 'hill' in the last case), a reference to some source of water appears to be at least implied. From the semantic angle the interpretation of **-pi-** as **-api** 'water' appears plausible. This interpretation seems to be strengthened by the existence of the name **Api Yakunha** 'spring near Yadnina (Balcanoona)', given by McEntee (1992:9), with the rare appearance of the word for 'water' at the beginning of the name. We would however have to admit that there is a higher degree of fusion than in the names ending in **-awi**, as can be seen by the contrast between **Thudupinha** and the name which was pronounced as **Muruawi** by Andrew Coulthard, but is recorded as **Mur-awi** 'Moorowie — Mount Chambers' by McEntee (1992:90) rather than **Muruwi*. The contrast is confirmed also by the different stress pattern: in my own experience, all names in **-awi** have the secondary stress on the first syllable of **-awi**,[12] whereas all three names in **-pi-nha** have the secondary stress on the syllable **-pi-**. Therefore, if we accept the interpretation of the three names as containing the variant form **-api** 'water', we must concede that the degree of fusion is so high as to have obscured that link probably even to the speakers. If all this is correct, it would suggest that these names, which are unlikely to be borrowings, are archaic forms.

There are some further names which at first sight might appear to fall into this category, but for which another explanation is readily available. The name **Wakarl-apina** 'Grindstone Range' is explained by McEntee as 'lit. mob of crows', and **Yarnmarr-apinha** 'hill west of Nepabunna' as 'lit. mob of steps' by McEntee, and **Yanmarri-apinha** 'Nepabunna: stepped hill' as 'several steps' by Tunbridge (1988:171).[13] **Warturlipinha** 'Angepena Station: hill' (Tunbridge, 1988:124) is explained by McEntee as **Warturl-ipi Yurru** 'Range north side of Angepena-Nepabunna road, Mount Constitution, lit. 'Ringneck parrot egg'. Tunbridge (1988:124) gives **warturli-vipi** 'ring-neck parrot's egg'. These interpretations appear quite plausible: in the present context they highlight the risks of misinterpretation.

---

[12] It will be noticed that Andrew Coulthard's pronunciation, as just reported, differs from that reported by McEntee in that Andrew Coulthard usually did not drop the last vowel of the word preceding **-awi**.

[13] It is interpreted as containing the plural marker, which is given as the 'suffix' **-apinha** by those two authors (also as **-aipinha/-aipina** by some present-day literate native speakers): it was rendered as the 'particle' **vapina/vapinha** by myself.

3.2 Names in -pa-nha

Names in **-pa-nha** appear to parallel those in **-pi-nha**, because of the stress pattern (secondary stress on the syllable **-pa-**), the presence of the suffix **-nha**,[14] and probably a certain degree of 'fusion'. The best known is **Nhipa-pa-nha** 'Nepabunna'. I would guess that at least some of these names are old diminutives, that is old compounds with the word **vapa** 'little, small'. In spite of the presence of variants such as **Akurrupanha** 'hill on NE side of **Ngurru Widlya**' (Tunbridge 1988:15) and **Akurrupanha** 'pad going through gap near Mount Serle' (McEntee 1992:2 and Tunbridge1988:161), we cannot be sure how far the fusion has gone.

Again we have names which at first may appear to belong here, but which clearly have a different explanation. Thus, McEntee's **Andupanha** 'Silver Gap' (1992:8) appears at first sight to belong here (with fusion); however, the variant form **Anduupanha**, which was recorded by Tunbridge as well as by myself, supports the literal translation 'white wallaby' given by several speakers from **andu** 'rock wallaby'+ **upa** 'white'. There are still other cases which clearly do not belong here: Tunbridge's explanation (1988:170) of the name **Yamuti Ardupanha** 'Wirrealpa: two ridges near Red Well' as '**Yamuti** married couple' is quite plausible, and the name **Yulupanha** 'Uliban **Yulupanha** Spring' (Tunbridge 1988:171) is no doubt a 'simple' name, derived from **yulhupa** 'for a long time'.

3.3 Names in -aka-nha

There are not many placenames in this group, the best known is 'Patsy Springs', given as **Mudlh-akanha** by McEntee, and as **Mudlhaakanha** by Tunbridge (1988:163). Austin et al. (1976) have suggested the translation 'fruit' for the word **aka**, but I had previously suggested 'little piece, bit'.[15] Words like **mayaaka** 'wild pear' seem to fit the 'fruit' interpretation well. The form **minaaka** 'eye', perhaps more particularly 'pupil' (McEntee 1992:84) besides **mina** 'eye', was the topic of the 1976 paper. There are other uses, which do not so obviously conform to the interpretation 'fruit', such as the expression **inda/indha aka** 'end of a rock', recorded by Andrew Coulthard. It was the existence of a particle **aka-nha**, that is **-aka** followed by the suffix **–nha**, which was the main reason for my suggesting a lexical meaning 'little piece, bit'; therefore, the meaning 'fruit' may be a secondary development. The particle **-aka-nha** is used in a way similar to English 'one', as in **inha aka-nha** 'this one'.[16]

---

[14] We cannot generalise to the point of declaring the suffix -nha obligatory in these names. Thus we find the variants **Wartapanha** 'Spring in Weetootla Gorge (Wortupa)' (McEntee 1992:108) and **Wartapa** 'Wortupa Spring' (Tunbridge (1988:169); for this name see §3.4 below). There is also, for instance, **Widap-awi** 'Spring in Bendieuta Creek' (McEntee) and **Widapa Awi** 'Widapawi Creek: waterhole (at Wirrealpa Station)' (Tunbridge 1988:170), and no variant with the suffix **-nha**.

[15] In Schebeck (1974:14–15, 47), this is called, somewhat infelicitously, a 'classifier'. I am leaving aside the use of a suffix -(a)ka- to distinguish female from male 'baby names', as in **marru** 'big; wide; fourth born boy', **marruana** 'fourth born boy', as against **marruka(nha)** 'fourth born girl', which I now do not directly link with the morpheme under discussion.

[16] McEntee distinguishes two suffixes -aka, one glossed as 'fruit (attached to nouns?)', and the other as 'piece, bit (attached to adjectives?)'; he also lists a suffix -akanha, which he glosses as 'that piece'. All these are distinguished in his dictionary from an archaic noun **aka**, glossed as 'head'.

3.4 Names in -wata-nha / -warta-nha

There is another 'classifier', which can be written **-warta-nha**, as in **inha warta-nha yakarti** 'this (one) child'. It is unclear how this is derived, but there is a word **warta**, glossed by McEntee (1992:108) as '1. bush; 2. large, big', and by Tunbridge (1988:122) as 'one, thing (or sometimes, big)'. The word does, indeed, occur in compounds with the meaning 'big' as in **warta vurdli** 'morning star', explained as 'big star' by Tunbridge. The interpretation as 'tree' is found in **igawarta** 'wild orange tree' given to me by Andrew Coulthard. It is not always clear which interpretation should be adopted. However, in several names the translation as 'big' (or even the translation 'one') will do. Perhaps the best-known name of this kind is **Mudlu-warta-nha/Mudluwartanha** 'Moolawatana' (McEntee 1992:88, Tunbridge 1988:163), which McEntee indeed explains as 'lit. big hip'. The explanations with 'big' appear to work mainly for names where the element in question is the second part. For, unlike in the previous cases, there are several examples where the morpheme is the first part of a name. Probably the best known is **Wartalyunha** 'Wertaloona', which is listed by both McEntee and Tunbridge without further explanation. In the case of the name **Wartapanha** (fn. 13) 'Spring in Weetootla Gorge (Wortupa)' (McEntee 1992:108) and **Wartapa** 'Wortupa Spring' (Tunbridge 1988:169), the translation as 'big' is much less convincing than in previous examples and the translation by 'bush, tree' appears at least equally acceptable. It is possible that this name does not belong here at all, but is simply the word **wartapa** 'shadow (of a cliff)', which was recorded by the author from A. Coulthard. The difficulties of interpretation are further illustrated by the example **Vityirliwartapanha** 'Sliding Rock near Beltana' (McEntee 1992:64), which appears to combine the morpheme meaning, among other things, 'big' and the diminutive form.[17]

3.5 A note on lexical elements

It is not an exaggeration to say that any word may occur in a placename. Of course, there are differences in frequency and, naturally, complete absences. For instance, names containing **u(r)dna** 'faeces' are common, whereas **umbu** 'urine' is almost absent. Besides words for natural features, those referring to natural species are particularly frequent. This and similar topics require further study.

3.6 A note on the construction of placenames

So far we have dealt mainly with compounds as placenames. A simple word, one which consists of a single lexical item, to which the suffix **-nha** can be added, may also function as a placename: a few examples were given in the preceding pages.[18] The statistical survey of 483

---

[17] Whereas the first part could be interpreted as **vityi** 'dish, coolamon' + suffix **-li** 'like' (not *-**rli**), 'like a dish'.

[18] The name pronounced as **Umburratanha** 'Umberatana' by Andrew Coulthard may for instance be an example; but Tunbridge gives the name as **Ngambadatanha**, making such an interpretation less likely.

Just for information the following further examples are listed without explanation: **Uldanha/Uldhanha** 'Wooltana' (McEntee 1992:33, explained as 'bough'), 'Wooltana Spring' (Tunbridge 1988:166); **Valdhanha** 'Beltana', **Yanggunha** (see Hercus & Potezny, 1999:168, with the explanation 'the left-handed one',

names showed that at the most about 10.5 per cent of the total are of this type. Furthermore, all attempts to find some sort of correlation between compounding and the suffixing of **-nha** have failed.

As is well known, whole phrases may function as placenames; the distinguishing feature is the presence of a verb form. In our rough survey, there are 14 such names. This type — interesting as it may be — is statistically insignificant. It is noteworthy that no less than 7 of these 14 names contain the intransitiviser **-i-**. Admittedly, a few of the verbs in question are intransitive, such as **winmii-** used by Andrew Coulthard when speaking to me, or **wirnmi-i-** (McEntee 1992:120) 'to whistle', but the majority of these examples may be interpreted as 'impersonal' rather than passive. Thus, the name **Wakarl-arrpa-indhanha/Wakarla Adpaindanha** 'Cave in **Wildu** legend' 'Angepena Station: rock with cave' (Tunbridge 1988:168) is explained by McEntee (McEntee 1992:107) as 'lit. painting of crows', i.e. 'the crows being painted up'. **Wabma Nambaindanha** 'Aroona valley area, hill' means '(the place where) the snakes (are) being covered up'. A striking feature of this type of name is the suffixing of **-nda-nha**, with, it seems, obligatory use of **-nha**. Hence Schebeck (1974:49) has interpreted these forms as verbal nouns. There is just one name which is in the 'verbal noun' form, but does not contain the 'intransitiviser' suffix **-i-**, namely **Yurlu Ngukandanha** 'lit. kingfisher going' (Tunbridge 1988:171). There is also one name which may be interpreted as 'participle' (that is, without the suffix **-nha**), namely **Arlarru Udu-udu-manda** 'Tea-tree Shadow', explained by McEntee (1988:13) as 'lit. tea-trees pulled down'. This is the only transitive example in this class,[19] and the question may be asked whether the forms in **-ndha-nha/-nda-nha** are obligatorily intransitive.

There are two examples of names with a finite verb form. The first might appear to belong to the previous subtype: but **Nguthunanga Mai Ambatanha** 'hill near Nepabunna' is explained by Tunbridge as 'the Spirits of the Dreaming cooking damper': the form is interpreted as an ergative and, therefore, the verb form must be interpreted as a present.[20] The other example is Tunbridge's **Akurra Avianggu** 'Akurra vomited', which is grammatically transparent.

The two remaining examples are grammatically unclear, as both appear to show an otherwise unknown construction involving a verb root. The first is the name **Arladu-vuyuvuyunha**, quoted in §2.2: this name is complicated by the fact that the actual verb root **vuyu-** (tr) 'blow with mouth (?); clean; smoke out; twist (?)' is marked as being uncertain in meaning in my vocabulary.[21] The other example appears in Tunbridge as **Yurndungarlpa** 'Mount Samuel' which is also unclear: it could be suggested that this is an abbreviated form of a name linked with the expression **yurndu ngarlpandana-thadi** 'west, lit. 'towards the sun setting' (McEntee 1992:102).

---

whereas Tunbridge (1988) explains it as 'looking over left shoulder' (1988:98)). This shows, yet again, that the literal meaning might not be enough to fully explain a name; **Yulupanha** 'Uliban [Yulupanha Spring]'.

[19] The verb, which McEntee (1992:13) gives as **udu-ma-** 'to pull down, to unpack', looks like a transitivised form, but no verb root ***udu-** has been recorded. For a discussion of transitivity see Schebeck (1976).

[20] McEntee (1992:45) reports also a name **Nguthuna Mai** 'Damper Hill', so one might be tempted to interpret the above as a locative construction. However, this interpretation is vitiated by the fact that the locative of this name would be **Nguthuna Mai-nga**, not * **Nguthuna-nga Mai**.

[21] McEntee (1992:73) lists only an intransitive verb **vuyu-vuyu-ri-** 'to be pretty and curly (as child's hair)'.

CONCLUSION

Without wishing to appear negative I have attempted to emphasise the risks that await too hastily assembled conclusions, and to counsel caution in what is a very worthwhile line of research. Research on names, and placenames in particular, is important from a practical viewpoint — perhaps with some implications. It is also important for more theoretical reasons. It is not only in Indo-European linguistics that the study of placenames allows us to peer just a little deeper into the history of the linguistic and hence the human geography of vast areas of the land.

REFERENCES

Austin, P., R. Ellis and L.A. Hercus, 1976, 'Fruit of the eyes' semantic diffusion in the Lakes languages of South Australia. *Papers in Australian Linguistics* 10, 57–77. Canberra: Pacific Linguistics.

Dixon, R.M.W., ed., 1976, *Grammatical Categories in Australian Languages*. Canberra: Australian Institute of Aboriginal Studies.

Hercus, Luise A. and Vlad Potezny, 1999, 'Finch' versus 'finch-water': A study of Aboriginal place-names in South Australia. *Records of the South Australian Museum* 31 (2):165–180.

McEntee, John and Pearl McKenzie, 1992, *Adna-mat-na English Dictionary*. Adelaide: the author.

Schebeck, B., 1974, *Texts on the Social System of the Atynyamathanha People with Grammatical Notes*. Canberra: Pacific Linguistics.

— 1976, Transitivity, ergativity and voice in Atjnjamathanha. In Dixon, ed., 1976:534–550.

Smith, Henry Quincy, 1879, The Nimbalda tribe (Far North). In Taplin, ed., 1879:87–89.

Taplin, G., 1879, *The Folklore, Manners, Customs and Languages of the South Australian Aborigines*. Adelaide: Government Printer.

Tunbridge, Dorothy, 1988, *Flinders Ranges Dreaming*. Canberra: Aboriginal Studies Press.

RECONSTRUCTING
PLACENAMES SYSTEMS

11 BLOWN TO WITEWITEKALK: PLACENAMES AND CULTURAL LANDSCAPES IN NORTH-WEST VICTORIA

Edward Ryan

Indigenous placenames of north-west Victoria originate in what has been described as the western Kulin group of languages which were traditionally spoken across the region.[1] More particularly in terms of this paper, Wergaia speakers occupied the northern Wimmera and much of the Mallee and were bounded on the north-east from around Lalbert Creek by speakers of Wemba Wemba. Current knowledge of the placenames of this area is limited by problems common to other areas of 'settled' Australia, involving a lack of knowledge of the dynamics of European naming processes, in addition to ignorance of that of the original inhabitants. As relevant surveyors' field notes do not appear to have survived, there are no data extant on the language affiliation of the 'native guides' necessary for early cartographic expeditions, and very little information on other indigenous sources for early placename compilations. Later problems experienced elsewhere across settled Australia, such as the migration of placenames, particularly during the creation of parish-based land divisions, have made the situation even more confusing.

An example from the region of such migration is seen in the case of the placename Carapugna, currently found between the towns of Birchip and Nullawil. In the division of the area into a parish system in the late nineteenth century, Carapugna moved 35 kilometres east of its earliest recorded position and changed from being the name of an outstation on the Morton Plains pastoral run, to being the name of a newly created parish. During the same period, Whirily became a parish placename located well to the north of its original location referring to a waterhole on the Wirrimbirchip pastoral run. More unusually, it appears to have metamorphosed into a form involving the Irish surname O'Reilly for a time, before resuming a supposedly indigenous form with the creation of parishes. The meaning of these placenames and indeed many others in the area remains opaque, despite the identification of original locations.

---

[1]   Hercus (1986).

L. Hercus, F. Hodges and J. Simpson, eds, *The Land is a Map: placenames of Indigenous origin in Australia*, 157–163. Canberra: Pandanus Books in association with Pacific Linguistics, 2002.
© Edward Ryan

A contributing factor to this lack of transparency is the absence of glosses accompanying maps. One gloss at least is extant, though, relating to a map of the area around the Tyrell and Lalbert Creeks and gives us further insight into the problems and potentialities involved in the use of cartographic material. Among the placenames listed is Cooroopajerup with 'news of water' given as a translation. No relevant data appear to exist in either the nineteenth century linguistic compilations or the work of Luise Hercus to translate this but the morpheme *jerup* is a recurring one in local placenames and is always found in relation to water. Besides the lake on the map in question, a Cherrup Swamp is recorded near Charlton, and a Cooroopajerup and Tcharrup Tcharrup Creek — both ephemeral and known only from oral sources — run from the Charlton area north to join the Tyrell Creek. White Jagur Swamp also appears on the map and is glossed as 'wattle blossom'. The Wergaia word for golden wattle, *wadj,* may be a relevant translation in the case of 'White'. This form is also found applied to a clan further west in Wergaia country, termed the 'White Cundie Bulluk', and may well refer to the ecology of the clan's territory.[2] The form *Jagor* is given in nineteenth-century sources as a word for swamp and thus a more direct translation could be 'wattle swamp' rather than wattle blossom. Another example from the list is 'Lar Mitiyen', glossed as 'moon camp', which in distinction to the cases above is a direct and accurate translation. The question remains though — what does a 'moon camp' mean? Even when names are linguistically transparent they can remain opaque in cultural terms. A final example here is 'Brimybill' which appears to be a transcription of 'Brambimbul', the name of the two brothers Bram, the central ancestral figures in the creation of much of the landscape of western Victoria. This is confirmed by the gloss of 'brothers lake'. There is something to be gained then from a more rigorous analysis of such sources despite their apparent shortcomings; but there is also the danger that such an examination when undertaken in the absence of greater knowledge of traditional society could result in the compilation of sterile word lists largely devoid of indigenous cultural meaning. Fortunately there is material in the ethnographic record which provides some semblance at least of indigenous conceptions of place.

Woodford Robinson, 'a native of the Loddon' whose language was Wemba Wemba, provided a clear statement of the origin of placenames in a story he told the Reverend John Mathew in 1909.[3] The story concerned is a Brambimbul narrative and relates an account of how the Brambimbul killed the spiders Werimbul and their dogs and then points out the location of the Brambimul in the sky. They are the two pointers of the Southern Cross and are standing near the *donggal*, which is the big black hole beneath the Southern Cross. This *donggal*, black hole, comprises two bunyips which the Brambimul killed near Koondrook at Gannawara on the Gunbower Creek, where there is a basin with deep water. After the Brambimbul killed the two bunyips, they cut them up to feed the people and the different peoples ate different parts of the bunyips. Some ate the flesh, some the fat and some even the mud that was on them and this caused the difference between the languages of the various tribes. We have contained in one story an account of how the Brambimbul shaped the landscape and its equivalent in the sky and created the different languages of the region through their actions. Thus the constituents of placenames, language, country and cultural practice are clearly derived from and linked to the Brambimbul. The importance of the ethnographic record as a means of deepening our understanding of traditional conceptions of place is also demonstrated.

---

[2] See Clark (1990:360).

[3] J. Mathew, MS 950, Australian Institute of Aboriginal and Torres Strait Islander Studies, Canberra.

It can be difficult to use such ethnographic material though because of the lack of attention by early ethnographers paid to the identities of the Aboriginal people involved. A.W. Howitt referred to his main informant for what he termed the Wotjoballuk tribe as 'Bobby', a name common to a number of men across the region at the time under nineteenth-century European naming practices.[4] Bobby can be identified, however, by association with other Aboriginal people and by association with country he describes in his stories. Specifically he narrates a tale of how, after transgressions by the tribe, the ancestor figure Kurn burned down the giant pine tree that linked the ancestors of the Richardson River people to the sky. When the tree fell some people were left in the sky and the depression, now known as Lake Buloke, was formed at Banyenong. The kernels of the pine tree fell out across the country to become the stones that lie across Morton Plains to Lake Tyrell. While the same elements of location in the sky and on the earth and the actions of ancestor figures shown in Woodford Robinson's story are again present, we can also deduce that the 'Bobby' narrating is the historical figure Morton Plains Bobby or King Robert. He was the second-last traditional head of the Bulugdja (or lake) clan of the Wergaia people whose country included those locations mentioned in the story. A clear picture of places of importance in Bobby's country can be seen on the map of the Morton Plains pastoral run, c.1850 (see Figure 1).[5]

While the story of Kurn and the ancestors of the Richardson River people explains the nature of local country, the Bram brothers' story cycle explains country right across the territory of Wergaia and other related peoples. In one of the main narratives from the cycle narrated by Bobby a range of places from the Lower Wimmera River in the south to Pine Plains in the north are mentioned. The location of some of the events of the narrative are given with their European equivalents and so the way the landscape was formed becomes obvious but at other times this is not the case. After a series of adventures with other beings the Brambimbul enter a contest with Gertuk the tree creeper during which Gertuk unleashes his bag of whirlwinds and the Bram brothers have to cling to trees for support. The elder brother held a strong pine and was safe but the younger held a small tree and was blown a great distance to 'Witewitekalk'. Witewitekalk translates as 'small tree' of which there are many in the Mallee and were many more in the 1880s when Howitt interviewed Bobby and when read out of cultural context this placename would remain an obscure piece of cartographic data. Turning to the map of Bobby's home country of Morton Plains though we can move beyond this as we find the placename Witchagulk — a cognate of Witewitekalk and presumably the place mentioned in the narrative given the highlighting of the normally inconsequential 'small tree' in both story and placename.

---

[4] A.W. Howitt, MS 759, National Museum of Victoria.

[5] Illustration and other cartographic information from Historical Maps Collection, Department of Natural Resources and Environment, Victoria.

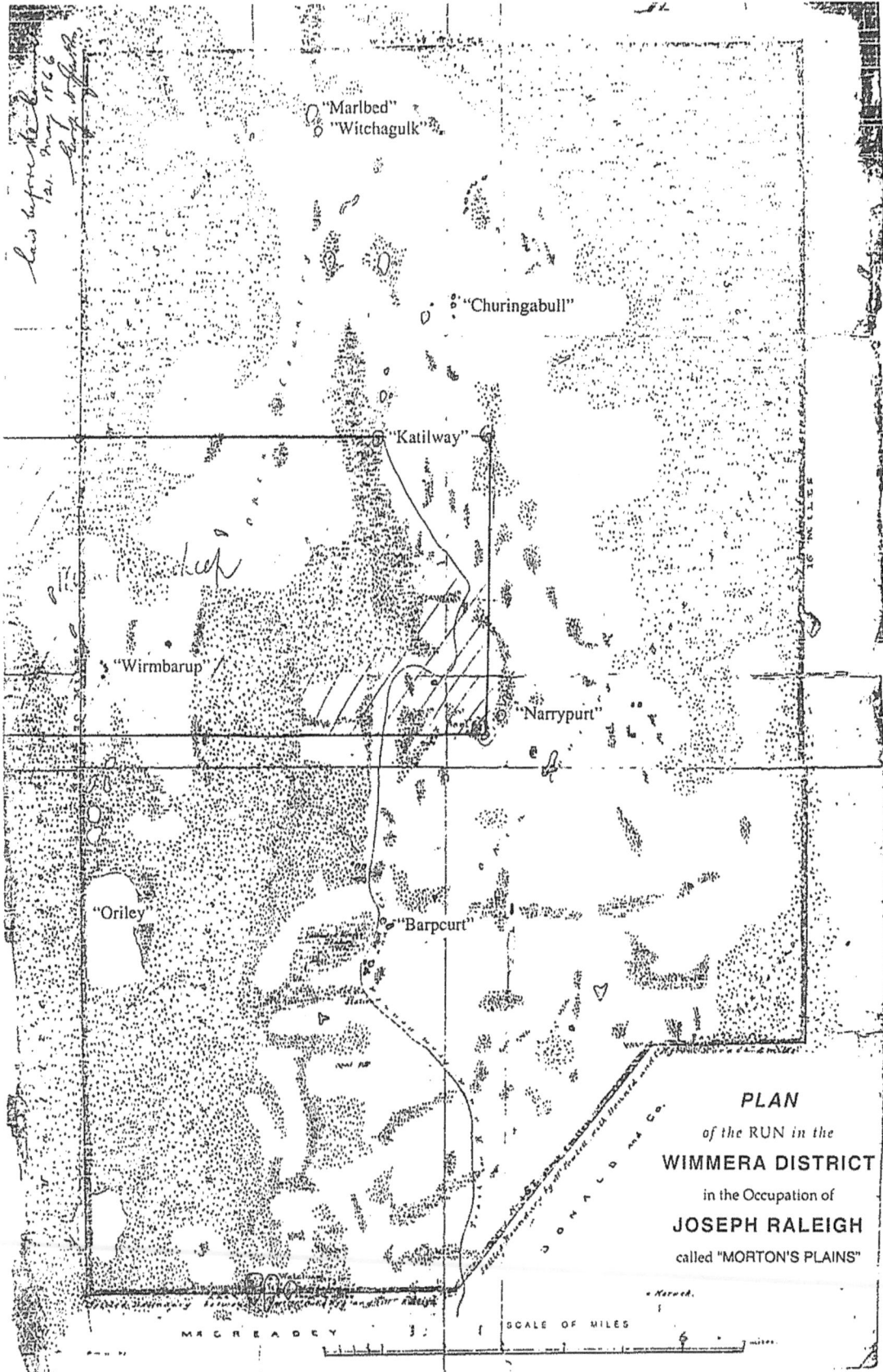

"Marlbed"
"Witchagulk"

"Churingabull"

"Katilway"

"Wirmbarup"

"Narrypurt"

"Oriley"

"Barpcurt"

PLAN
of the RUN *in the*
WIMMERA DISTRICT
in the Occupation of
JOSEPH RALEIGH
called "MORTON'S PLAINS"

SCALE OF MILES

MACREADEY

Figure 1: Map of the Morton Plains pastoral run, c.1850

Narrative cycles are naturally not the only basis for placenames in the area, as exemplified by Watchem on the southern boundary of the Morton Plains pastoral run. This name is generally translated as 'silver wattle' which may be an accurate rendering given the presence of the morpheme *wadj*, meaning 'golden wattle' as noted above. King Anthony Anderson, last traditional head of the Bulugdja clan, stated to European settlers that this was the case and strong stands of wattle originally grew near the lake at Watchem. An alternative origin of the name was given by 'Syntax'.[6] 'Blackfella lie alonga lake to watchem wild cattle come for water; then blackfella spear cattle'.[7] While this story gives us a good insight into the use of humour and irony by Aboriginal people when questioned on country by Europeans, it also illustrates the importance of the lake to its traditional owners and incoming settlers as one of few sources of permanent water in the area. However, not all placenames in the area refer to water sources. A range of documentary sources refer to a King Barney of Brabcut but in doing so illustrate a mis-hearing of Indigenous words and a misunderstanding of traditional authority systems. Brabcut is more accurately rendered *barpcurt* in the earliest sources including the 1850 pastoral map of Morton Plains. The initial morpheme of the name is transparent in meaning and represents *burb* — the Wergaia word for head or hill, which is also found in the nearby placename Wycheproof — *widji burb* meaning 'basket grass hill'. Barpcurt was the site of the Morton Plains head station, which occupied a rise in a generally flat area. In terms of a mis-reading of traditional authority systems King Barney of Brabcut was in fact a Djadjawurrung man from near St Arnaud but was free to much of Wergaia country because his mother came from Pine Plains and he spent much time on Morton Plains, the country of his wife Maria of the Bulugdja clan of the Wergaia. Rather than being a traditional head man of the Wergaia, his authority among Aboriginal people of the area came from him being a *bangal*, that is a 'Doctor' or 'Clever Man'.

Another individual named Syntax was also titled King of Brabcut, possibly again through being a *bangal*. Regardless of his place in the traditional authority system, his relation to place warrants further scrutiny. There is little evidence of King Syntax, otherwise known as Godfrey Syntax, leaving his immediate clan territory, unlike many other clan members. Following the relocation of the last of the old people of the clan including Morton Plains Bobby to the Ebenezer Mission on the Wimmera River in 1884, King Syntax appears to have lived as the sole permanent traditional inhabitant of the area until his death aged 50 in December 1885. In an area rapidly being taken up for closer settlement by incoming selectors he chose to live in his camp at Lake Marlbed. While this was still part of a pastoral run and so was less environmentally changed, it was at the northern end of the Bulugdja traditional country and well away from permanent reliable water sources to the south such as those at Watchem and Lake Buloke. As this site was of great inconvenience in utilitarian terms, its significance as a location for him then must be sought in the cultural realm. Lake Marlbed home station, as can be seen from the map, is also the location of Witchagulk, the site where the younger of the two Bram brothers was blown by Gertuk's whirlwinds in Bobby's Brambimbul narrative.

---

[6]   Either Syntax Harrison or Godfrey Syntax.

[7]   Donald History and Natural History Group (1989:41).

Figure 2: King Anthony Anderson, photograph courtesy of the State Library of Victoria.

Evidence exists of a similar choice being made by King Anthony Anderson. After travelling over Wergaia and other country to which he was 'free' and well beyond for some years, King Anthony settled back on his clan country as he grew older in the 1890s. Though drawing Protection Board rations from the new township of Birchip, he too chose to camp in drier country well to the north at points such as Gould's Lagoon and Marlbed, each 15 kilometres from town. Ethnographic sources again contain material which may help explain this 'illogical' choice through shedding more light on the place of Witchagulk in Bobby's Bram brothers narrative. Sergeant Major, a son of King Barney who had also spent much time in Bulugdja country, explained to A.W. Howitt that the Bram brothers were of the Patjingal or pelican moiety subsection. While Patjingal was a subsection name common to both white cockatoo and black cockatoo moieties, Sergeant Major also stated that the mother of the Bram brothers, Tuk the bullfrog, was of the black cockatoo moiety.[8] Working on the basis of matrilineal descent of moiety affiliation, the Bram brothers must then have been black cockatoo men — as indeed was King Anthony Anderson.

There is an evident need to consider the narrative cycles associated with the ancestral beings and indeed a range of other cultural and historical material when considering Indigenous placenames in settled south-eastern Australia. In some ways this may be of even more importance than in areas where language and traditional culture are stronger as there are less linguistic data available to attempt to bring some degree of transparency to placenames. As can be seen in the case of north-west Victoria, a discussion based on wider cultural considerations can not only contribute to the attempted translation of opaque placenames, but also provide transparency to the actions of known historical figures to whom those placenames had associations of great importance.

In adding cultural knowledge to what we know of the lives of the people of the north-west, and in comparing the actions of the ancestor beings with the actions of the Wergaia and associated peoples, we can begin to approach an understanding of the deeper meanings of the Indigenous placenames of the region. In doing so, we can retrieve placenames from the obscurity of antiquarian maps and begin to place them in the cultural landscapes in which they belong. We can move beyond Indigenous placenames taking the role of collectables without context as mere names on a page and indeed move beyond Witewitekalk.

REFERENCES

Clark, I.D., 1990, *Aboriginal Languages and Clans: an historical atlas of western and central Victoria, 1800–1900*. Clayton: Monash Publications in Geography.

Donald History and Natural History Group, 1989, *Past & Present: 'The Donald Times' jubilee souvenir*. Donald.

Hercus, L.A., 1986, *Victorian Languages: a late survey*. Canberra: Pacific Linguistics.

Historical Maps Collection, Department of Natural Resources and Environment, Victoria.

Howitt, A.W., MS 759, National Museum of Victoria.

— MS 9356, La Trobe Library, State Library of Victoria.

Mathew, J., MS 950, Australian Institute of Aboriginal and Torres Strait Islander Studies, Canberra.

---

[8] A.W. Howitt, MS 9356, LaTrobe Library, State Library of Victoria.

12   WEEDING OUT SPURIOUS ETYMOLOGIES: TOPONYMS ON THE ADELAIDE PLAINS

Rob Amery

---

1 ETYMOLOGY OF KAURNA NAMES[1]

The etymology of Kaurna placenames is often very difficult to determine. With the passage of more than 160 years since colonisation and several generations since the death of the last fluent Kaurna speakers, it is difficult to be certain about anything. Folk etymology is rampant. About 60[2] different Kaurna placenames are recorded in the vocabulary sources. Teichelmann and Schürmann (1840) (henceforth referred to as T&S), Williams (1840), Piesse (1840), Wyatt (1879) and Black (1920) all recorded Kaurna placenames.

Apart from the recognised sources of Kaurna language, placenames are recorded on a variety of maps. They appear in journals, government records, local histories, newspaper articles and especially in letters to the editor over the years. Others may be preserved in property names.[3] Much of this material has been collated and published in several popular books on South Australian placenames, notably Praite and Tolley (1970), Manning (1986) and Cockburn (1908; 1984; 1990).

---

[1]   Abbreviations used in this paper are: LOC locative; PL plural.

[2]   This is a far cry from the two to three thousand placenames known by any given adult in the Wik region of Cape York as discussed by Sutton (this volume), indicating the extent of language loss in the Adelaide area.

[3]   It should be noted, however, that many property names of apparent or obvious Aboriginal origins are drawn from other regions and other languages.

L. Hercus, F. Hodges and J. Simpson, eds, *The Land is a Map: placenames of Indigenous origin in Australia*, 165–180. Canberra: Pandanus Books in association with Pacific Linguistics, 2002.

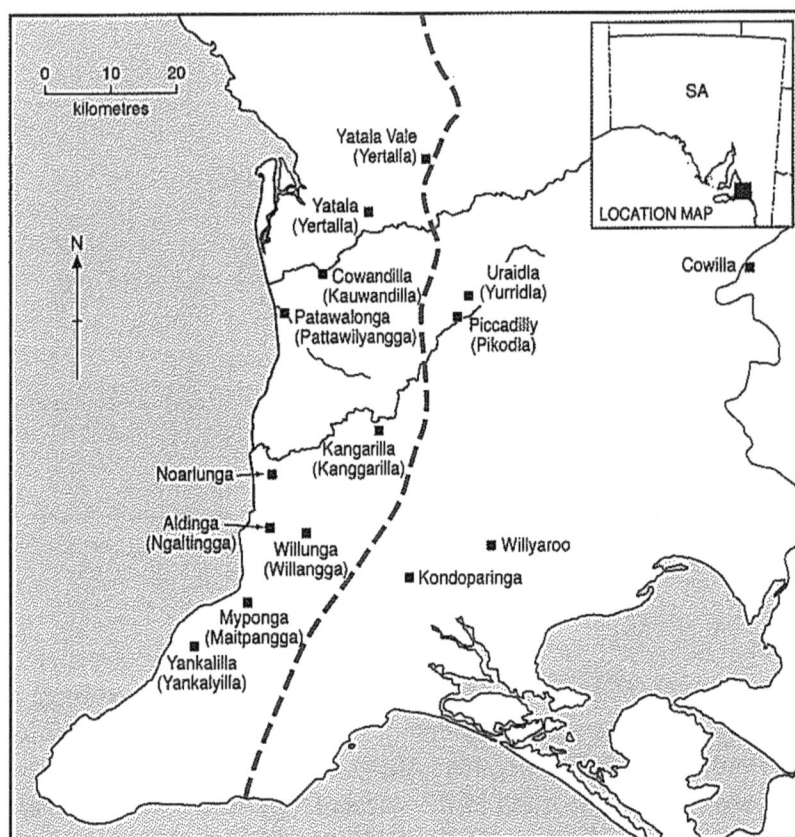

Map 1: Placenames of Aboriginal origin on the Adelaide Plains

Nineteenth-century sources on the Kaurna language, with the exception of Wyatt's (1879:5) explanation of Yurre illa 'two ears' and T&S's observation that Noarlunga is derived from *nurlo* 'corner' (discussed later), did not try to provide etymologies for placenames on the Adelaide Plains. Black (1920) does attempt to do this for the 11 placenames he records, but often this is simply on the basis of T&S. However, in the case of Ngangkipari (= Onkaparinga) he does appeal to his source, Ivaritji, for an explanation. He notes in the case of Pattawilya that Ivaritji was unable to explain the name, but on the basis of T&S he ascribes the meaning 'gumtree foliage' to the name. We are left with the fact that, in almost all cases, meanings ascribed to placenames of Indigenous origins in the Adelaide area have been the result of speculation on the part of Black, Tindale, Webb and numerous others, including myself.

In some cases, such as Witongga 'The Reedbeds'[4] in the Fulham area, the meaning is quite apparent. The word for 'reed' is *wito* and reeds are prolific in that locality. So there is little doubt that 'reed place' is in fact its rightful meaning and no competing etymology has been noted in the sources. In other cases, the meaning is not so readily apparent.

The popular placename books include some fanciful etymologies. In many cases we can say little about the true etymologies of these names, though we can use a knowledge of linguistics to call into question or exclude certain suggestions. A distinction needs to be

---

[4]   This is one of the few cases where the English name, The Reedbeds, is a translation (at least in part) of the Kaurna name.

drawn between *literal* and *connotative* meaning, the difference between say *wirra* literally 'forest', though the connotative meaning might be 'hunting ground' or 'good country' or even 'wild', 'rugged' or 'untamed' depending on one's cultural background. The longer and more florid the description is, the less likely it is a literal meaning of the placename, though we cannot discount the possibility that it may be related in some way to the particular name through connotative or affective meanings, or perhaps by a story linked to the place. For instance, the Kaurna name for Morphett Vale is Parnanga, from *parna* 'autumn star' + *-ngga* 'LOC'. Therefore, Parnanga literally means 'autumn star place'. We do not know the story behind Parnanga, but from what we know of Kaurna cosmology, it is likely that Parnanga is the place from where the Ancestral Being Parna ascended into the heavens, or where he or she performed some other feat. Manning (1986:141) says it means 'autumn rain', in which case it is a connotative meaning.

Consider the following additional examples:

1.1 Aldinga

T&S record the name as *Ngalti-ngga,* yet do not provide a vocabulary entry for *ngalti.* Manning (1986:3) notes that various meanings have been suggested for Aldinga, including 'much water'; 'good place for meat'; 'open wide plain'; and 'tree district'. In addition, Cockburn (1990:7) adds 'battle or burial ground'. None of the meanings noted by Manning or Cockburn make any sense in terms of the Kaurna words documented in the historical sources.

Now the suffix *-ngga* could be translated as 'place', and *ngalti* almost certainly consists of a single morpheme. Yet a number of the etymologies above suggest that a compound would be required to encode the notions. It could not possibly literally mean 'good place for meat', even though historical sources suggest that the Aldinga area was used as a kind of 'fattening paddock', because that would require at least two morphemes in addition to the *-ngga* suffix. Furthermore, the known Kaurna words for meat *(paru),* water *(kauwe),* plain *(womma)* etc. bear no resemblance whatsoever to *ngalti.*

Perhaps there was no meaning for Aldinga in the minds of Kaurna people alive back in the 1830s. Ngaltingga could have been just a name with no known meaning, just as Indulkana is just a name for Pitjantjatjara/Yankunytjatjara people today. If Aldinga had meant something, there is a good possibility that T&S would have listed *ngalti* in their vocabulary, after having noted and listed the placename Ngalti-ngga.

1.2 Willyaroo

Willyaroo, close to Strathalbyn and a little to the east of Kaurna territory, is said by both Cockburn (1990:240) and Manning (1986:229) to mean 'to invoke a good harvest', though Praite and Tolley (1970:195) say it means 'scrubby plain'. *Wilyaru* is given by T&S as 'one who has gone through all the initiatory ceremonies; a fully grown-up man'. *Wilyaru* also appears in Hercus's (1992) Nukunu dictionary as 'name of the final stage of initiation involving cicatrization' and in other languages to the north. Similar words are not found in Ngarrindjeri (Meyer 1843; Taplin 1879) or Ngayawang (Moorhouse 1846).

The meaning 'to invoke a good harvest' is most unlikely. It does not sound like the kind of notion encoded within an Indigenous language and in any case would require several morphemes and should look like a verb, whereas it bears no resemblance to one in either Ngarrindjeri or Kaurna, which take -*un* and -*ndi* endings respectively. The reference to a stage of initiation, as found in T&S, is a strong possibility. This practice may be associated with success in hunting or fertility rites, but there is certainly nothing in the name itself to suggest that it means 'to invoke a good harvest'.

1.3 Kondoparinga

An even better example is that of Kondoparinga, the name given by Governor Robe to a Hundred[5] in the Willunga area in 1846. According to Cockburn (1908:76) it is a 'native word meaning "long, winding water, breeding crawfish between steep banks"', a notion which by my estimation would require at least seven morphemes to encode literally. It is almost certain that Kondoparinga consists of just three morphemes, possibly *kundo* 'chest' + *parri* 'river' + -*ngga* 'LOC' (i.e. 'chest river place').

1.4 Noarlunga

Cockburn (1990:160) and Manning (1986:152) assert that Noarlunga means 'fishing place' while Praite and Tolley (1970:129) say that it means 'the place with a hill'. Praite and Tolley are obviously drawing on the fact that Ramindjeri has *ngurle* 'hill' and this, combined with the Kaurna suffix -*ngga*, results in Noarlunga.

However, it is far more likely that Noarlunga derives from Kaurna *nurlo* 'curvature; corner' and referred to Horseshoe Bend, on the Onkaparinga River, where the town was first established. Certainly, as we observed earlier, T&S recognised this.

1.5 Pattawilya

In his earliest publication, Cockburn (1908:50) gives the meaning of Patawilya as 'cloggy, green place', an etymology that also appears in Martin (1943), though Martin gives Patawalonga as 'Aboriginal for boggy and bushy stretch with fish'. In like fashion, later, Cockburn (1984; 1990:87) claims that Pattawilya, the Indigenous name for Glenelg or Holdfast Bay, means 'swampy and bushy, with fish'. Here again, such a notion would require at least three and probably four distinct morphemes. As most root morphemes in Kaurna are disyllabic, the word *pattawilya* almost certainly consists of just two morphemes. Furthermore, no part of the word bears any resemblance to *kuya* 'fish'. The name is in fact transparently composed of *patta* 'a species of gum tree' (possibly the swamp gum) and *wilya* 'foliage; young branches; brushwood'. So 'swampy and bushy, with fish' is fanciful to say the least, as a supposed literal meaning of Pattawilya.

---

[5] 'Hundred' means a major division of land, the purpose of which is to provide unique identities for properties in conjunction with the section numbers. The actual size varies from one to another (Bill Watt pers. comm.).

1.6 Cowandilla

All three sources (Praite & Tolley 1970:32; Manning 1986:48; and Cockburn 1990:52) state that Cowandilla means 'water', or in Manning's case 'locality of the waters'. Further, Manning records the name as 'kaunenna-dlla'. While Manning does not specifically name Webb as the source, it is fairly clear that the latter is responsible for this etymology. Webb (1936–37:302) says that

> certain waterholes near Glenelg were called 'Kauwenna'. 'Kauwe' is water. 'Kauwenna' means the Waters. 'Kauwenna-dlla', the name of the locality survives in the pretty placename 'Cowandilla', which means the locality of the waters.

Now, 'locality of the waters' should appear as *kauwe* + *nna* + *illa* 'water+PL+LOC' resulting in *kauwinnilla*. It is actually questionable whether a plural morpheme would appear in such a Kaurna word. *Kauwingga* may be more likely. In any event, there is no way of accounting for the presence of the 'd' in such an etymology. Therefore, this etymology is most unlikely.

Tindale (card files) also claims that Cowandilla means literally 'in the waters' and breaks down the word as *kauwe* 'water' + *an* 'PL' + *til:a* 'in' or 'among', though it is unclear where he gets *til:a* from.

An alternative etymology is far more likely. The word for 'north' is *kauanda* and upon attaching the location suffix *-illa*, as it is a trisyllabic root, the result would be *kauandilla* 'in the north', corresponding exactly to the form of the name as it is known today. This contrasts with Patpangga, the name for Rapid Bay, which means 'in the south'.[6]

In this case, I am suggesting that *all* the numerous popular accounts are false. Undoubtedly there were springs in the area, and Cowandilla may indeed have been the rightful name for one or more of these springs, but I assert that it did not literally mean 'water' or 'spring' or anything related to water. *Kauwe* 'water' was one of the few Kaurna words with which many colonists would have been familiar, and I would suggest that people jumped to conclusions.

1.7 Uraidla

It was noted earlier that Uraidla was one of just two Kaurna placenames for which nineteenth-century sources provided a meaning. It is derived from *yurridla* 'ear-DUAL'. However, Webb writing in the *Advertiser* on 3 December 1927 claimed that it meant 'place of the ear' and that *-dlla* was the location suffix. Here Webb is confused. The location suffix on trisyllabic roots is *-(i)lla* and the 'd' cannot be accounted for using this etymology. Cockburn (1990:225) also suggests that, on the basis of T&S, Uraidla is derived from *yurre* 'ear'+ *idla* 'whelp'. Further he quotes a letter from Thomas Playford written in 1872 claiming that Uraidla meant 'spring, or source of many waters'.

---

[6] It is not clear what the point of reference is for Patpangga and Kauandilla. At the *Indigenous Placenames Colloquium* on 8 April 2000, Jane Simpson suggested that these places may be relative to the country of the Kaurna men who served as the main informants for T&S, Wyatt and others. Mullawirraburka's and Ityamaiitpinna's country lay roughly mid-way between these two locations. However, Patpangga is a Kaurna name which was clearly in use before the arrival of the colonists, as it was used by Kalloongoo in her interview with George Augustus Robinson in June 1837 and she had been kidnapped from the Rapid Bay–Yankalilla area many years before that (see Amery 1996).

1.8 Piccadilly

Cockburn gives three different accounts of Piccadilly having been named by different people after Piccadilly in London:

1. It was named by Mrs John Young after her birthplace in London in about 1853.

2. It was named in about 1862 or 1863 by Mary Johns, an English cook who worked for Arthur Hardy, one of the earliest residents of the district.

3. The gardener at Piccadilly, H. Curtis, disputed the second account and asserted that Samuel Tonkinson bestowed the name.

Who should we believe? The second possibility given by Cockburn is accompanied by a lengthy story contained in a letter published in a newspaper in 1928:

> Mrs Caulfield Barton, daughter of Arthur Hardy, one of the earliest residents of Mount Lofty district wrote to the *Advertiser:* 'My mother had for us an English nurse, Miss Everitt and when we were living at Mount Lofty House — about 1862 or 1863 — she also had a young English cook named Mary Johns. Right and lively girl, and as my father's house was at the time the only one on Mount Lofty proper, it was naturally dull for the maid. One day she said to Miss Everett, "Come for a walk" and on going out Miss Everett asked, "Where shall we go?" "Let's take a walk down Piccadilly", said Mary Johns, and they followed the track to the bottom of the gully, where at the time there were only charcoal burners and no houses at all. So Miss Everett told my mother of the joke Mary had made and the name stuck to it — in fun at first — and afterwards it was accepted by everyone. The very fact of its unlikeness to London Piccadilly was the cause of its getting the name.' (in Cockburn 1990:175)

I have no reason to doubt that Mary Johns and Miss Everett did go for a walk and that the conversation as quoted in the letter did take place. But why would Mary Johns have said 'Let's take a walk down Piccadilly'? She may well have been familiar with the Indigenous name of the place, which reminded her of Piccadilly. After all, nothing else about the location bore the least resemblance to Piccadilly in London as the writer of the letter points out.

In fact, Manning (1986:168) tells us that the name derives from piccoddla 'the locality of the eyebrow'. This is confirmed by T&S who give *piko* 'eyebrow'. The dual form would be *pikodla* (i.e. 'two eyebrows'), just as Uraidla or *yurreidla* is the 'two ears'.

I would suggest that a number of patently English names were given to localities because they bore resemblances like this to the Indigenous name. Another good example of this is the town of Gawler, named after South Australia's second governor Colonel George Gawler. The Kaurna name is recorded as Kaleeya or Kaleteeya. The close resemblance is probably not a matter of sheer coincidence. It could also be that the recorded Kaurna names were a Kaurna pronunciation or corruption of Gawler.

There are even more striking cases of conflicting etymologies than these. Consider the following examples:

1.9 Yankalilla

Yankalilla has been the subject of much discussion and fantasy as to its origins. Some observers have even suggested that Yankalilla is not an Aboriginal word. Note Cockburn's entry:

Yankalilla has always been a puzzle in nomenclature and probably always will be. Undoubtedly it is an Aboriginal word, but although it is known that the terminal 'illa' indicates locality, either at, on, or near, no satisfactory record of the meaning of the prefix has ever been established. The orthography of the word went through an extraordinary range of variations before settling down to its present form.

Teichelmann and Schürmann's Vocabulary, published in 1840, gave it as Yankalya-illa, but quite a dozen different spellings could be quoted. The first time that Colonel Light mentioned the name in his journal, he made it Yankalilla, but on one of his plans he wrote it Yanky-lilly, together with Yanky Point. This gave rise to the diehard theory that the appellative was derived from the circumstance that an American whaler, who had a daughter called Lily, used to frequent the coast near Yankalilla; a variation of the story being, that an American whaler known as the *Lilly* was wrecked thereabouts. But there are plenty of Aboriginal words with the suffix '-illa' — for example Tunkalilla district and beach in the same region as Yankalilla. In one of the dark recesses at the gorge of the later <sic> place there is a never-failing spring of water and the precise locality was a favourite camping place of the natives. It is highly probably [sic] that the name has reference to that fact. (Cockburn 1990:247)

Governor Hindmarsh, in a dispatch to Angas in 1837, wrote:

There is a spot near Cape Jervis with one of the sweetest sounding names I ever heard, pronounced Yoongalilla, which, like all their other words, is extremely liquid in its sound. Colonel Light, I am sorry to say, in his letter to the Commissioners, has adopted the Kangaroo Island Whalers cognomen of Yanky-lilly. I thought it was some place that had been frequented by the American Whalers. (quoted in Williams 1985:6)

Even more compelling is the fact that Yankalilla is referred to in an interview given in 1837 by Sarah or Kalloongoo, a Kaurna woman kidnapped by sealers and taken to Kangaroo Island. Kalloongoo was kidnapped most likely in the early 1820s, taken to Kangaroo Island where she remained for some time, then taken to Portland and later to Bass Strait (see Amery 1996). She told George Augustus Robinson on 2 June 1837 that she had been kidnapped from the Yankalilla–Rapid Bay area:

Said the country where she came from was called BAT.BUN.GER [Patpangga = Rapid Bay] YANG.GAL.LALE.LAR [Yankalilla]. It is situate [sic] at the west point of St Vincents Gulf. (Plomley 1987:446)

So, if the origins of Yankalilla are from English, it must have been very early, prior to the establishment of the colony of South Australia, and the name must have been known, remembered and used by a Kaurna woman kidnapped from there many years previous. This is most unlikely. Thus the non-Aboriginal origins of Yankalilla can be readily dismissed.

However, that is not the end of the matter, for we have conflicting Indigenous etymologies for the word, including two totally different explanations put forward by the same author. On the one hand, Tindale provides a Kaurna etymology for Yankalilla which appeals to the well-known Tjilbruke Dreaming story. Manning (1986:237) writes:

A difference of opinion regarding the origin of the name [Yankalilla] prompted a spate of letters to the Register, each writer giving different versions. (see 10, 13, 16, 17, 20 & 25 Feb 1928). Professor N.B. Tindale says 'it is derived from the Aboriginal word *jankalan,* meaning "falling", from an incident in the myth of Tjilbruke, whose sister's [sic] mummified body began to fall into pieces here, as he was carrying it from Brighton to Cape Jervis for burial'.

However, Tindale's Kaurna card file, held by the South Australian Museum, provides an entirely different etymology for the name as follows:

Jangkaljil:a 'Yankalilla' Lit. 'Upon the hill' Deriv. ['jangkalja] 'hill' + [íl:a] 'at or upon'

Jangkaljawangk 'Yankalilla' Deriv: ['jangkalja] 'hill' + ['wangk] 'upon'

This etymology is based on Ramindjeri and Kaurna. Meyer (1843) lists the Ramindjeri word *yangaiãke* 'hill', which Tindale has combined with the Kaurna locative suffix *-illa*.

R.F. Williams (1985) in his volume on the local history of the Yankalilla area has made even more far-reaching blunders in confusing Kaurna with Ramindjeri. He writes:

> A mission was established in the territory of the Ramindjeri people at Encounter Bay. Teaching commenced and a study of the language, sufficient to publish a vocabulary in the native tongue, was made by 1840.
>
> The missionaries gave the locality name as Yankalya-illa. The vocabulary records that nganka is Ramindjeri for woman, alya for tragedy and illa for place terminal. It is not a far step from Nganka-alya-illa to Yankalya-illa. (Williams 1985:6–7)

Unfortunately this statement is wrong in several details. The published vocabulary Williams refers to is that of Teichelmann and Schürmann (1840). Meyer's Ramindjeri vocabulary was not published till 1843. The Ramindjeri word for 'woman' is *mimine* (Meyer 1843:80), a word still used today. No such word *nganka* is listed by Meyer. Williams must have mis-transcribed the Kaurna word *ngangki* 'woman'. Meyer (1843:50) lists Yangkallyaw-angk for Yangkalilla. I contacted Roy Williams by phone in 1995. He was unable to locate his sources and seemed to fail to understand that Ramindjeri and Kaurna were different languages. Unfortunately, a Royal Geographic Society publication has perpetuated Williams' errors:

> The meaning of this word is still a puzzle. Light wrote of it in its present form, but later referred to it as Yanky-lilly.
>
> Missionaries at Encounter Bay in 1849 who published a vocabulary in the native tongue, in 1840 referred to it as Yanky-aly-illa. Ngangka is Ramindjari for woman, alya for tragedy and illa for place, hence Nganga-alya-illa to Yank-aly-illa. (RGSASA 1989:12–13)

Now Teichelmann and Schürmann (1840:21, 76) clearly identify Yankalilla as a Kaurna name. It is given as an example of the location suffix *-illa* affixed to trisyllabic roots — '*Yangkalyilla*, in *Yangkalya*'. It derives from *yernka-*, the root of the verb *yernkandi* 'to hang down, on; to join; impart; infect' + *-lya* 'CONTINUOUS' + *-illa* 'LOC'. Thus it translates literally as 'the place where [it] kept falling apart', a meaning entirely consistent with that given by Manning above on the basis of information supplied by Norman Tindale. Yankalilla might be thought of as 'the place of the fallen bits'. Tindale's other etymology which relates Yankalilla to a Ngarrindjeri word is unlikely to be the 'true' etymology, though it was undoubtedly believed to be so by Tindale's Ngarrindjeri informants, such as Milerum.

1.10 Yatala

Yatala is another clear example where Tindale seems to have gone astray. According to Tindale's Card File:

> 'Jatala 'Yatala; Dry Creek'
> Probably post-contact name arising from the presence of a white man's prison. The name seems to be linked with the verb ['jat:un] 'to steal'.

However, Tindale's ['jat:un] is clearly a Ngarrindjeri word. Meyer (1843) lists *yart-in* 'stretching out the hand to receive' while Taplin (1879:136) gives *yartin* 'reaching out the hand to receive'. Yatala is a placename that has been used by non-Aboriginal people since 1836. In 1846 the Hundred of Yatala was proclaimed, long before the establishment of the Yatala Labour Prison, which was known as the Stockade of Dry Creek in 1854. The prison was re-named Yatala Labour Prison some time between 1854 and 1860 (phone enquiry, Yatala Labour Prison 21 November 1997). Thus in this case, Tindale's etymology is clearly falsifiable. The prison authorities now understand Yatala to mean 'meeting place' or 'by the water'.

Yatala most likely derives from *yertalla* 'water running by the side of a river; inundation; cascade'. As Manning (1986:238) observes 'in winter when water flowed from the hills, over the plains, the Dry Creek area became a morass'.

1.11 Kangarilla

Kangarilla almost certainly derives from *kanggari-* 'to bring forth' + *-(i)lla* 'LOC'. According to Manning (1986:100), Tindale says 'it is derived from the Aboriginal word Kanggarila, which may mean birthplace', but we have no information about the context. Perhaps it referred to the encampment of a shepherd who first moved into the area. Note that T&S give *sheepi kanggallanggalla* 'a shepherd', derived from *kanggandi* 'to lead; conduct; accompany; to bear a child "bring forth"'.

Cockburn (1908:76) claims that Kangarilla is derived from Kangooarinilla, 'native for "place where sheep mother sits down" — "kanga", a ewe with lamb; "illa" the place of sitting' and notes that it was formerly called Eyre's Flat, on account of it once being the property of Edward John Eyre, the explorer. In Cockburn (1984; 1990:112) this etymology has been replaced by a reference to Dutton's (1846) *South Australia and its Mines* which claims that Kangarilla is derived from Kangowirranilla, meaning the 'place for kangaroo and water'. However, Cockburn himself thinks that it is 'more likely to be kangaroo and timber', no doubt having noticed the presence of *wirra* 'forest, bush' in the word.

However, Manning also cites an entirely different etymology to both his own and Cockburn's as follows (1986: 100):

> In *Kangarilla Historical Records* the compiler says — 'the Reverend Gordon Rowe of the Aborigines Friends' Association obtained the following information from Mr David Unaipon, an eighty-two year old full blood member of the Tailem Bend tribe. His definition of the meaning of the origin of the name is — "Kang means two; Ra'mulia means outflow or water flowing ..." When first approached on the matter Mr Unaipon at once asked if there were two waterholes. Upon enquiry it was found that there were two ...'

This entry is extremely revealing of the ways in which etymologies arise, particularly many of those in Tindale's materials (1987; Card files; assorted papers) which were mostly obtained from Ngarrindjeri sources. In this case, Kangarilla was not Unaipon's country. Having been asked about it, of course he tries to make sense of the name in his own language and comes up with *kang + ra'mulia*. First, Kangra'mulia is not particularly close phonetically to Kangarilla. In fact, it is difficult to see how the two forms could refer to the same name. Further, we have seen that Kangarilla already has a perfectly good etymology in Kaurna, the language of the territory of the place itself. This Kaurna etymology is based on morphology which conforms perfectly to expected Kaurna patterns. Unaipon's suggestion is clearly a false folk etymology.

1.12 Willunga

The town of Willunga, laid out in 1839, is another interesting case. There are in fact three competing etymologies for Willunga, all from within the Kaurna language, and all of them plausible.

1. Willunga is from *willa* 'dust' + *-ngga* 'LOC'. Williams (1840) records *wil-lah* 'dust'.

2. Willunga is from *wilya* 'foliage' + *-ngga* 'LOC'. This is certainly the etymology that the Willunga Council ascribes to. Manning (1986) says it means 'a place of green trees'.

3. Willunga is from *willi* 'the chest of a kangaroo or other animal' + *-ngga* 'LOC'. This is the etymology put forward by Webb when he links the place with Uraidla and other locations referring to parts of a giant kangaroo laid out across the landscape.

The first etymology is in fact the most likely of the three, though perhaps least favoured. T&S record the name as Willa-ngga, though they do not record *willa* in their vocabulary. If it had been derived from *wilya* or *willi*, most likely they would have recorded the name as Wilya-ngga or Willi-ngga. Perhaps Willa-ngga means something entirely different. There are four distinct laterals (or 'l' sounds) in Kaurna. T&S often wrote three of these, namely an interdental [lh], alveolar [l] and retroflex [rl], with a double 'll'. Nor did Williams distinguish between these three 'l' sounds. So T&S's Willa-ngga may bear no relationship to Williams' *wil-lah* 'dust'.

1.13 Warriparri

Similarly, three meanings for Warriparri, the Sturt River, are to be found:

1. According to Cockburn (1990:209) Warriparri means a 'creek fringed with trees'. No morphological analysis is suggested however.

2. Webb (1936–37:308) claims that Warriparri means 'the throat river' and suggests that it is derived from *warra* 'throat' and *parri* 'river'. The existence of the suburb Warradale in the vicinity, which itself was named after a property in the area, lends some credence to this suggestion.

3. Most sources say that Warriparri means 'windy river' and is thus derived from *warri* 'wind'. This is entirely consistent with T&S's spelling of the word and with other historical sources: War-rey par-rey (Williams 1840) and Wari pari (Black 1920). They all agree on the vowel being 'i' rather than 'a'. Thus this is the most likely etymology. This is also the meaning favoured by literature produced by the Kaurna Aboriginal Community and Heritage Association (KACHA Inc.) and the Marion City Council in relation to the Warriparinga heritage site and planned Interpretive Centre (Warriparinga brochure, October 1997).

2 BODY PARTS OF THE GIANT KANGAROO

Noel Augustin Webb spent considerable effort puzzling over Kaurna placenames. He published an article 'The Place Names of the Adelaide Tribe' for several years running. Webb had refined and expanded this article with successive publications.

Webb suggests that no less than 11 Kaurna names, some of which have been discussed already, are related to body parts of a giant kangaroo stretching from Gumeracha, north-east of Adelaide, to Encounter Bay and Kangaroo Island. According to Webb (1936–37), these parts are as follows:

Uraidla	*yurridla* = *yurre* 'ear' + *-dla* 'DUAL' (i.e. two ears)
Piccadilly	*pikodla* = *piko* 'eyebrow' + *-dla* 'DUAL' (i.e. two eyebrows)
Gumeracha	*ngarrumuka* 'brain'
Mudlangga (Le Fevre Peninsula)	*mudla* 'nose' + *-ngga* 'LOC'
Warripari (Sturt River)	*warra* 'throat' + *parri* 'river'
Marino	*marra* 'hand' + *-nna* 'PL' (i.e. paws)
Willunga	*willi* 'chest' + *-ngga* 'LOC'
Kundoparinga (near Willunga)	*kundo* 'chest' + *parri* 'river' +*-ngga* 'LOC'
Murtapari (Inman River)	*murta* 'excrement' + *parri* 'river'
Yerltoworti (Hindmarsh River)	*yerta* 'land' + *worti* 'tail'
Karta (Kangaroo Island)	*karta* 'lap'

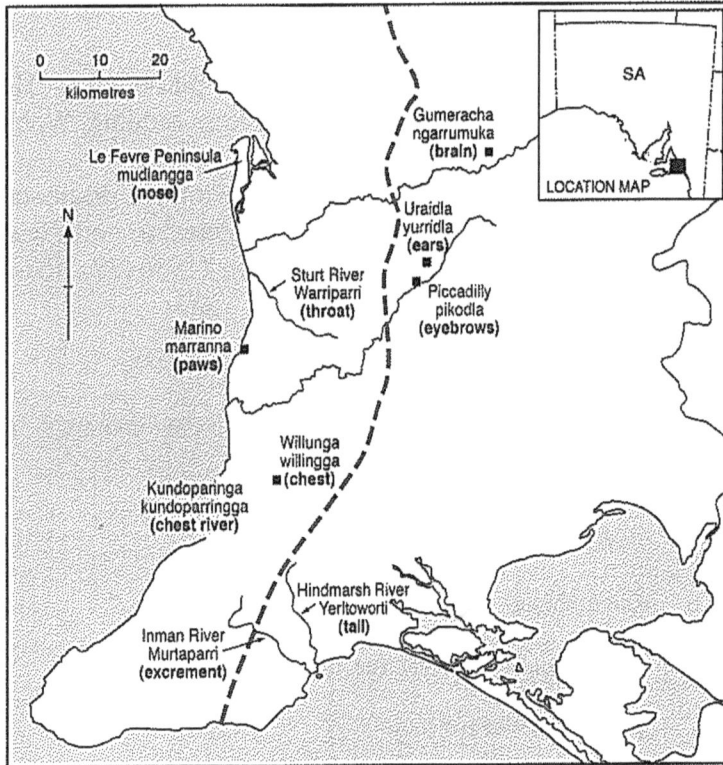

Map 2: Placenames related to body parts of a kangaroo (according to Webb 1936)

As discussed above, there is little doubt that the two peaks, Mount Lofty and Mount Bonython named Yurreidla 'two ears', are likened to the ears of this being. There is also a strong possibility that Piccadilly, an Anglicisation of *pikodla* 'two eyebrows' is a part of the same being, though Simpson (1995) suggests that it might derive from *piko+rti+dla* 'eyebrow-without-DUAL'. Mudlangga 'Le Fevre Peninsula' is undoubtedly 'nose place', but whether or not it refers to the nose of this same kangaroo being is another matter. The word for 'nose' is used by other Australian languages to refer to peninsulas, as in Yolngu Matha *ngurru* 'nose; tip; front; peninsula; cape; point (of land, spear)' (Zorc 1986:233). Several of the locations suggested by Webb, while plausible, have other competing etymologies. Willunga and Warriparri have been discussed above. Webb's etymology for Marino as *maranna* 'hands' is suspect. If the name did in fact refer to the front paws, we would expect *mararla* with the dual suffix. A hind leg is probably referred to as *tidna* 'foot' rather than *marra* 'hand'. According to Cockburn (1990:137), Marino was named after San Marino, while Praite and Tolley (1970:109) say that Marino was 'originally Marina, which is the Italian word for "seashore"'. It is also highly unlikely that *yerlto* in Yerltoworti as transcribed by T&S is a variant or mis-transcription of *yerta* 'land'. T&S are generally consistent in their transcriptions. Had they meant *yerta*, they would have written it as *yerta*. Deriving Gumeracha from *ngarrumuka* 'brain' is also perhaps drawing a long bow. It would involve metathesis or transposition of sounds within the word (i.e. swapping of the 'rr' and 'm'), transcription of the initial velar nasal with 'g' and the transcription of the high back vowel as 'er' and 'a'. All these are possible, but in the light of other questionable etymologies this one must be suspect too. It would appear that Webb did have an opportunity to meet with Ivaritji, and it is possible that some of these etymologies came from her. However, Webb does not clarify the source of his etymologies and one is left wondering whether Webb himself is extending a metaphor and a naming process well beyond its actual application.

It is evident from some of the above examples and the following that Webb does not have a good grasp of Kaurna morphology. He provides these two false etymologies elsewhere in his article:

Cowandilla	from *kauwennadlla* = *kauwe* 'water' + *-nna* 'PL' + *-dlla* 'LOC' <sic>
	<cf *kawandilla* = *kawanda* 'north' + *-illa* 'LOC'>
Tarndanyangga	from *tarndannangga* = *tarnda* 'red kangaroo' + *-nna* 'PL' + *-ngga* 'LOC'
	<cf *tarndanyangga* = *tarnda* 'red kangaroo' + *kanya* 'rock' + *-ngga* 'LOC'>

So, on linguistic grounds we are forced to question the validity of some of his etymologies, in the absence of the identification of his sources and further information. However, the etymologies suggested by Webb seem to have gained some acceptance within the Kaurna community and make sense to them.

3 COWILLA — BEWARE OF JUMPING TO CONCLUSIONS

Several years ago I saw the name Cowilla at the Port Elliot show. We might be tempted to suggest that the name means 'water place', derived from *kauwe* 'water' + *-illa* 'LOC'. A basic word in Kaurna and other languages to the north is *kauwe*. It appears in many placenames in the Flinders Ranges, mid-north and Yorke Peninsula, whilst *-illa* is found in a number of placenames nearby (such as Yankalilla, Tunkalilla and Kangarilla). However, *-illa* is suffixed only to trisyllabic roots. If the name were indeed 'water place', then we would expect Kauwingga (or Cowinga or some other variant spelling). Cowilla is in fact the name of a dairy farm, and for a while I thought that it might derive from English 'cow' with the Kaurna location suffix attached, in the same way that the names of the Adelaide suburbs of Glenunga, Glenalta, Warradale and Paraville have been formed. Upon making enquiries, however, it turns out that the dairy stud is located at Mannum, a considerable distance from Kaurna country adjacent to Cowirra Swamp (meaning unknown). The name Cowirra was already in use, so the owner of the dairy substituted *ll*'s for the *rr*'s because they sounded all right (Steven Vivienne pers. comm., 6 April 2000). So the name is not related to a Kaurna word at all.

4 SUMMARY OF LINGUISTIC PRINCIPLES AND METHODS FOR WEEDING OUT SPURIOUS ETYMOLOGIES

In the examples above, I have demonstrated that linguistic principles and a detailed knowledge of the Kaurna language can be used to call into question and exclude certain etymologies put forward in the literature for placenames on the Adelaide Plains. The general principles outlined here can be applied to other languages, though of course the specifics will vary in accordance with the particular language.

For any given name, the following procedures can be applied:

1. Assemble all known spelling variants and meanings, together with their sources.

2. How reliable are the respective sources? What other kinds of errors did the writer make? (e.g. Webb; Tindale) What were the ultimate sources of the information? How well did these sources know the country? (e.g. Ngarrindjeri men Unaipon and Milerum talking about Kaurna places)

3. Identify known morphemes (e.g. *-ngga* 'LOC'; *-(i)lla* 'LOC'; *-rla ~ -dla* 'DUAL'; *parri* 'river'; *wirra* 'forest' etc.) often found in placenames of the area.

4. Check for similar forms in recognised sources (e.g. T&S; Wyatt etc.).

5. Calculate the number of morphemes required to encode the notion conveyed in the English gloss and check against the likely number of morphemes present in the Indigenous word. In Kaurna there are very few monosyllabic roots. Most are disyllabic, while some are trisyllabic. So a word like Myponga (Maitpangga) is likely to consist of a disyllabic root *maitpa* plus the location suffix.

6. Identify morphophonemic irregularities. For example, *-(i)lla* only ever appears on trisyllabic roots. A final *la* syllable on a disyllabic root is more likely to be the dual suffix *-rla ~ -dla* 'DUAL' than the location suffix *-(i)lla* 'LOC'.

7. Check for semantic consistency. For example, the paws of a kangaroo should appear as *marrarla* 'hand-DUAL' rather than *marranna* 'hand-PL' since the hind feet are referred to as *tidna* 'foot'.

8. Beware of meanings like 'many springs', 'plenty of water' or 'good water' etc. in the absence of identifiable words for water (e.g. *pudna* 'spring', *kauwe* 'water', *parri* 'river'). There is a strong possibility that such a word may have indeed been the name of a spring, but did not literally mean 'spring' etc.

CONCLUSIONS

In this short paper I have discussed a range of Kaurna placenames where linguistics can be brought to bear, in the absence of speakers of the language, to question and weed out certain etymologies recorded in the literature. A knowledge of Kaurna morphology and phonology helps to narrow down allowable possibilities. In the final analysis, however, in most cases we simply cannot be certain of the 'true' etymology.

REFERENCES

Amery, Rob, 1996, Kaurna in Tasmania: a case of mistaken identity. *Aboriginal History* 20: 24–50.

Black, J.M., 1920, Vocabularies of four South Australian languages, Adelaide, Narrunga, Kukata, and Narrinyeri, with special reference to their speech sounds. *Transactions of the Royal Society of South Australia* 44:76–93.

Cockburn, Rodney, 1908, *Nomenclature of South Australia*. Adelaide: W.K. Thomas & Co.

— 1990, *South Australia. What's in a name?* Adelaide: Axiom Publishing (Revised edition. First published 1984).

Dutton, Francis, 1846, *South Australia and its Mines*. London: T. & W. Boone.

Hercus, Luise, 1992, *Nukunu Dictionary*. Canberra: the author.

Manning, Geoffrey H., 1986, *The Romance of Place Names of South Australia*. Adelaide: the author; Gillingham Printers.

Martin, A.E., 1943, *Twelve Hundred and More Place Names in South Australia, Western Australia and the Northern Territory*. Sydney: NSW Bookstall Co.

Meyer, H.A.E., 1843, *Vocabulary of the Language Spoken by the Aborigines of the Southern and Eastern Portions of the Settled Districts of South Australia*. Adelaide: James Allen.

Moorhouse, Matthew, 1846, *Vocabulary and Outline of the Grammatical Structure of the Murray River Language*. Adelaide: printed by Andrew Murray.

Piesse, Louis, 1840, Letter to *South Australian Colonist,* July 1840.

Plomley, N.J.B., ed., 1987, *Weep in Silence: a history of the Flinders Island Aboriginal settlement*. Hobart: Blubber Head Press.

Praite, R. and J.C. Tolley, 1970, *Place Names of South Australia*. Adelaide, Sydney, Melbourne, Brisbane and Perth: Rigby.

Royal Geographical Society of Australasia (SA Branch), 1989, *Captain Collet Barker Field Day*. Adelaide: RGSASA.

Simpson, Jane, 1995, CUP Adelaide placenames reduced electronic data file (last updated 11 April 1995).

Taplin, George, 1879, *The Folklore, Manners, Customs, and Languages of the South Australian Aborigines*. Adelaide: Government Printer.

Teichelmann, C.G. and C.W. Schürmann, 1840, *Outlines of a Grammar, Vocabulary, and Phraseology, of the Aboriginal Language of South Australia, Spoken by the Natives in and for Some Distance around Adelaide*. Adelaide: published by the authors at the native location. Facsimile edition 1962, State Library of South Australia. Facsimile edition 1982, Tjintu Books, Adelaide. A copy annotated by Teichelmann was sent to Grey in 1858 and is held in the Sir George Grey Collection, South African Public Library, Cape Town.

Tindale, Norman B., [Assorted Papers] held in the Tindale Collection, Anthropology Section, South Australian Museum, Adelaide.

— Kaurna place names card file held in the Anthropology Section, South Australian Museum, Adelaide.

— 1987, The wanderings of Tjirbruki: a tale of the Kaurna people of Adelaide. *Records of the South Australian Museum,* vol. 20, May 1987:5–13.

Webb, Noel Augustin, 1936–37, Place names of the Adelaide tribe. In *Municipal Year Book, City of Adelaide*, 302–310. Adelaide: *The Advertiser*.

Williams, Roy F., 1985, *To Find the Way: Yankalilla and district 1836–1986*. Yankalilla and District Historical Society Inc., South Australia.

Williams, William, 1840, The language of the natives of South Australia. *The South Australian Colonist* 1(19):295–296.

Wyatt, William, 1879, Some account of the manners and superstitions of the Adelaide and Encounter Bay tribes. In J.D. Woods, ed., *The Native Tribes of South Australia*, 157–181. Adelaide: Government Printer. (Original manuscript with corrections in BSL Special Collection)

Zorc, David, 1986, *Yolngu-Matha Dictionary.* Batchelor, NT: School of Australian Linguistics.

13 PLACENAMES IN YUWAALARAAY, YUWAALIYAAY AND GAMILARAAY LANGUAGES OF NORTH-WEST NEW SOUTH WALES

Anna Ash

1 INTRODUCTION[1]

Yuwaalaraay, Yuwaaliyaay and Gamilaraay are closely related languages that cover a large area of north-west New South Wales, from the New South Wales–Queensland border down to the Tamworth area, and from the edge of the Tablelands, west to beyond Walgett. Cognacy rates of around 60–80 per cent (Williams 1980:1) and comparable grammars mean that the three languages are dialects. The names of these languages have two parts, the first part is the word for 'no', and the second is the comitative suffix meaning 'having'. So Yuwaalaraay has *yuwaal* (actually *waal*) 'no', and Gamilaraay has *gamil* 'no'. This is a fairly common way of naming Aboriginal languages in this area; compare, for example, Yugambal of the Inverell area which has *yuga* 'no', and Wiradjuri from central-southern NSW which has *wirray* 'no'.[2]

During 1999 work began on the Yuwaalaraay, Yuwaaliyaay and Gamilaraay dictionary database. The project involves creating a dictionary with and for the Yuwaalaraay, Yuwaaliyaay and Gamilaraay people. The author is compiling data from historical sources,[3] from Yuwaalaraay, Yuwaaliyaay and Gamilaraay people, and from the checking of historical

---

[1] This report comes from a work in progress, and while some information has been published by other people (as referenced), other information is still subject to further community consultation and revision. I would like to acknowledge the input of Uncle Ted Fields, Aunty Pearl Trindall and Uncle Jo Trindall, John Giacon, Peter Thompson, Meg Leathart and Tamsin Donaldson. Of course, any errors rest with the author.

[2] Abbreviations used in this paper are: COM comitative; DIM diminutive; LOC locative; NONFUT non future; PRIV privative; REDUP reduplication.

[3] Sources include tapes recorded by Corinne Williams and Janet Mathews (Yry, Yyy) and Stephen Wurm (Gry); Ian Sim's word list (Yyy); and Peter Austin's Web Dictionary (Gry). These are supplemented by Giacon (1998), Giacon and Sim (1998), Williams (1980) and many historical documents by people such as R.H. Mathews, Revs W. Ridley and C. Greenway, K. Langloh-Parker and M.J. Cain.

L. Hercus, F. Hodges and J. Simpson, eds, *The Land is a Map: placenames of Indigenous origin in Australia*, 181–185. Canberra: Pandanus Books in association with Pacific Linguistics, 2002.

information with the people. The content and final form of publication[4] will be determined by local Aboriginal people.

The current dictionary project is just one aspect of Yuwaalaraay (Yry), Yuwaaliyaay (Yyy) and Gamilaraay (Gry) language reclamation that also includes many school and community-based activities. The languages are taught in schools and Technical and Further Education (TAFE) courses for adults; Yuwaalaraay, Yuwaaliyaay and Gamilaraay language conferences are regularly held, bringing together Elders, teachers, community members and linguists to use language, share ideas and plan for further language reclamation. Local Aboriginal languages are increasingly being used in speeches, songs, naming, creative writing, and memorials, such as the recently opened Memorial to the Myall Creek Massacre of 10 June 1838.

In the course of this work, names (so far, about 70) for places in the region have been incorporated into the database. These come from historical sources, other linguistic work and from Elders of the region, such as Uncle Ted Fields of Walgett who have a lot of valuable local knowledge. Analysis has begun on many other placenames that are potentially of Yuwaalaraay, Yuwaaliyaay or Gamilaraay origin; these are still to be checked with informants. Understandably, culturally sensitive information, such as placenames cited in Dreaming stories, are the subject of discussion as to whether or not they should be included in the dictionary.

It must be emphasised that this is a report on one section of a work in progress, the primary aim of which is to produce a dictionary that is useful across a wide region. The first step has been to concentrate on entering data from historical sources; while there has been some community consultation, the bulk of this is still to occur. The historical data therefore provide a base on which to build the current language knowledge, which may include significant variety across the region, and some words borrowed from languages outside the Yuwaalaraay, Yuwaaliyaay and Gamilaraay regions.

2 FINDINGS[5]

Around 70 placenames have been analysed so far: of the four types described by Hercus and Simpson (this volume), the placenames range from those that describe aspects of topography and environment (such as landforms, trees, and animals) to those that refer to the activities of Ancestral Beings. In the following analysis reference will also be made to Hercus's classification (this volume) of placenames as 'generic', 'intermediate' and (perhaps) 'silly'.

Yuwaalaraay, Yuwaaliyaay and Gamilaraay placenames, as in many areas, often refer to events from dreamtime stories. Thus *Bumaygarriya* (Yyy) is a place on the Narran Lake, where *Baayama* 'the Creator' ambushed and killed the *garriya* 'crocodile'. The water gushing out of the body created Narran Lake (Giacon & Sim 1998:26:App. 8). Note that syntactically it consists of a verb and its object:

---

[4] The immediate goal is to produce a hard copy dictionary from the database, for use in local language programs. A FileMaker Pro database is being used, so that we can later produce other computer-based or on-line resources if the Yyr, Yyy and Gry communities so desire.

[5] In this paper I have used the following conventions: when a placename is discussed as a name, it is given in italic; when a placename is mentioned as a place, it is given in plain type.

Buma-y   *garriya*
kill-NONFUT   crocodile
'Bumaygarriya'

Without the knowledge of Elders, placename analysis can be guesswork (albeit linguistically informed guesswork); the full meaning of a placename, for example, whether it refers to a landscape feature or a creation event may never be known. In north-west New South Wales, there are several Elders who have contributed knowledge to enable a fuller understanding of certain placenames. 'Cumborah Knob' is a flat-topped hill that used to be known as *Babarraa* probably 'red and yellow snake' (Ted Fields pers. comm.), but perhaps from *babarr-a* 'snake species-LOC' (author's analysis). Uncle Ted Fields tells the story of how it was originally a peaked hill, but two enormous snakes fought there; one threw its boomerang and sliced off the top of the hill. A place on the Narran River, near Angledool — *maluwil* 'human shadow, spirit' (Yyy) — is so called 'because of the shadows there' (Giacon & Sim 1998:24). In this case we may not have access to the full meaning of the name, for example, whether or not the site was connected to an Ancestral Being's journey.

Many of the current Yuwaalaraay, Yuwaaliyaay and Gamilaraay placenames based on flora, fauna and topography may be of the type that Hercus (this volume) classifies as 'generic' (that is, the use of a generic term where there may have been a specific term) or 'intermediate ... or descriptive'. Examples of generic placenames are Warrumbal (River) from *Warrumbal* (Yyy) 'watercourse' (but also a name for the Milky Way); and Burrul Gungan (Yyy) 'big water', which Sim was given as the name for Narran Lake (Giacon & Sim 1998:App. 9).

Names that refer to topographic features include, in Gamilaraay: Boggabri, *bagaay-baraay* 'creek-COM', and Boggabila, *Bagaay-bil-a* 'creek-having many-LOC' (Austin 1993:2) and Milbulah, *Mil-bil-a* 'eye/hole-having many-LOC'; and Dandara, *Dhandarr-a* 'frost/ice-LOC' (author's analysis). In Yuwaaliyaay, Sim (1998) mentions Cowal, *Gaawul* 'creek, lagoon'; and two other as yet unknown locations: Garrabila, *Garra-bilaa* 'cracks-parallel' and Garradhuul, *Garra-dhuul* 'cracks-DIM'.

Many placenames refer to vegetation; in Gamilaraay there is Collarenabri, *Galariin-baraay* 'coolabah blossom-COM'; Tarilarai, *Dhariil-araay* 'reed-COM'; Drilldool, *Dhariil-duul* 'reed-DIM' (Austin 1993:2); and Gurley, *gurralay* 'river wattle' (O'Rourke 1995:95). Yuwaaliyaay placenames based on flora may include: Ngamanbirrabaa, *Ngamanbirra-baa* 'wild plum-LOC'; Brewarrina, possibly *Birraa-warra-nhi* 'whitewood tree-stand-NONFUT' (author's analysis); and Yerranbaa, *Yaran-baa* 'a type of acacia-LOC' (Ted Fields pers. comm.). Many of these placenames referring to vegetation may be classified as Hercus's 'intermediate' names.

Several placenames have been recorded that reflect the significance of the type of soil and rock of that country. Yuwaaliyaay provides the following examples: Nee-Nee, *Nhii-Nhii* 'charcoal-REDUP', reduplication often forms an adjective from a noun; and Goonoo, *Gunu* 'lime gypsum' (Giacon & Sim 1998). Again, Sim's work with Yuwaaliyaay people in the 1950s provides some insight: 'Boggy Ridge' which was once known as Buggy Ridge, is perhaps from *bagi* 'white pipe clay' (Giacon & Sim 1998:19).

Fauna are the source of several placenames in the region. In Gamilaraay there is Bundarra, *Bandaarr-a* 'kangaroo-LOC' (Austin 1993:4). From Yuwaaliyaay we have Ballone, *baluun* 'great egret' and Bollonbillion, *Baluun-bilyan* 'great egret-waterhole' (author's analysis). Dirrinbandi, *Dhurrun-banda-y* 'hairy caterpillars-going along in a line-NONFUT' (Giacon & Sim 1998:12); and Coorigel, *Guligal* 'beehive debris' (Giacon & Sim 1998:11). However, as yet, it is not possible to say whether these names have a purely descriptive reference or are associated with Creation beliefs.

Still other placenames show the effect of colonisation: Yarraman 'horse' and Timbumburi (Creek) *Dhimba-m-baraay* 'sheep-epenthetic /m/-COM' (author's analysis). The English source of *dhimba* is thought by some to be jumbuck, however it could be a borrowing of 'sheep' with adaptation into Gamilaraay phonology and epenthetic /m/. This is in accordance with the fact that the word for sheep in some languages is *dhimbak* (Simpson pers. comm.) which would go to *dhimba* in Gamilaraay, as the language does not allow final /k/.

Coonabarabran is given the meaning 'inquisitive person' in Mary Jane Cain's 1920 (Gry) wordlist; this may be an example of Hercus's 'silly names' or, more precisely, the informant's way of avoiding giving a translation. It is not known whether 'inquisitive person' was Mrs Cain's response, or the response of an informant; however, there is nothing in the literature that supports this translation. It is more likely to be someone's polite way of not giving the true meaning which probably includes *guna* 'faeces' and perhaps also *barabin* 'semen'; compare also Coonamble *Guna-m-bil* 'faeces-/m/-having lots' (Gry, author's analysis).

There are several placenames which relate to body parts; although at this stage it is not certain whether they are metaphorical (compare English: Hungry Head, Hat Head) or whether they refer to Ancestral Beings. Giacon and Sim (1998:4) provide the following derivations for some Yuwaaliyaay placenames: Nullawa, *Nguluu-wawul* 'forehead/face-narrow', as in a narrow point of land going into a river; and Terewah (a branch of the Narran River), *Dharra-wawul* 'leg-narrow'. There is also Angledool, *Yanggal-dhuul* 'vagina-small' (Giacon pers. comm.).

3 RELATED ISSUES

Placenames can contribute to knowledge about the boundaries between languages. For example, Gamilaraay does not permit final /ng/ (velar nasal), whereas a neighbouring language, Wiradjuri, does. Therefore places such as 'Wallumburrawang', 'Gowang Mountain' and 'Windurong' are likely to be outside Gamilaraay country. Similarly, places ending in the Wangaaybuwan and Wayilwan comitative suffix *-buwan*, such as 'Gulargambone', 'Quambone' and 'Mumblebone' are likely to be outside Gamilaraay country. Donaldson (1985:77) provides the full etymology of Gulargambone *kilaampuwan* as 'with young galahs'. Of course this kind of evidence is not firm proof of language boundaries, as the situation is complicated by borrowed or imported placenames.

An interesting example of the Elders' knowledge and historical records working together is the case of Narran (Lakes, River, Plains and Narrandool). There was nothing in recent recordings to hint at the meaning, and people said that they did not know the meaning of 'narran'. However a Gamilaraay Elder, Aunty Pearl Trindall, recalls that they used to call very skinny people 'narran gutted'; this is supported by Greenway (1911:85): 'Nerang or Noorong: small or nearer to, as opposed to the Coolgoa: not going so far round. The Aboriginal name of the river now called Narran'. We also have Williams' (Yry) recording of *ngaarrigulay* 'over here, this way', and Austin's (Gry) reconstruction from written sources, *ngaariyalana* 'this side of'. This is obviously not conclusive; there is some evidence for an interpretation of Narran (*Nharran* or perhaps *Nharrang*) as 'small', and some support for a kind of deictic 'here/this side'. This may be resolved if further evidence is found, otherwise it may be appropriate that Elders make a recommendation in the course of language reclamation.

Finally, a single placename may occur in more than one country or language area; for example, there are several places named *Wiidhalibaa* 'wood/fire-PRIV'. In Yuwaalaraay country there is Weetaliba Waterholes (north-west of Lightning Ridge). In Gamilaraay country there is Weetaliba Station (south of Coonabarabran), and Weetalibah Creek (north of Coonamble). East of the Tablelands in Ngarrabul country there is also Wytalibah locality. (It is also the name of the station in *Coonardoo* by Katharine Susannah Prichard!) This is fairly common and may result from several factors including: high cognacy rate within the dialects; shared placenaming strategies across languages; and the non-Aboriginal tendency to take names with them when they move.

It is hoped that further research and consultation will fill some of the gaps in our current understanding of placenames in north-west New South Wales. While some derivations may remain 'best guesses', it is important that wherever possible, this linguistic analysis of placenames is informed by and checked with the Elders of the Gamilaraay, Yuwaalaraay and Yuwaaliyaay peoples.

REFERENCES

Austin, Peter, 1993, *A Reference Dictionary of Gamilaraay, Northern New South Wales*. Melbourne: La Trobe University, Department of Linguistics.

Austin, Peter and David Nathan, Kamilaroi/Gamilaraay web dictionary. On the internet at: http://coombs.anu.edu.au/WWWVLPages/AborigPages/LANG/GAMDICT/GAMDICT.HTM #2001.

Cain, Mary Jane, 1920, Names of places and their meanings in the native language. Unpublished word list.

Donaldson, Tamsin, 1985, Hearing the first Australians. In Ian Donaldson and Tamsin Donaldson, eds, *Seeing the First Australians*, 76–91. Sydney: George Allen and Unwin.

Giacon, John, ed., 1998, *Yuwaalaraay/Gamilaraay Wordlist*. Walgett: Walgett High School Yuwaalaraay–Gamilaraay Language Program.

Giacon, John, ed., and Ian Sim (recorder), 1998, *Yuwaalayaay: the language of the Narran River*. Walgett: Giacon.

Greenway, Rev. C.C., 1911, Aboriginal place names in the county of Benarba (cont.). Kamilari tribe. *Science of Man* 13(4):85.

O'Rourke, Michael, 1995, *The Kamilaroi Lands: north-central New South Wales in the early 19th century*. Griffith, ACT: the author.

Williams, Corinne, 1980, *A Grammar of Yuwaalaraay*. Canberra: Pacific Linguistics.

14 NAMING THE DEAD HEART: HILLIER'S MAP AND REUTHER'S GAZETTEER OF 2,468 PLACENAMES IN NORTH-EASTERN SOUTH AUSTRALIA

Philip Jones

This paper is about the Hillier Map of Aboriginal placenames of north-eastern South Australia. By 1904, when this map was drawn (Fig. 1), the region east and north of Lake Eyre had been largely explored and surveyed, its principal features named and the country had been divided and then subdivided for pastoral purposes. A couple of years earlier at the height of a severe drought, the last Aboriginal people living beyond the bounds of European influence had chosen to leave their Simpson Desert home.[1] Their destinations were the remote cattle stations such as Alton Downs, Cowarie and Macumba, established on the fringes of the Desert. These isolated settlements, run by Europeans who came to rely almost entirely upon Aboriginal knowledge and expertise in this harsh country, became the foci of subsistence and sociability for remnants of Wangkangurru, Ngamini, Yarluyendi and related groups. Their choice contrasted to that of the greater number of Aboriginal people from the region who had progressively 'come in' to the Killalpaninna mission station on the Cooper since its foundation in 1866. It is the descendants of those small and vulnerable mixed communities, comprising representatives of different language groups, who have maintained an unbroken presence in the northern part of the region until the present day and who preserve the greater store of knowledge about Aboriginal placenames.[2]

In contrast, the corporate identity of the much larger and apparently more homogeneous community of mainly Diyari speakers at Killalpaninna Mission — where the Hillier Map was made — is considerably less evident.

---

[1]   Hercus (1985; 1986).

[2]   Jones, (1991); see also, for example, Hercus and Potezny (1990).

L. Hercus, F. Hodges and J. Simpson, eds, *The Land is a Map: placenames of Indigenous origin in Australia*, 187–200. Canberra: Pandanus Books in association with Pacific Linguistics, 2002.

Figure 1: The Hillier Map, drawn in 1904 by schoolteacher Henry Hillier, at Killalpaninna Mission, Cooper Creek, South Australia. South Australian Museum

The Hillier Map is an extraordinary document, as much for what it excludes as for its dense profusion of named Aboriginal sites. It contains approximately 2,500 placenames — probably more than the 2,468 listed in the accompanying gazetteer compiled by Pastor J.G. Reuther as volume VII of his Diyari manuscript.[3] These names are inscribed on silk in Henry Hillier's minute ink letters over his tracing of an 1890s 1:500 000 pastoral plan.[4] But while this tracing includes the colony's boundaries, the delicate dotted lines of the Birdsville and Strzelecki Tracks and some mail routes to Lake Hope, Cowarie and other stations, these traces of European influence are exceptional. The town of 'Birdsville' constitutes the map's sole

---

[3] Reuther, J.G. (1981).

[4] Hillier (1904).

European placename. The larger town of Marree is marked only by its placename *Marina*, Innamincka by *Jidniminka*, and Lake Eyre itself by the larger inscription, *KATITANTA*. Although Hillier appears to have retained European trig points (a scattering of small triangles across the extent of the region) — probably as references for plotting placenames — crucial landscape features are barely evident. The drainage system which dominates European maps has all but disappeared under the density of Aboriginal placenames.

The Hillier Map more than negates the image of emptiness and desolation conveyed by European maps of the region. That perception of a desert void, generated by Edward Eyre, Charles Sturt, Peter Warburton and others, was subsequently reinforced by twentieth-century explorers such as Cecil Madigan whose 'Dead Heart' appellation is now entrenched in popular consciousness. The most promising vistas observed by Charles Sturt's party during his epic struggle to penetrate this region during 1844–45 were revealed as mirages. Marooned at its edge by a withering drought, Sturt neither dignified his forced camp with a European proper noun nor its Aboriginal name. At Depot Glen his party was, in his words, 'bound by the heat as fast as if we were amidst the eternal Ice at the Poles'.[5] This site seemed to him a non-place, the nineteenth-century explorer's equivalent of an anonymous airport lounge; Sturt waiting for a connection. The expedition's boat was abandoned unused at the edge of the Simpson Desert and fittingly it was the inland sea that provided the most telling metaphor for this wasteland environment. Sturt and subsequent explorers conceived of the desert as an ocean; a 'perfect sea of dunes', 'trackless', 'as far as the eye can see', were phrases used in their descriptions.

But like a sea, the Simpson Desert was both barrier and means of communication. Different language groups and differing kinship systems characterised groups bordering it to the east and west, north and south. And yet these groups were not self-contained entities. People married in and out and traded goods and ceremonies with each other. To varying degrees they also spoke each other's languages and used each other's sites and placenames. They followed defined routes across the Desert, navigating from *mikiri* to *mikiri* well as though from island to island. Lines of mythology and of genealogy crossed the Desert, binding individuals and groups together across hundreds of kilometres. Here the analogy ends, for Aboriginal people knew the north-eastern deserts intimately as their home rather than as the forbidding landscape perceived by Europeans. As Luise Hercus put it,

> it was the home of people who were familiar with every dune, every claypan and swamp: they lived there on a permanent basis. They did not just travel from mikiri to mikiri, every salt lake, all the bigger sandhills and claypans had names. It was simply their home.
> (L. Hercus, pers. comm.)

There is no doubt that Sturt's attitude towards this region as a series of stony and sandy deserts separated by dry waterholes and creek-beds, which had to be traversed in order to reach his desired goal, meant that it never became an object of investigation in its own right. It was an impediment, country to be passed over. Its Aboriginal inhabitants were interrogated not about their own country and its named features, but only about the country beyond. In this sense, Sturt wasted most of his opportunities with Aboriginal people visiting his camps and with those encountered during his party's dead-end reconnaissance expeditions. He did not obtain a single Aboriginal placename for sites on the Strzelecki, Cooper, Diamantina or Eyre

---

[5] Sturt (1844–46).

Creeks and made no apparent effort to come to grips with Aboriginal languages of the region once his own Aboriginal guides (from the Darling) moved beyond their linguistic range.

Neither Eyre, on his expedition of 1841, nor Goyder (1857), Burke and Wills (1861–62), Warburton (1866) or Lewis (1874) added more than a handful of additional Aboriginal names. Of these explorers Lewis was perhaps the most dependent upon Aboriginal guidance. He used at least two interpreters who knew sufficient English to communicate placenames accurately, but his 1874–75 map is a sad reflection of his capacity to absorb this information.[6] Punctuating his route are 'Christmas Water', 'Gardiner's Waterhole', 'Camel Water', 'Yellow Waterhole' or simply 'Waterhole' and several examples of 'Native Well'. The site 'Tommy's Well', named after the Ngamini or Wangkangurru guide who had 'a good smattering of English having been with the telegraph parties for some time and was very useful as a guide and interpreter', was the closest Lewis came to applying an Aboriginal name during his traverse to the north of Lake Eyre (Lewis 1876).

The surveyors and pastoralists who followed Lewis made the greatest advance in recording placenames for this north-eastern region. Most notable was William Cornish, whose 1879 survey of the Warburton and Kallakoopah Creeks covered some of Lewis's ground and added more than a dozen new Aboriginal names to maps. Cornish does not mention it in his report to the Surveyor General, but it is probable that he relied partly upon the first owner of Cowarie Station, a German, August Helling. Later, both Cornish and another German based at Helling's Cowarie Station, W.J. Paull, contributed Aboriginal vocabularies from the Warburton to E.M. Curr's 1886 volume (Cornish 1886; Paull 1886). By this time Helling's knowledge of the region's Aboriginal placenames was supplemented through his role in pioneering the mail run to Birdsville (Litchfield 1983:5). In fact, Helling emerges as a possible important source for the Hillier Map; his German background led to close contact with the Killalpaninna Mission during the 1870s and 1880s, if not beyond. Writing in 1934 from his store at Mulka on the Birdsville Track, George Aiston acknowledged that Helling was 'an independent observer and was a reliable authority [who] lived in the heart of Wongkonguru country', further maintaining that Reuther's manuscript records of the Aboriginal peoples of the Cooper and Warburton were 'nearly all Helling'.[7]

The Killalpaninna missionaries had the incentive and means to augment European maps of the Aboriginal landscape surrounding them from the time of their arrival at Bucaltaninna, Kopperamanna and Killalpaninna in 1866. Their efforts to 'enter the mental world of the heathen' could succeed only through an understanding of the various languages spoken in the region and by orienting themselves in relation to those complexes of sites which distracted Aboriginal attention, away from the mission. While this broader project did not take final shape until Reuther's time (accompanying his ambitious investigations into the range of beliefs relating to the mythological landscape), an early priority was a meaningful map of the 'mission block itself'. Two versions of such a map apparently exist, dating from the early 1870s [fig. 2].[8] The northern section of a poor copy of one version is reproduced below,

[6] Lewis (1876).

[7] G. Aiston to W.D. Gill, 26 June 1934. Aiston, G. (1920s–1940s) MS. Correspondence with W.H. Gill. A2535–A2537, Mitchell Library, Sydney. Note that Helling was in fact living in traditional Ngamini, not Wangkangurru, country.

[8] One map has been sighted in the Lutheran Archives (J. McEntee, pers. comm.), another is in the Anthropology Archives, South Australian Museum (AA266).

showing the way in which the Aboriginal landscape was becoming merged with a European, mission landscape. The Killalpaninna lake and mission station are evident at the top of the map. Wells, horse-paddocks, sheds and roads are marked together with a concentration of Aboriginal sites along the Cooper, prefiguring the density of placenames on the Hillier Map.

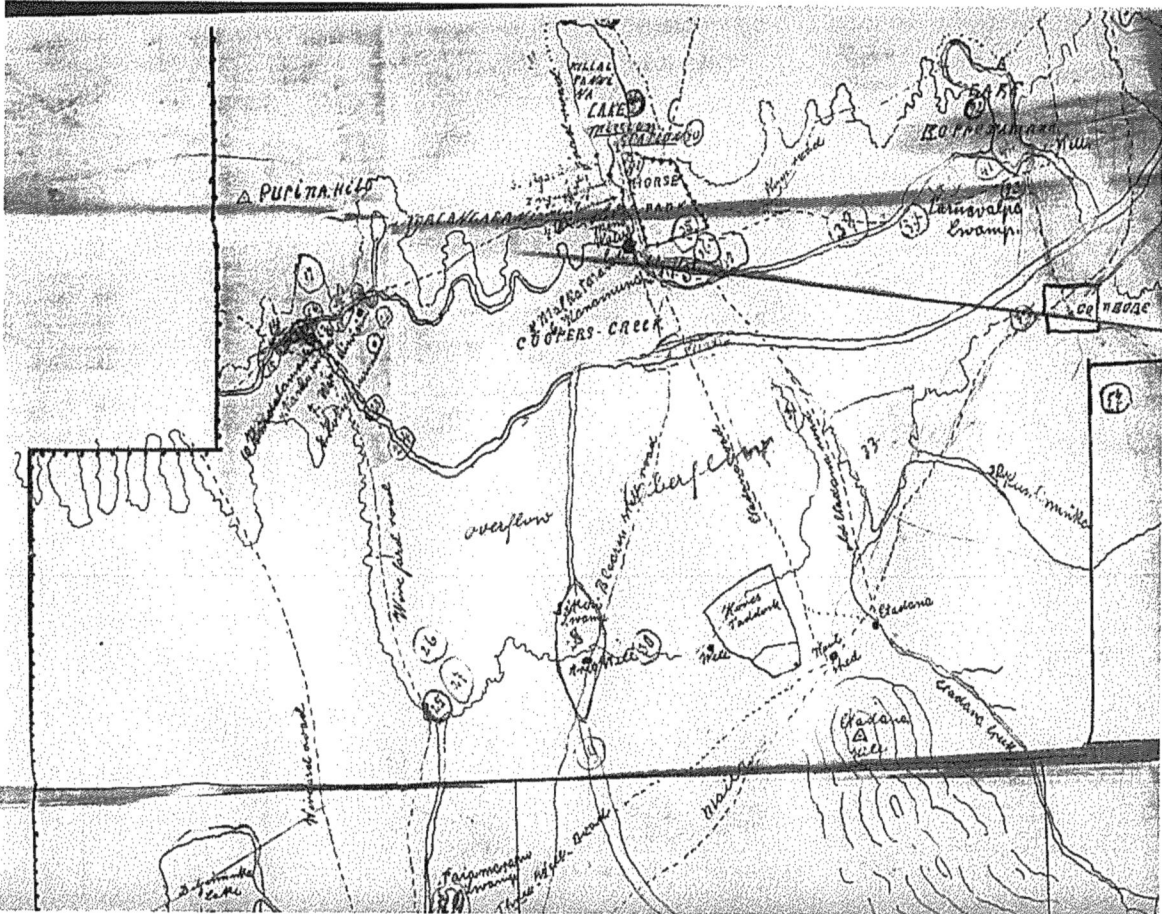

Figure 2: Detail from Gason's map of the Killalpaninna 'mission block', c.1870. Anthropology Archives, South Australian Museum

We can assume that the ringed figures on this map refer to a list of Aboriginal placenames, perhaps with expanded text. Intriguingly, while the map obviously refers to the territory of the mission and includes some German text, the bulk of the annotations seem to be in the hand of Samuel Gason, the police-trooper who had preceded the Lutherans in the region, based consecutively at Lake Hope and Kopperamanna until posted to Barrow Creek in 1873. Gason's subsequent correspondence with Alfred Howitt suggests that, like the missionaries, Gason had elevated the Diyari as a corporate entity, subsuming separate references to other language groups at the Lutheran mission under that general heading.[9]

By the 1880s even recently arrived pastoralists in the Cooper and Warburton region were well aware of the density of Aboriginal placenames, even though Lewis's map and the first

---

[9]   Gason to Howitt, 1870s. A.W. Howitt Correspondence, Australian Institute of Aboriginal and Torres Strait Islander Studies, Canberra.

general pastoral plans remained the only readily available maps. The unpublished record of the 1882 and 1883 horseback tours of inspection by pastoralists K. Robinson and a Mr Taylor along the Warburton Creek west of Cowarie shows that a reliable set of placenames was becoming essential when establishing sources of water and feed for stock. In contrast to Lewis's bland and meaningless naming, these aspiring pastoralists could not afford to abandon the precision and relevance of Aboriginal names.[10] Again, it is likely that Helling was the source of these names used by Taylor and Robertson. The Taylor journals offer possibilities for confirming particular placenames on the Hillier Map. Taylor documents two expeditions by horseback from Cowarie Station along the Warburton and notes details of each halt, a mileage, and the Aboriginal names of waterholes and springs. Where these coincide with the names on the Hillier Map, it should be possible to localise those names, particularly where other known reference points are included. The highlighted names on Figure 3 are those Hillier sites along the eastern section of the Warburton Creek visited by Taylor and Robertson, providing an indication of the value of interrogating archival records for corroborative placename detail.

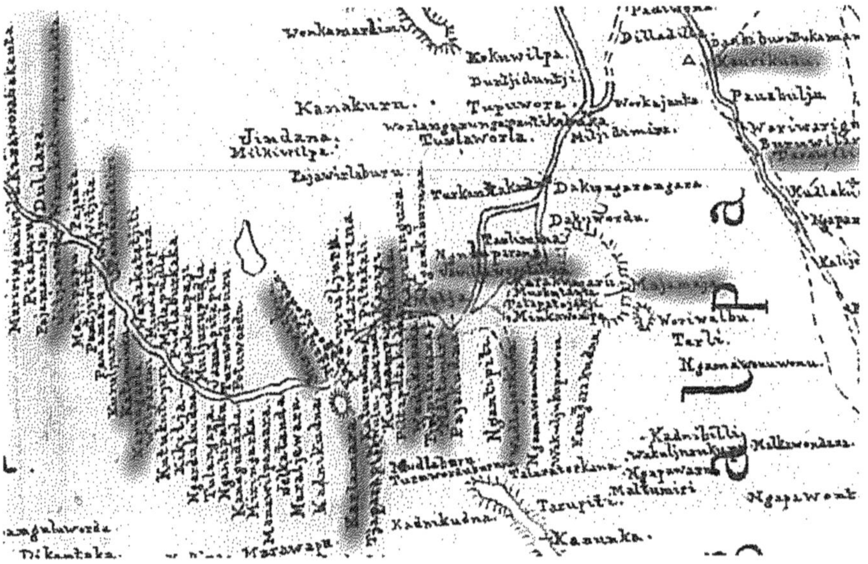

Figure 3: Annotated detail of Hillier Map along eastern Warburton Creek, with placenames identified from the 1882–1883 Taylor journals

An inevitable corollary of this European reliance upon the Aboriginal landscape was that these names became Anglicised (or perhaps 'Germanised') to a degree. Such changes occurred through constant use by European speakers rather than through any concerted alteration. The Lutherans' own attempt to change the name of Killalpaninna (*Kirla-wilpa-ni-nha* = vagina) to the benign Hebrew 'Bethesda' ('healing pool') was a conspicuous failure, except in genteel Lutheran circles.

---

[10] An accurate transcription has not yet been made of this journal which is in the possession of a family member. It contains several Aboriginal placenames along the Warburton which are otherwise known only from the Hillier Map. Copy in possession of author.

THE HILLIER MAP AS A RESEARCH TOOL

Until recently the Hillier Map has operated most effectively at a symbolic rather than an informational level. The fact that the map contains 2,500 Aboriginal placenames, and seems to have resulted from a remarkable if mysterious partnership, has overshadowed the information that these names carry or the issue of their exact location. Despite that, two quite different lines of research have focused upon the map, gleaning valuable information from it. The first, and most productive of these has been the meticulous site-recording work of Luise Hercus and Vlad Potezny, undertaken particularly during the 1980s. Their work has undoubtedly illuminated more of the map's deficiencies than its virtues, but has shown that the map can function as a source of information which complements and informs other sources, including first-hand data from Aboriginal informants. Hercus's 1987 paper, 'Just one Toa', also drew attention to the particular, if inconsistent, relationship between the Hillier Map, the Reuther placename volume which describes the names on it, the Reuther toas which refer to and symbolise these places, and the other Reuther manuscript volumes (Hercus 1987). Hercus's search for the important Wangkangurru ceremonial site of *MaRaru* was ultimately successful in spite of, rather than because of, the Hillier Map. The site appears on the map as one entry in an inaccurate and unlocalised list of sites, placed almost at random in the Simpson Desert.

Hercus and Potezny's 1990 paper, dealing with the section of the Hillier Map south of Birdsville along the Diamantina and the Mulligan channels, provides even more salutary reading for those who might imagine it possible to 'simply go out and follow the map, use the information from the [Reuther] volumes and locate all the important Aboriginal sites in the north-east of South Australia' (Hercus & Potezny 1990:139). In fact, as the authors make clear, the map's primary value lies in its corroborative importance when used in conjunction with other historical sources, evidence provided by Aboriginal informants, and field research. Their paper gives several comprehensive examples of how this blend of careful research can supply meaning and exact locations to the names on the Hillier Map.

Hercus and Potezny's work confirms that the Hillier Map was largely a desk-based exercise, probably undertaken by both Reuther and Hillier working together to fix the generalised localities of sites in one region after another. But the fact that the numbered sequences of sites in Reuther's volume VII often appear on the map in a similar sequence suggests not only the ethnographers' working practice; it also indicates that Aboriginal informants were supplying their own sequences of sites. Whether those sequences correspond to the course of particular Dreamings, or to an imagined journey through familiar country, may be tested through further analysis. Limited research into this question suggests that Hillier and/or Reuther used existing mapped topographic references, such as water features and trig points to make cartographic sense of the data being supplied to them by their informants. In areas of the base map where such topographic markers did not exist, their cartographic rendering of this information is barely distinguishable from a list of names superimposed on a blank. This process is exposed most clearly in the remotest areas from the mission itself, particularly in the Simpson Desert and along and adjacent to the Diamantina and Mulligan channels. Further, it is often difficult to be sure whether the indicated site is at the beginning or the end of a placename, as Hillier used locative dots for a minority of sites. His practice of finishing each name with a tiny dot is a trap for the unwary. Where the names clearly refer to sites along a watercourse or around a lake, this is less of a problem. Elsewhere the task is much more difficult: as Roderick Wilson observed (1981:6), a single Hillier placename may cover 15 kilometres of country.

Even on the Warburton and Cooper channels, probably well known to mission staff, Hillier seems to have paid as much attention to fitting the names neatly on the map in geographical sequence without overlapping, as to fixing their exact location. For example, in that part of the Warburton that bristles most thickly with his names, Hillier allowed just enough space for each name to be read individually (see Fig. 3). But while this inevitably imparts a distortion to the map, the Taylor journals' record of site distribution along the north and south banks of the Warburton matches that of Hillier's.

Nevertheless, in these stretches of the Warburton and Cooper, where the names are so thick that the meanderings of the rivers themselves are often obscured, Hillier was placing a name on every twist and turn. This itself suggests possibilities for a process of retrieving lost placenames by informed guesswork, even without the possibility of confirming those guesses through Aboriginal informants. An occasional correlation with named Aboriginal sites on other maps — particularly the 1890s pastoral plans, but also including the recent 1:250 000 series — may provide the basis for firmer attributions from the Hillier Map. The accompanying descriptions in the Reuther placename volume may provide an additional corroboration, although these can rarely be interpreted as site-specific. This approach has been attempted by this author and John McEntee on the Warburton Creek west of Cowarie, for example, using a combination of the Hillier Map, the Taylor journal and the Helling and Cornish maps, and at Lake Hope on the Cooper. This will be the only possible approach in the future, given that opportunities for carefully fitting all the jigsaw pieces of song-verse, Aboriginal knowledge and memory, field observation, archival fragment and map-name together have already vanished.

The Hillier Map has potential research value of another order. This line of enquiry was first explored during the 1920s by Norman Tindale and more recently by the geographer Roderick Wilson (Wilson 1981). It involves treating the Hillier Map as an artefact in its own right, revealing a pattern of information which, despite its shortcomings as an accurate linguistic and ethnographic source, nevertheless illuminates the historical context of the map's creation. Norman Tindale was the first to realise that each of the Reuther placenames could be coded according to the tribal affiliations documented in volume VII of his manuscript (Reuther 1981). By concentrating upon this association, rather than upon any exact plotting or analysis of the placename itself, an impression of the broader mosaic of language-group affiliations might be gained. Tindale attempted this by allotting a colour to each tribal group and underlining each placename in ink of that colour on a large photo-mosaic of the map. The result, rather tattered and illegible now after seven decades, gives a remarkable insight into the map as a cultural artefact of Aboriginal language group affiliations in the eastern Lake Eyre region. Of course, as Luise Hercus has pointed out, this pattern of affiliation may well be flawed, as it relies upon the language group attributions made by Reuther and/or Hillier. We cannot be certain, for instance, that their attributions related to the actual site's affiliation, or to the language group of their particular informants, who may have been describing sites belonging to other groups. It is worth noting, though, that while the sites associated with toa names, recorded in volumes XII and XIII as well as in volume VII, often differ fundamentally in their Dreaming ancestor affiliations and in other details (suggesting different informants), their language group affiliations do not vary.

The recent digitisation of volume VII, supervised by Bill Watt of the Geographical Names Board, has enabled the most numerous language group affiliations of the Hillier placenames to be viewed as a table (see Fig. 4).

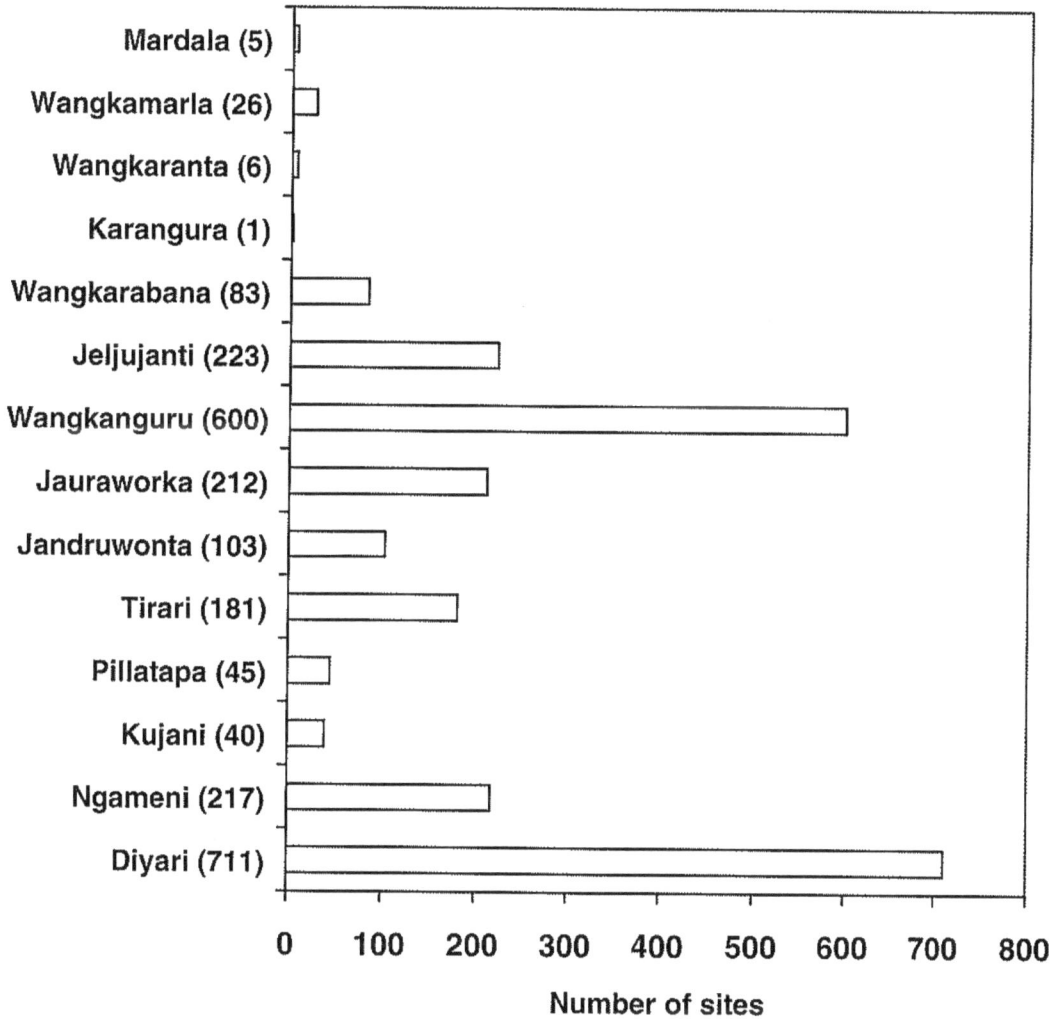

Figure 4: Numbers of placenames on the Hillier Map by language-group affiliation (Reuther's orthography)

The data are striking in one immediate sense: we are reminded that despite the Lutheran missionaries' commitment to the Diyari and Reuther's characterisation of his ethnographic work as a study of that single group, the reality was far different. The tally of Diyari placenames is almost equalled by Wangkangurru, and both together represent only slightly more than half of the recorded total. Despite its location in Diyari heartland, by the 1890s Killalpaninna was a cosmopolitan place, a focus for most of the major language groups of the region. It is reasonable to assume that the men and women contributing data to the Hillier

Map were drawn from this range of groups, and that they were attributing their own language group names to their own sites. Luise Hercus has drawn attention to errors in this record, or to sites that may have had shared affiliations of which only the lesser was recorded by Reuther (the *MaRaru* site is an example, see Hercus 1987). But broadly speaking, the coloured mosaic which can be drawn of the Hillier names in digital format should reflect the ethnographic reality of the accepted divisions — and, perhaps most interestingly, interpenetrations — of the region's language groups. This mosaic of language group/site distribution has been encoded in the Hillier Map and the Reuther manuscript since these documents were created, but it is unlikely that either the German missionary or the English schoolteacher was aware of the extent to which this 'pattern in the carpet' might be revealed.

Figure 5 shows a Hillier Map section of the Coongie Lakes region, incorporating border territory of the Diyari (south-western), Yawarrawarka (northern) and Yadliyawarra (south-eastern) groups; the superimposed 'boundaries' mark the relatively sharp divisions between language group-affiliated sites. The interpenetration of site affiliations here is minor compared to the region just a few kilometres to the west, where a mixture of Yarluyendi, Ngamini, Yawarrawarka, Diyari and even Wangkangurru-affiliated sites is recorded by Reuther and Hillier. The map contains several contrasting vignettes of this kind — each requiring detailed analysis, particularly by linguists.

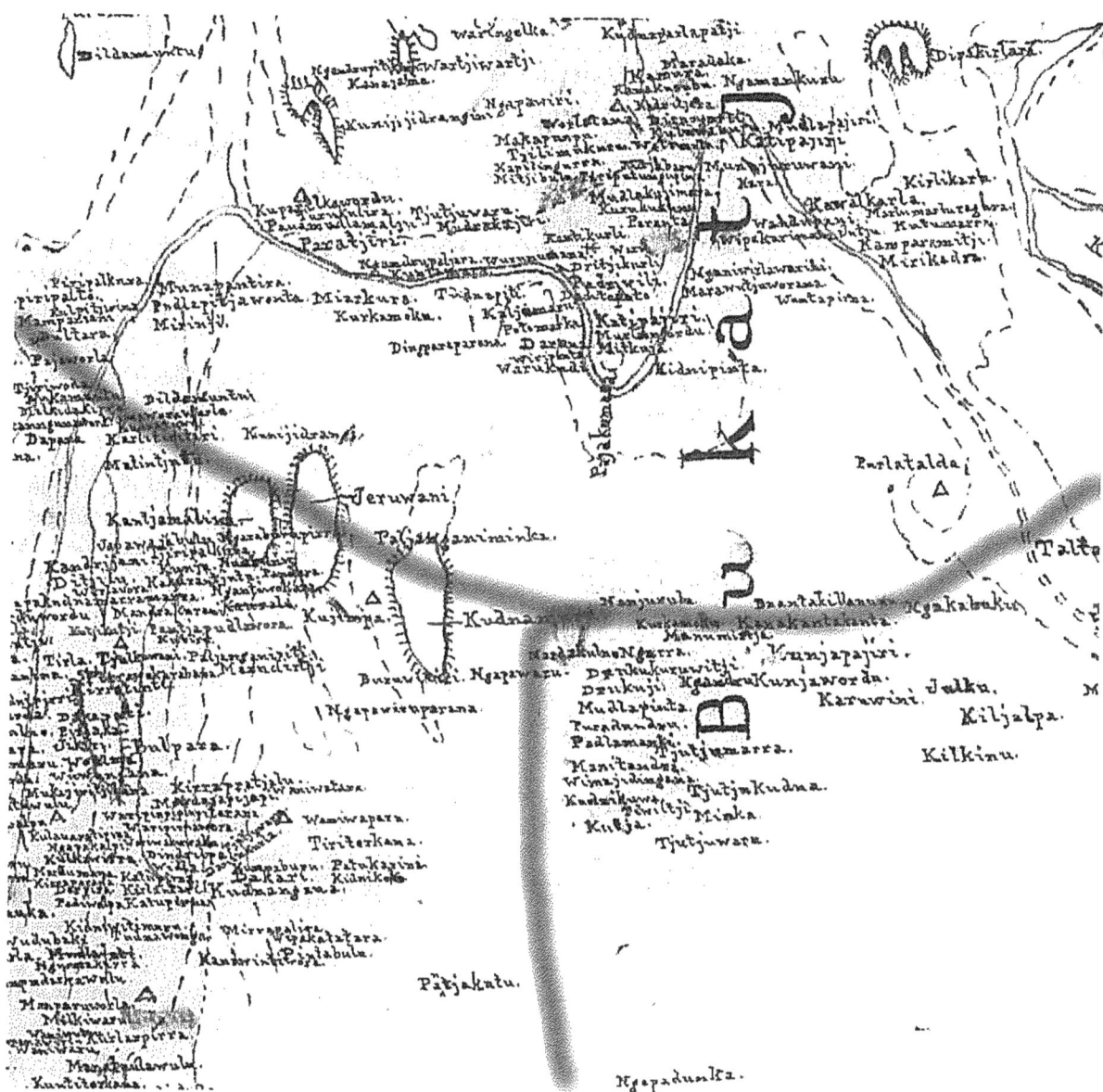

Figure 5: Section of Hillier Map showing abutting 'territories' associated with Yawarrawarka (northern), Yandruwantha (south-eastern) and Diyari (south-western) placenames. Boundaries superimposed.

Wilson's 1981 thesis indicates what may be gleaned from the Hillier Map at other, broader levels of analysis. Again, his plotting of Hillier placenames affiliated to particular *mura-mura* or Dreaming Ancestors reflects Reuther and Hillier's data, rather than ethnographic reality. But following his lead, the digitisation of the Hillier Map will eventually enable a more accurate analysis of these site 'sets' associated with the various *mura-mura*.

A further problem to be faced here is that of the uneven fit between the sets of *mura-mura* mentioned in Reuther's volume VII, his toa volumes (vols XII and XIII) and two other volumes concerning religion and legends (vols X and XI). Several *mura-mura* discussed in the toa volumes do not appear in the other volumes, and vice versa. It appears that Reuther may not have understood that the same Ancestor could be referred to in several ways, and by various cryptic terms. Apparent discontinuities between the data relating to the *mura-mura* recorded in these different volumes also suggest that he may have been working with different informants at particular times, and so was obtaining varying accounts of the same Dreaming from different perspectives. It is therefore not a straightforward matter to simply digitise Reuther's data and match it to the mapped placenames. Even so, with the eventual digital coding of all Hillier's placenames according to the Ancestral and language group affiliations ascribed to them in volume VII, it may be possible to interrogate data contained in the other volumes in an effort to identify some meaningful, previously hidden correlations. In the process, as John McEntee's research has revealed (pers. comm.), these other volumes contain a quantity of other placenames not referred to in volume VII, and possibly not recorded on the map.

One productive line of research may be to highlight (if not to clarify) the relationship between the Hillier Map, the placenames volume and volumes XII and XIV, which both concern the toas. Each of these three volumes refers directly to named sites appearing on the Hillier Map, but in different terms. The variation between the descriptions in volumes XII and XIII is usually so slight as to suggest that this may be accounted for by the process of transcription itself, especially if Reuther was attempting to improve his written expression, in order to promote his collection for sale. But the contrast between these accounts of toa sites and the corresponding placename descriptions in volume VII is marked. Most strikingly, the Ancestor associated with the toa site is rarely described as the same Ancestor for that site in volume VII. This aspect requires more detailed research, but it is worth noting that the preponderant 'set' of toa Ancestors does not mirror the most numerous set of volume VII site Ancestors.

These differences may simply reflect Reuther's reliance upon different groups of informants for his site and toa data. The range of language group and mythological alignments among the Aboriginal community based at Killalpaninna may well account for such consistently different emphases and explanations, and once the etymology of the Ancestor names is unravelled, these differences may also make sense. Nevertheless, this strong inconsistency reminds us of the controversial origins of the toas themselves and the likelihood that they emerged as innovative elements of a mission-based material culture responding to intense demand for elaborated artefact forms (see Jones & Sutton 1986). That said, the toas' references to particular sites and Dreamings are indisputably authentic. In fact, these objects can be regarded as 'sculptural placenames', not merely in the sense that they often operate logographically as symbolic embodiments of named sites or the ancestral action taking place there, but because each toa represents a site on the Hillier Map.

The Reuther toas are known to have been made on Killalpaninna Mission itself, probably during late 1903 or early 1904. This mapped distribution of 'toa sites' is a further reminder that the Aboriginal individuals involved in the creation of the toas were drawing on the wide extent of their territorial knowledge, despite a growing dependence upon the single site of the Lutheran mission. But more importantly, with the insight that they provide into the heroic episodes of eastern Lake Eyre Histories or Dreamings, the toas offer a useful reminder of the distinction between the Hillier Map and a conventional European map of the Australian desert (see also Sutton 1998:383–384). There is a deceptive uniformity to the Hillier placenames, a sense that all are on a level and weighted to the same degree of meaning, unlike a European map where the features are differentiated, separating destinations from impediments, centres from outliers, hills from valleys. It is the Reuther manuscript that provides that key for differentiating the Hillier placenames, enabling us to see that the dynamic which underlies and invigorates the placenames is that of Ancestral action, multiple imprints left on the landscape by a legion of *mura-mura* pursuing their individual and interlocking destinies. In 1904, even if some of this meaning and the Aboriginal taxonomy underlying it was apparent to Reuther and Hillier, it was clearly impossible for them to represent it graphically. A century later, using the clues embedded in the Reuther manuscript and the essential research contribution of linguists and Aboriginal people, it may be possible to redraw Hillier's Map with that in mind.

REFERENCES

Cornish, William H., 1886, Cooper's Creek & Warburton River. In E.M. Curr, ed., *The Australian Race*. Vol. II: 22–23; 24–25; 28–29. Melbourne, John Ferres.

Hercus, Luise, 1985, Leaving the Simpson Desert. *Aboriginal History* 9(1):22–43.

— 1986, The end of the Mindiri people. In L.A. Hercus and P.J. Sutton, eds, *This is What Happened*, 182–192. Canberra, Australian Institute of Aboriginal Studies.

— 1987, Just one Toa. *Records of the South Australian Museum* 20:59–69.

Hercus, Luise and Vlad Potezny, 1990, Locating Aboriginal sites: a note on J.G. Reuther and the Hillier Map of 1904. *Records of the South Australian Museum* 24(2):139–151.

Hillier, Harry, 1904, Map of 2,468 place-names north and east of Lake Eyre. AA266, South Australian Museum.

Jones, Philip, 1991, Ngapamanha: a case study in the population history of north-eastern South Australia. In P. Austin et al., eds, *Language and History: essays in honour of Luise A. Hercus,* 157–173. Canberra: Pacific Linguistics.

Jones, Philip and Peter Sutton, 1986, *Art and Land: Aboriginal sculptures of the Lake Eyre region*. Adelaide: South Australian Museum.

Lewis, John W., 1876, Journal of Mr Lewis's Lake Eyre expedition, 1874–75. *South Australian Parliamentary Papers* 19:1–42.

Litchfield, Lois, 1983, *Marree and the Tracks Beyond in Black and White*. Adelaide, the author.

Paull, W.J., 1886, Warburton River. In E.M. Curr, ed., *The Australian Race*. Vol: II, 18–21. Melbourne: John Ferres.

Reuther, Johann G., 1981, *The Diari* vols 1–13. Translated by Philipp A. Scherer. Vol. 5 translated by T. Schwarzschild and L.A. Hercus. AIAS Microfiche No. 2, Canberra: Australian Institute of Aboriginal Studies. Original volumes in South Australian Museum, AA266.

Sturt, Charles, Journal of 1844–46 expedition, entry for 13 January 1845. Microfilm, Mortlock Library of South Australiana, Adelaide.

Sutton, Peter, 1998, Icons of country: topographic representations in classical Aboriginal traditions. In D. Woodward and G.M. Lewis, eds, *Cartography in the Traditional African, American, Arctic, Australian and Pacific Societies. The History of Cartography.* Vol. 2, Book 3: 353–386. Chicago: University of Chicago Press.

Wilson, Roderick S., 1981, Geography and the totemic landscape — the Dieri case: a study of Dieri social organisation including territorial organisation. BA Hons thesis, University of Queensland.

INTERACTION BETWEEN THE
TWO PLACENAMES SYSTEMS

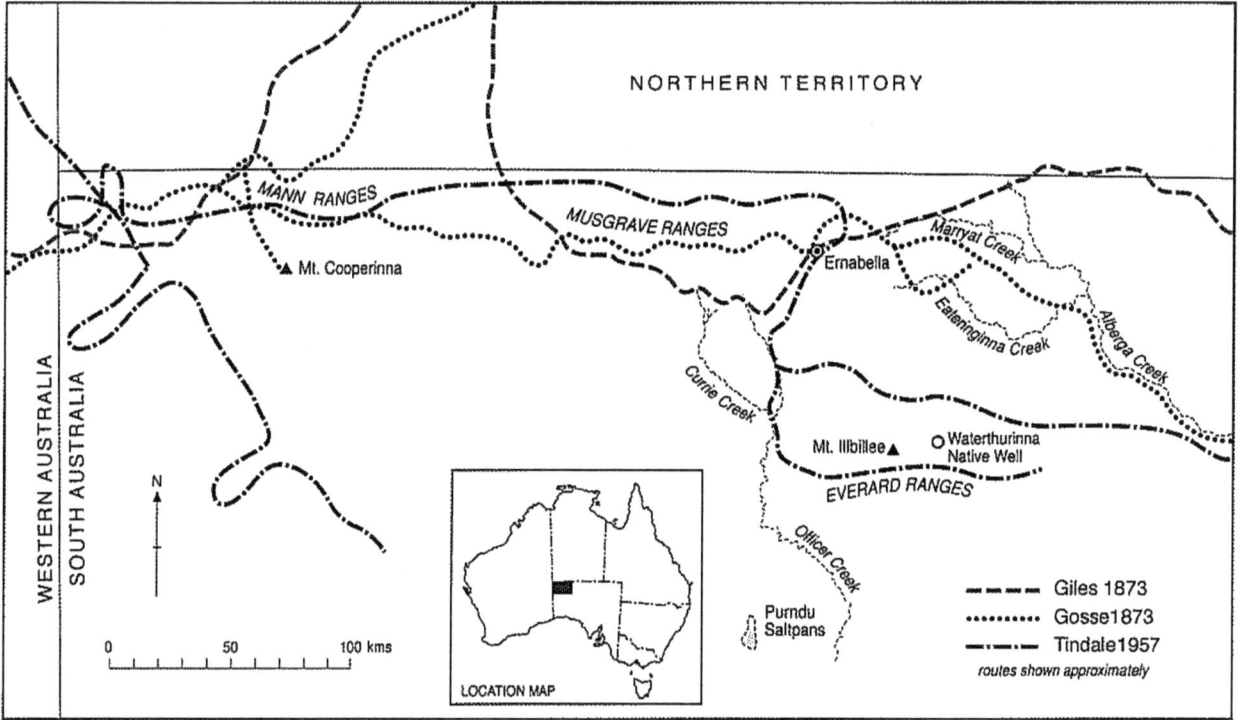

Map 1: Placenames in the Western Desert of South Australia

15 'WHAT NAME?': THE RECORDING OF INDIGENOUS PLACENAMES IN THE WESTERN DESERT OF SOUTH AUSTRALIA

Paul Monaghan

I would like to make a few brief comments on an issue arising from my current research on placenames in the north-west of South Australia (see Map 1).[1] While the research considers both Indigenous placenaming practices and those names imposed by colonisers, an early problem is suggested by the fact that nearly all Indigenous placenames on record have been recorded by non-Indigenous people. Who recorded the names and under what conditions, what sorts of features had their names recorded, and what types of inaccuracies occurred? These are the main questions I will now briefly address.

(i)   The first major explorations to pass through the north-west region were conducted in the 1870s by Ernest Giles, John Forrest and W.C. Gosse. In general, Indigenous placenames were overlooked — either because contact with locals was minimal or non-existent, or colonial naming practices in English, commemorating an influential colonial figure, expedition member, relative or friend, predominated. A few names, however, were recorded by Gosse towards the end of his expedition, largely because of increased contact with locals. As can be seen from the following example though, the Indigenous name remained the marked case:

> Charlie ... showed me a hole of water thirty yards long, six wide, and four deep, which I have called by its native name, Appatinna. (1874:22)

Charlie, described as a blackfellow 'who could speak a few words of English' (1874:21–22), was not attached to the expedition but was met within the vicinity of the waterhole. This type of incidental meeting, with communication in Pidgin English, was fairly typical of what

---

[1]   My thanks go to Rob Amery who made some helpful suggestions to a draft version of this paper. Thanks also to Jane Simpson for editorial comments on a later draft. Of course, any remaining errors are my own.

This research has arisen as part of PhD I am currently undertaking at the University of Adelaide. At the present stage of its genesis, my thesis examines the wider cultural contexts in which Indigenous and non-Indigenous placenames have been recorded and used in the north-west of South Australia. It is particularly concerned with the linguistic 'construction' of land–language relations and how this impacts on current Native Title issues.

L. Hercus, F. Hodges and J. Simpson, eds, *The Land is a Map: placenames of Indigenous origin in Australia*, 202–206.
Canberra: Pandanus Books in association with Pacific Linguistics, 2002.
© Paul Monaghan

was to follow, with names of water-related features figuring prominently. Nearly all of the names recorded by Herbert Basedow in 1903, for example, are in this category (see Basedow 1914).

(ii)   Under the instruction to 'name all the hills piled by you where not already named, preserving as far as possible the native names' (PRG 10/6), the surveyor John Carruthers set out to triangulate the north-west region in 1888. His survey plan from the years 1888–1892 (Adelaide Surveyor General's Office, n.d.[1893?]) contains a wealth of Indigenous names for a broad range of geographical features, many of which were no doubt recorded by Carruthers himself. But how were these names recorded? A journal entry for 3 July 1889 states:

> Built stone pile on 'Cooperinna' hill ... Met some Natives this morning and through our interpreter Tommy Carrunda obtained the Native name of this hill. (PRG 10/3)

But this approach was not always successful, as the following account reveals:

> Our interpreter Tommy Carrunda had some difficulty in understanding the meaning of a lot of their words ... Unfortunately Tommy could not make them understand that we wanted their Native names for our hills and other prominent features. (PRG 10/7)

Apart from the use of interpreters, Carruthers used Pidgin English to gather the names directly from Indigenous people, if the following account by Basedow is anything to go by:

> Mr. W.R. Murray, one of J. Carruthers' party, tells me that an appropriate native name was wanted for an imposing, bare, granite mound in the heart of the Everard Ranges. An old warrior was consequently addressed as follows: 'What name that one hill, Billy?' The old man looked up smirkingly at the white men and, as they usually do, repeated the last syllables of the query as near to the original as he could. 'Ill Billy' came the guttural reply. 'No, that one hill, Billy,' shouted Carruthers. 'Ill Billy' shouted the old chap with a pitiful smile and a peculiar gesture with his hands. 'Well, if it must be,' said Carruthers, 'we shall name the hill Illbillee.' (1914:233)

And so it appeared on the survey plan, and on subsequent maps. While the humour of this account rests on a pun, it also points to the degree of uncertainty or ambiguity that could arise during the name recording process. While it is possible to trace the likely meaning of 'Illbillee'(between Papunya and Kintore is a place Ilpilli, where ti-trees are a prominent feature in the landscape (David Nash pers. comm.); Cliff Goddard (1992) gives 'ilpili' as 'ti-tree'), it is not so easy to determine whether the recording of this name rests on a misunderstanding or merely a humorous aside based on a coincidence. Perhaps we will never know for certain, but it is likely that an element of both was involved.

(iii) An interesting question is raised by David Lindsay's recording of 'Purndu Saltpans' (1893:12) in 1891, while leading the Elder Expedition. Upon arriving at a saltpan with Billy, an Everard Range local who had been 'following' the party, Lindsay learned that it was 'purndu' and recorded the name of the feature as Purndu Saltpans. In Pitjantjatjara/Yankunytjatjara (P/Y), however, pantu is a general term for saltpan; so it seems we have the case of a geographical feature in the Indigenous language being recorded as its placename. At present an alternative Indigenous name for this feature is not known to me, and one does not appear on the South Australian Gazetteer, so the question remains: 'Should this be taken as an Indigenous name, a name imposed by colonists, or in some sense a joint English–P/Y construction?' Of course, the possibility that it is both general term and placename should not be ruled out at this stage (that is, before fieldwork is undertaken).

(iv) Probably more in the category of inaccuracy rather than error, most names recorded up to this point were missing a number of retroflex and nasal sounds that were not picked up by the native English speakers. A 'system of orthography for native names of places' put together by the Royal Geographical Society in London was available to some travellers, and this did seem to regularise some of the orthographic practices (see 'Handbook of Instructions for ... the Elder Scientific Exploring Expedition', 1891:35–36). Largely though, the names recorded were influenced by English writing practices. On Carruthers' survey plan we find for instance:

*Eat*eringinna

*Water*thurinna

*Cooper*inna

(v) After the explorers, surveyors, prospectors and scientists came the anthropologists, and it is with their arrival that we find a marked increase in the recording of placenames. Moreover, the work was carried out, particularly by Norman Tindale, in what seems a more systematic and rigorous fashion. Tindale set out to record as many names as he could, employed a more detailed orthography (although he did struggle with it), and often revised the spelling of names previously recorded by others. During field trips in 1933 and 1957 he gathered names in the standard way of asking informants while at a location, but he also broadened this to include names heard while talking about, as he puts it, 'totems, ancestral legends & placenames' (1933:59), and from the crayon drawings also collected. As his journals show, he was careful to record the names and cross-check them later while on a peak or high point:

The native pointed out various native places of interest & helped locate places of which I had previously only vague ideas as to direction. Thus list of native names of places is growing. (Tindale 1933:341)

There are many other examples I could have chosen to illustrate this point.

Part of my task with Tindale's names is to locate any blind spots in his recording as well as to consider his reliability given the difficulties he faced as an early researcher in this area. At one point he corrects a name on Carruthers' survey plan, noting that the recording Pinundinna Hill, south of the Musgrave Ranges, is 'in error' and that it is in actual fact 'Iŋanja'. While this may have been an error on Carruthers' part (which I have not been able to establish yet), it could also simply be an alternative name — an important aspect of Indigenous naming practices that Tindale does not seem to be aware of at this point. In general he seems to have been content with one name and to move on to the next feature.

(vi) Finally, I should mention the work of Bill Watt from the South Australian Geographical Names Advisory Committee. In the late 1980s Bill conducted fieldwork in the north-west region, recording placenames and verifying those already on the South Australian Gazetteer. Importantly he was helped with the orthographic aspects of this work by Cliff Goddard, a linguist who was working at that time in the north-west region. This approach, introducing GPS (Global Positioning System) technology and a culture/language specialist into the official recording process, obviously stands in stark contrast to the first attempts of early colonists to record the locations and names of 'native places'.

In conclusion it should be evident that the recording of placenames in the north-west has developed from incidental and haphazard to well organised and thorough over time. Nevertheless, questions of accuracy still remain and work needs to be done in checking the names already recorded.

REFERENCES

Basedow, Herbert, 1914, Journal of the Government north-west expedition (1903). *Royal Geographical Society of Australia, SA Branch Proceedings* 15:57–242.

Carruthers, John, (1888–1890 MS), Journal of John Carruthers. State Library of South Australia, Mortlock Library of South Australiana: PRG:10/3.

— (MS), Letters received from the Surveyor General, 1882–1895. State Library of South Australia, Mortlock Library of South Australiana: PRG:10/6.

— (MS), Papers of John Carruthers. State Library of South Australia, Mortlock Library of South Australiana: PRG:10/7.

Goddard, Cliff, 1992, *Pitjantjatjara/Yankunytjatjara to English Dictionary*. 2nd edition. Alice Springs: IAD.

Gosse, W.C., 1874, Report and diary of Mr W.C. Gosse's central and western exploring expedition, 1873. *South Australian Parliamentary Paper* No. 48.

Handbook of Instructions for the Guidance of the Officers of the Elder Scientific Exploring Expedition to the Unknown Portions of Australia, 1891. Royal Geographical Society of Australia, South Australian Branch. Adelaide: W.K. Thomas & Co.

Lindsay, David, 1893, Journal of the Elder exploring expedition, 1891. *South Australian Parliamentary Paper* No. 45.

Surveyor General's Office, n.d.[1893?], Map of country in the north-west portion of the province triangulated by J. Carruthers during 1888–1892. Adelaide.

Tindale, Norman, 1933, Journal of an anthropological expedition to the Mann and Musgrave Ranges, north west of South Australia, May–July 1933 and a personal record of the anthropological expedition to Ernabella, Aug 1933. MS. South Australian Museum: AA 338/1/9.

16 'WHAT THEY CALL THAT IN THE WHITES?': NGIYAMPAA AND OTHER PLACENAMES IN A NEW SOUTH WALES *NGURRAMPAA*

Tamsin Donaldson

1 INTRODUCTION[1]

'The world as we know it is in the last resort the words through which we imagine and name it.' David Malouf puts these words into the head of a fictional nineteenth-century lexicographer recalling his encounter with 'The Only Speaker of His Tongue' (Malouf 1985:69).

When I wrote a grammar of the Wangaaypuwan variety of Ngiyampaa, I began with a map of what my teachers called their *ngurram-paa*[2], their 'camp-world', captioned 'the area in which surviving Ngiyampaa speakers were born and for which they know placenames' (Donaldson 1980:xxix, see Map 1).

---

[1] Many Ngiyampaa people helped to make this record of their placename heritage possible besides those whose contributions are mentioned in the text, either by sharing their time and their stories with me or by encouraging me to write and publish this record in this form, or both. Jeremy Beckett made his 1957 fieldnotes available. I am grateful to him and also to Ben Donaldson and John Stowell for comments on earlier drafts and to Jeannette Hope for detailed and fruitful discussion, and practical help resulting in the inclusion of Maps 3 and 4. Paul Stillwell kindly tested and enabled me to improve the representations of pronunciations. Winifred Mumford prepared the base maps which made Maps 2 and 3 possible and Margaret Tyrie computerised the 1980s tables. Dymphna Clark made the facilities of Manning Clark House available and Rod Burstall generously contributed to the costs.

[2] Except in quotations from sources which may have used different conventions, all indigenous language words in italics, including placenames, have been written according to the spelling and pronunciation system outlined at §2.1.

L. Hercus, F. Hodges and J. Simpson, eds, *The Land is a Map: placenames of Indigenous origin in Australia*, 207–238. Canberra: Pandanus Books in association with Pacific Linguistics, 2002.
© T. Donaldson

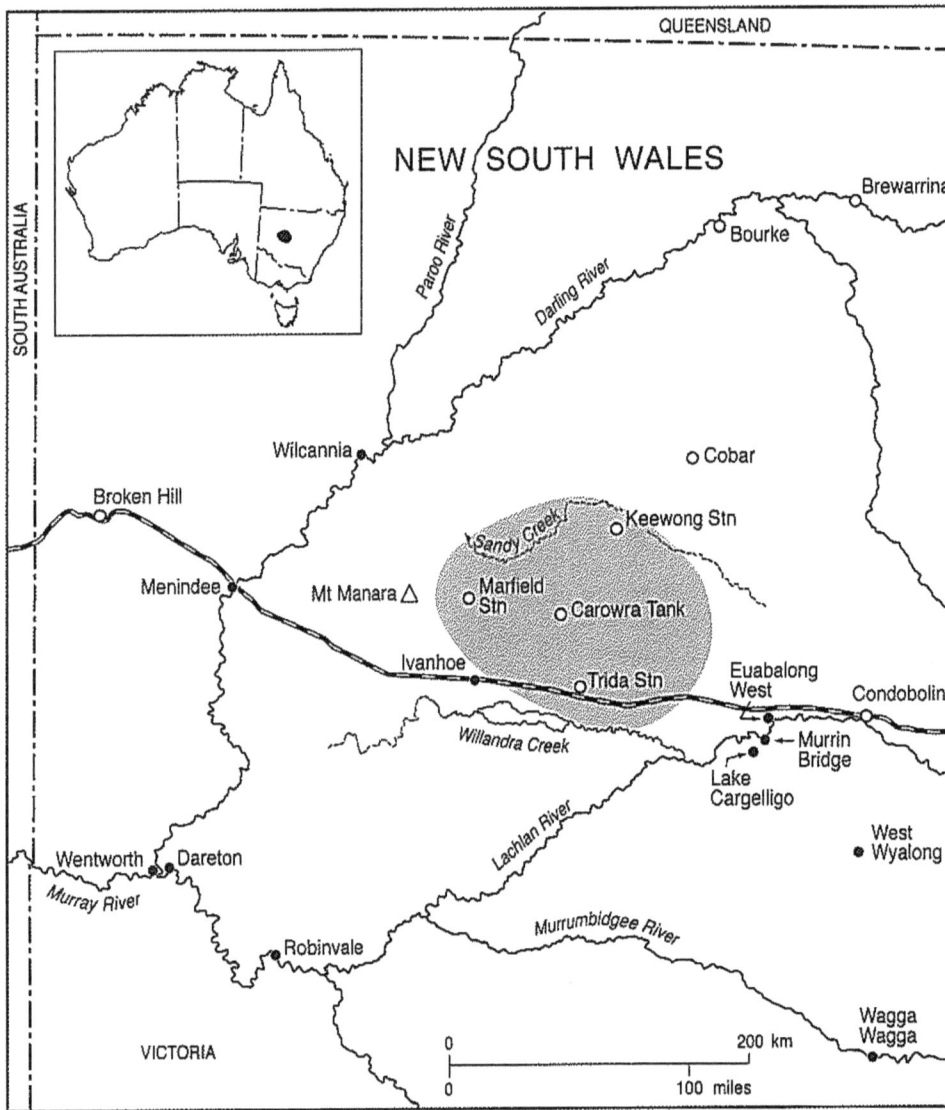

Map 1: The *ngurrampaa*, the 'campworld' in which the Ngiyampaa speakers who gave the placenames at Tables 1–4 were born, c.1900–1920. Based on Map 1, Donaldson (1997:iv). *Cartographer: Kay Dancey*

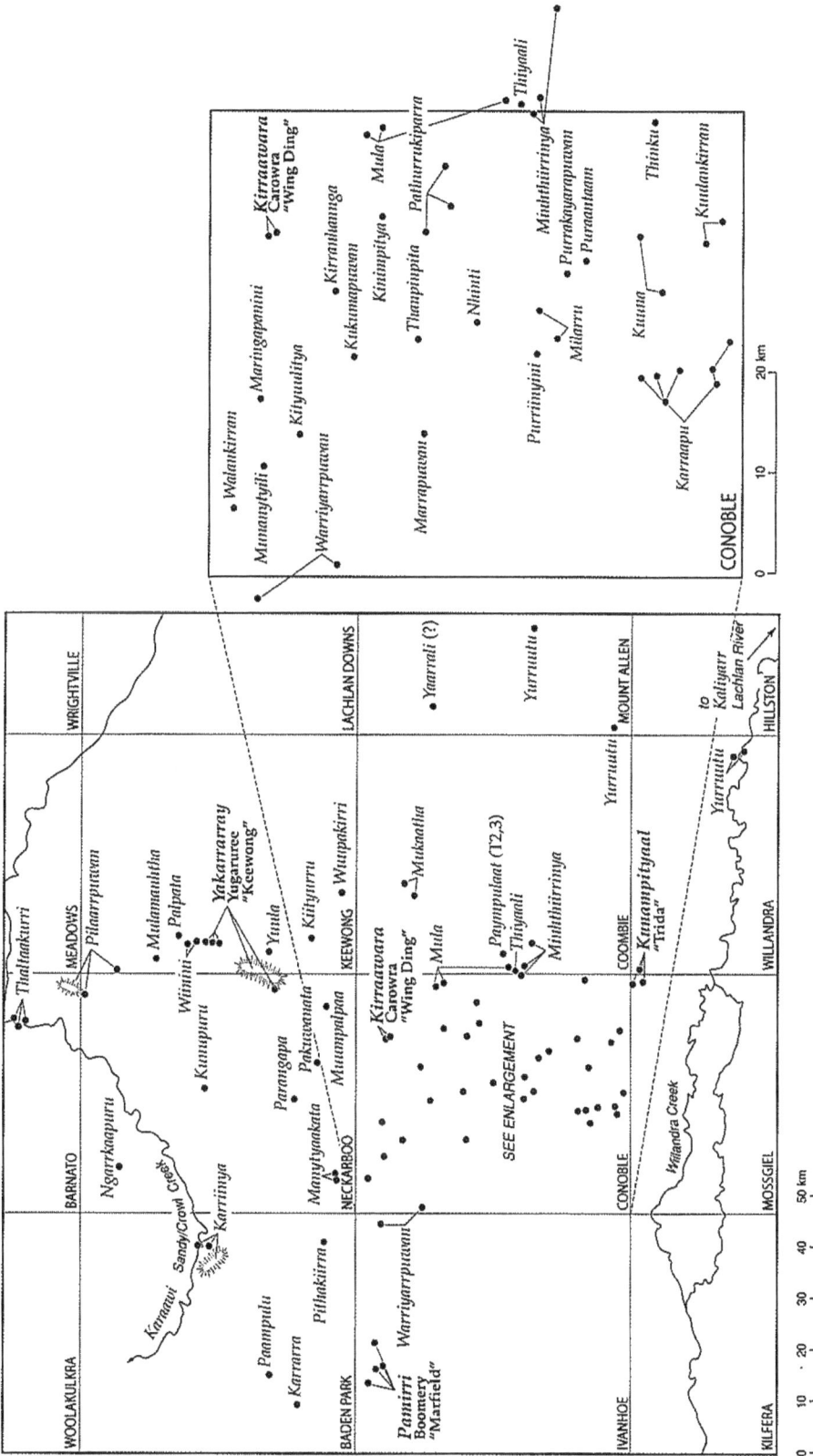

Map 2: Close-up of the *ngurrampaa* at Map 1. Ngiyampaa names are attached to physical features whose 1975 map names were recognised by my 1984 teachers as versions or replacements of Ngiyampaa names (see Table 1). *Cartographer: Kay Dancey*

WOOLAKULKRA   BARNATO   MEADOWS   WRIGHTVILLE

Karraawi   Sandy/Crow Creek   Thaltaakurri   Pilaarrputawn

Ngarrkaapuru   Karrinya   Mulamaalutha   Palpala

Paampula   Pithakiirra   Kunupuru   Wiinini   *Yakarrarray Yugaruree "Keewong"*   Yuula

Karrarra   Parangapa   Pakuawnata   Kiityurru   Wuupakirri

BADEN PARK   Warriyarrputawn   Manytyaakata   Muumpalpaa   KEEWONG   NECKARBOO

Pantirri Boomery "Marfield"   *Kirraawara Carowra "Wing Ding"*   Mula   Mukaatha

SEE ENLARGEMENT   Paympulaat (T2,3)   Thiyaali   Miuhthiirrinya

CONOBLE   *Kunampityaal "Trida"*   Yurruutu   Yurruutu   LACHLAN DOWNS   Yaarrali (?)   Yurruutu

IVANHOE   MOUNT ALLEN   Witandra Creek   to Kaliyarr Lachlan River   COOMBIE   WILLANDRA

KILFERA   MOSSGIEL   HILLSTON

0 10 20 30 40 50 km

CONOBLE enlargement:

Walankirran   Munanytyili   Maringapanini   Kityuulitya   Warriyarrputawn   *Kirraawara Carowra "Wing Ding"*   Kirranhannga   Kuktunaputawn   Mula   Thiyaali   Kininpitya   Thanpinpita   Pathurrukiparra   Miuhthiirrinya   Purrakayarruputawn   Purnaantaam

Marrapuawn   Niiinti   Purrtinyiii   Milarru   Kuuna   Thinku   Karraapu   Kuulankirran

0 10 20 km

This paper presents these names and shows how they both reflected and shaped Ngiyampaa speakers' inherited 'camp-world' — and their sense of themselves. With respect to country, Aboriginal people often put the imaginary lexicographer's idea the other way round: the country does the speaking and the naming. While I was preparing this paper an Aboriginal interviewee from Queensland said on the ABC 'My country tells me who I am'. For the Ngiyampaa people of my teachers' generation, mostly born around 1900, your birthplace in the *ngurrampaa* named you, and your relations followed suit: Liza Kennedy said that her Uncle Red Tank called her 'nothing but' *Wirrpinyi*, the name of the place where she was born (see Table 3, 26).

My teachers learned placenames first of all through growing up 'camping about' in their *ngurrampaa*, walking from place to place, shifting camp, and 'taking notice' of their Elders. They had paid attention to conversations, songs, stories and performances, including the placenames that were 'sung out at the *purrpa*', the ceremonies held in 1914 to make boys into men. (For one life story, see Kennedy & Donaldson 1982.) By the time I was learning the names from them, they were living mainly in the places marked with black dots on Map 1.

The paper also shows which Ngiyampaa placenames survived colonisation and in what forms, and some of the other re-naming strategies used by the newcomers. The changes were not all one way. Some of the new names were taken back into Ngiyampaa.

Nearly 90 Ngiyampaa placenames are presented in four tables followed by discussion. The paper concludes with some general questions about the relations between the Ngiyampaa names and naming practices and the names and naming practices of colonists and mapmakers (§3–4). These are explored by reference to the tables (§2) and to material presented in Maps 1–4 and Figure 1.

2 PRESENTATION OF THE NAMES IN TABLES

The names were recorded in 1984 or earlier from my Ngiyampaa teachers, *Pilaarrkiyalu*, 'Belar tree people', Ngiyampaa from the Keewong and Trida mobs, especially Liza Kennedy (Keewong), Archie King, Sarah Johnson and Mamie King (Trida). (The Marfield mob were *Nhiilyikiyalu*, 'Nelia tree people'.) Other people's particular contributions are acknowledged in the text and tables. I had been learning placenames incidentally, in the course of linguistic and related work, since 1972. In 1984, as well as reviewing names and information about them that I already knew, I systematically worked through all the names on all the 1975 Commonwealth Mapping Authority (CMA) 1:100 000 maps covering the *ngurrampaa* and surrounding country with Liza Kennedy and through some of them with others, usually at her suggestion. Though I had already travelled with some of my teachers in parts of the *ngurrampaa*, I was not able to visit the places named at the tables in their company in 1984 to enrich and refine our work, because of their increasing frailty.

By the end of 1984 Ngiyampaa people of the generations born around 1900 up until the end of the First World War had given equivalents 'in the whites' for more than 50 of the Ngiyampaa names they remembered. These names and their equivalents are presented at Tables 1 and 2. The proportion of anglicised versions of the Ngiyampaa names to introduced names among the equivalents was about two to three. The equivalents for around 50 of the names (sometimes more than one per name) could be found on CMA 1:100 000 maps published in 1975. Map 2 locates the uses of these equivalents on the 1975 maps.

Ngiyampaa people did not know of any equivalents 'in the whites' for 30 more of their placenames, roughly two out of every five. They located the places they referred to as near or between other places named at Table 1, or not at all. These names are presented at Tables 2 and 3.

In a few cases (not listed), the reverse was the case, as with the 'Red Tank' after which Liza Kennedy's uncle was named. They knew that there *was* a Ngiyampaa name for a place but, in the absence of the previous generation to remind them of the name, they were now able to refer to it only by the English name they had grown accustomed to using instead.

In addition, they did not recognise any Ngiyampaa origin for many of the apparently Aboriginal names used on the 1975 maps. Figure 1 illustrates this in relation to a listing of *all* the names on one of the 1975 maps, Neckarboo. The 'Neckarboo' section of Map 2 shows the Ngiyampaa names that correspond to those underlined at Figure 1.

Material from the tables is referred to in the text by table and row number, for example T3, 26. Contents and conventions for the columns of Tables 1–4 are as follows:

2.1 Ngiyampaa name

Column 1 gives the Ngiyampaa name (Tables 1–3) or asterisked reconstructed Ngiyampaa name (Table 4). The spelling system provides a guide to pronunciation. Each distinct sound in Ngiyampaa is written with a single letter or pair of letters. This is the full list of sounds:

a, aa(a:), i, ii(i:), k(g), l, m, n, ng(ŋ), nh, ny(nj), p(b), r(ɽ), rr(r), t(d), th(dh), ty(dj), u, uu, w, y

This is the system used since Johnson et al. (1982). Where different letters or symbols were used in Donaldson (1980), they follow their current equivalents in parentheses.

Words begin with these sounds: *k, m, nh, ng, th, ty, w, y*; and finish with these sounds: *a, aa, i, ii, l, n, rr, u, uu, y*.

The following pronunciation aids are used in Donaldson (1997:x):

VOWEL SOUNDS, AND HOW TO SAY THEM:

a   like the underlined part of c<u>u</u>t, l<u>o</u>ve, t<u>ou</u>ch

aa   the same as *a* but longer, like gal<u>ah</u>

i   like t<u>i</u>n, pr<u>e</u>tty

ii   the same as *i* but longer, like s<u>ee</u>, l<u>ea</u>f, sk<u>i</u>

u   like p<u>u</u>t, g<u>oo</u>d

uu   the same as *u* but longer, like m<u>oo</u>n

CONSONANT SOUNDS AND HOW TO SAY THEM:

k, p, t always like s<u>ch</u>ool, s<u>p</u>it, s<u>t</u>op, or ki<u>ck</u>, pu<u>p</u>, too<u>t</u>, never like <u>c</u>ool, <u>k</u>ick, <u>p</u>it, <u>p</u>up, <u>t</u>op or <u>t</u>oot

ng   always like si<u>ng</u>er, never like fi<u>ng</u>er

nh   like <u>n</u>ip, but put your tongue between your teeth as for <u>th</u>in

th   like <u>t</u>in, but put your tongue between your teeth as for <u>th</u>in

nhth like u<u>nt</u>ie, but keep your tongue between your teeth for both sounds *nh-th*

ny   always like ca<u>ny</u>on, never like ma<u>ny</u>

ty   like pea<u>ch</u>, but flatten your tongue against the roof of your mouth

nyty   like in<u>ch</u>, but keep your tongue against the roof of your mouth for both sounds *ny-ty* (imagine in<u>ych</u>)

rr   like a Scottish pronunciation of Ma<u>r</u>y, ma<u>rr</u>y

Say *l, m, n, p, r, t, w, y,* as you would in English words.

For phonological and phonetic details see Donaldson (1980). Stress is on the first syllable unless the second vowel is long as in *Kirraawara* at Table 1 (T1, 7). Hyphens separate analysable parts of names in this column, except at T1, 32 where the name is extra long.

Information on two names, *Thipara* (see Comment at T1, 10) and *Muunkatha* (see Comment at T3, 8), was found too late for them to be listed as Ngiyampaa names in the first columns of these tables.

Some well-remembered Wongaaypuwan Ngiyampaa placenames, such as *Kuparr* meaning 'raddle' for Cobar and *Kaliyarr*, the Lachlan River, are not included in the tables, as being beyond the densely named *ngurrampaa* of the generation of my *Pilaarrkiyalu* teachers and the *Nhiilyikiyalu* they grew up with as neighbours. For discussion of the 'camp-worlds' of their birth (where their birthplaces are known) and of their 'drylander' childhoods spent 'camping about' in comparison to those of earlier generations, see §3.4.

Mount Manara (Map 1) was not remembered by my teachers as being called *Manaara, Manaarra* or *Manara*. This may be because, as an important story place (as Mount Manara is for both Niyampaa and Paakantyi further west), it was not often appropriate to mention it by name. Fred Biggs, a generation older than my teachers, told Jeremy Beckett that it was not appropriate to mention the Being whose cave is at Mount Manara by name. He is now more openly named by people dealing with anglicised versions of his name, with spellings such as Baiami. Fred Biggs also told a story 'in language' 'from the very beginning' (*marrathal-pu*) about a journey 'to Mount Manara'. Jeremy Beckett thinks that the name, followed by *-ku* meaning 'to', was said with its English pronunciation [mə'narə] (pers. comm. 2000). Therefore, Mount Manara, earlier spelt Manaro (Crotton n.d.) and Muneru (Town and Country Journal September 1871:304), is not included in the *ngurrampaa*, as defined for this study, at Map 1, nor is *Manaara* included at Table 1.

The representation of the *ngurrampaa* at Map 1 was made by shading so as to include all the physical features marked on 1975 maps named with recognised equivalents of Ngiyampaa names, not by mapping Ngiyampaa referents for Ngiyampaa names (see the end of §3.1). A further 30-odd remembered names (from Tables 2 and 3) could not be mapped even in this roundabout fashion. Some of them may have referred to places outside this shaded area: *Puru pinytyilanhi* (T3, 18) is described as a 'tank on Keewong or Kaleno', where Keewong relations of Liza Kennedy camped during the first decade of the twentieth century, much further north (Map 3), and closer to Cobar (Donaldson 1985a:129).

2.2 Literal meaning

Names and parts of names which Ngiyampaa people recognised as meaningful are glossed in the columns headed 'Literal meaning' in Tables 1–4. No meanings differed from current meanings for the same forms. *Thantaay pumanaarra* (T3, 20) could have been translated 'frog(s) getting hit' or 'getting killed'. Only the translation 'frogs getting killed' was given.

Grammatically, the analysable names consist of nominal words and phrases, and/or phrases of other kinds, including fully explicit sentences (T3, 18). Nominal suffixes include ones deriving stems with the same reference as the root, such as *-kirran* 'big', but much more frequently ones deriving stems with a different reference, such as *-pityaal* 'with big', most commonly the plain comitative *-puwan* 'with' or 'having'. The comitative is the most frequently and readily recognisable placename element in today's anglicised map versions of placenames throughout the countries whose languages' names are formed by attaching the comitative of that language to its word for 'no!' — which include both Wangaaypuwan and Wayilwan varieties of Ngiyampaa and Wiradjuri (Donaldson 1985b) and the languages discussed by Ash (this volume). Early placename etymologists tend to record this suffix as meaning not 'with' or 'having' but 'place of' when it forms part of a placename.

Nominal roots refer most commonly to plants and their parts and products, body parts and products, and creatures.

2.3 Pronunciation of equivalent name

My teachers were able to answer each other's questions 'What they call that in the whites?' in respect of all the Ngiyampaa names at Table 1, sometimes with more than one name. Sometimes these names are anglicised versions of the Ngiyampaa name, sometimes they are quite different names. The 'pronunciation' columns in Tables 1, 2 and 4 give a guide to the way Ngiyampaa people said that the names based on Ngiyampaa names are pronounced 'in the whites'. The columns also give a guide to the pronunciation of other names not obvious from the spelling, and of any names that have no English spelling known to me. I follow the conventions set out at the front of the *Macquarie Dictionary* (1981:46) for using symbols of the International Phonetic Alphabet to represent the phonemes of Australian English.

There are occasionally specifically Aboriginal characteristics in these pronunciations such as an initial /h/ (T1, 36) or omitted final /s/ (T1, 13), but otherwise the pronunciations appear to be the same as those of anyone else with long local experience. (A spot rather than comprehensive check was done in the area of Ivanhoe in 1984.) If two pronunciations are given, the more specifically Aboriginal one comes first (T1, 36). The pronunciation given at T2, 1 contains the consonant cluster /nk/, fine midword in Ngiyampaa, but in most dialects of English only when a word is formed from two consecutive words (in 'turnkey' as opposed to 'monkey'). I was unable to find a map spelling for this name, or someone other than my teachers to pronounce it for me.

Using the *Macquarie Dictionary* conventions in 1984 (not always adequate to the task I was setting them), I recorded the first vowel of Boomery (*Pamirri* T1, 30) from Liza Kennedy as pronounced as in 'put', by contrast with that of Moolah (*Mula* T1, 21) as in 'pool'. I have not re-checked the Boomery pronunciation with either a Ngiyampaa or non-Ngiyampaa local.

2.4 Spelling of equivalent name

The spellings given in these columns in Tables 1, 2 and 4 are those used on the 1975 CMA 1:100 000 map sheets named at Map 2. Sometimes there is more than one name, or more than one spelling of the same name on the maps. (Additional spellings from other, usually earlier, maps or sources are given in the 'Comment and information' columns.) Some of these spellings are used for more than one feature on the maps. No feature specifications are included in this column, except where the distinguishing part of the name plays an otherwise unintelligible adjectival role, as in Blue Tank (T1, 23), Pine Flat (T2, 3) and so on. For the reason why, see §3.1.

Figure 1 shows the full range of named physical features on the map sheet 'Neckarboo'. On some of the other map sheets represented at Map 2, there are additional kinds of physical features, some named with equivalents of Ngiyampaa names, especially sidings etc. associated with the railway built to Roto (T1, 48) and as far as Trida (T1, 11) by 1919.

2.5 Location

Table 2 has a column giving Ngiyampaa people's locations for places whose equivalent names 'in the whites' they know, but which are not represented on the 1975 map sheets at Map 2. Table 3 has a column giving Ngiyampaa people's locations for the places for which they know only Ngiyampaa names. These are usually given in terms of their position in relation to other places whose names do have mapped English equivalents, as listed at Table 1. (Pineflat Tank was found on 'Coombie' map sheet too late for *Paympulaat* (T2, 3) to be transferred to Table 1, but it appears at Map 2.)

2.6 Comment and information

The present tense is used in this column in Tables 1, 3 and 4 to attribute comments made in 1984. Placename information is included from A.L.P. Cameron (1899). He lived locally during the last couple of decades of the nineteenth century, first outside the *ngurrampaa*, at Moolbong, south of Willandra Creek, then at Murrumbong (*Kuuna* T1, 14) (Hope, Donaldson & Hercus 1986:128–132 and 109 (Figure 8.1, partly reproduced here as Map 3)). He is interestingly clear about the limits of his language knowledge, but not about those of 'this part' (he writes from Murrumbong) and he did not attempt to map the names.

2.7 Significance of place

This column in Tables 1–3 gives examples of a range of kinds of Ngiyampaa associations with particular Ngiyampaa places. They include references, especially about swamp names and swamp 'owners', to Jeremy Beckett's fieldnotes (1957). He was recording mainly from people born in the 1880s, in particular from Fred Biggs. Beckett gives this account of Biggs' information about swamp 'ownership':

> Every man 'owned' a series of (adjacent) swamps ... He shouted their names as he came onto the ceremonial ground and he might sometimes be addressed by the name of the most important one. He was not the sole 'owner' but he had the right to hunt in them and to give others the permission to do so, whereas hunting in another man's swamp necessitated giving the owner half the kill. In the only two cases Biggs could cite ... the swamps had been acquired from the father-in-law; however (they) ... could be acquired from one's father or mother's brother. (Beckett 1959:206)

There are also examples of songs 'in language' that include Ngiyampaa placenames. It was part of the song-making tradition to locate scenes or incidents celebrated in songs, often by reference to placenames, either in the songs themselves or in the stories told to enable people to appreciate the subtleties of their often cryptically allusive style. As long as such songs continued to be made in or travel into the *ngurrampaa*, placenames appeared in them, including new names (Dardanelles-*ku*, (war-*ku* 'to war'), Ivanhoe (the races, and an indirect allusion to someone met there), Maitland (jail)). Since songs are often remembered with fondness by people who do not otherwise speak a great deal of their language, some placenames without mnemonic equivalents 'in the whites' have been preserved this way (*Punpirrpuwan* T3, 17). One song, which Mamie King taped for me in

April 1981 at Murrin Bridge for deposit at the Australian Institute of Aboriginal and Torres Strait Islander Studies (AIATSIS), records two names: *Kuluwarra* (T3, 4) and *Warriyarrpuwan* (T1, 43). It was sung by Moolbong Johnson, a Wiradjuri clever man and song-maker who spent time at Carowra, but may have belonged to another, Scottie Lanky, who died before Mamie was born. (An unfamiliar Wiradjuri word may underlie *tyingkatyi*, which neither Mamie nor her elder sister Sarah Johnson was able to 'straighten out'.)

Kuluwarra-ku=thu   manapi-y-aka.
K.-DAT=1NOM   hunt-CM-IRR
I'll go hunting to *Kuluwarra*.

Ngathi =tyu   yuwa-nga-y-aka   mukaa   Warriyarrpuwan-ti.
In that (place)=1NOM   sleep-EVENING-CM-IRR asleep   W.-CIRC
That's where I might camp in the evening, at *Warriyarrpuwan*. (lit. lie asleep in W.)

Ka-yili-nya   minkiyan.
Be-ULT.FOCUS-PRES   degree.adverb
(I) am very selfish. (lit. (I) am focusing on others more, i.e. on preventing them joining me.)

Ngathu-para   good man.
1+NOM-CATEG.ASSERT   good man
I (am) truly a good man.

Wirrka-nha-kalay=tyu   tyingkatyi.
limp-PRES-ONLY=1NOM   [foot-CIRC?]
I just limp [(lame) in the foot?].

Ngaku=thu   yana-nha.
to.that.(place)=1NOM   go-PRES
That's where I'm going.

Shilling a day-ku   Warriyarrpuwan-ku.
Shilling a day-DAT   W.-DAT
For a shilling a day, to *Warriyarrpuwan*. [the shilling probably paid for rabbiting]

Abbreviations used here and elsewhere in the paper include CATEG.ASSERT categorical assertion; CIRC circumstantive; CM conjugation marker; CONT continued action; DAT dative; DU dual; IGNOR ignorative; IRR irrealis; LING.EVID linguistic evidence; NOM nominative; OBL oblique; PRES present; PURP purposive; ULT.FOCUS ulterior focus; 1 first person. Donaldson (1980:xxiii, 343) indexes the functions of all forms with upper case glosses.

Table 1: Ngiyampaa names with known equivalents which appear on 1975 1:100 000 maps

	Ngiyampaa name	Literal meaning	Pronunciation of equivalent name	Spelling of equivalent name	Comment and information	Significance of place
1	*Karaawi*		/kraʊl/	Sandy Creek	Crowl Creek on many maps Ngiyampaa prefer 'Sandy Creek'	Marks approximate northern extent of my teachers' *ngurrampaa* as defined here
2	*Karraa-pu*	don't-at all (Wait on!)	/kə'noʊbəl/ /kə'napʊ/	Conoble	Earlier spelt Canoble. The Ngiyampaa name comes from Manny Johnson. Liza Kennedy does not know it, but recalls a local Chinaman's pronunciation, also given here. see T4,4	
3	*Karrarra*		/kər'arə/	Karrara		NB Not Carowra
4	*Karriinya*		/kər'ɪɲə/	Corinya		
5	*Kiityurru*		/gɪdʒər'u/	Gidgeroo	A government tank	
6	*Kinimpitya*			East McDonalds Tank		
7	*Kirraa-wara*	leaves-often with	/kər'aʊrə/	Carowra	Earlier spelt Corowra. Main government tank in *ngurrampaa*. Present homestead is Wing Ding	'Old Parkes' [Fred Biggs' father-in-law's] swamp' (Beckett 1957:75). Site of Carowra Tank Aboriginal Reserve.
8	*Kirranhannga*			West McDonalds Tank		Place of 1914 *purrpa*
9	*Kityuulitya*		/kədʒ'ulɪdʒɪ/, /kədʒ'uligə/	Kajuligah	Liza Kennedy (not herself a reader) recalls people saying 'How do you spell this? It's very hard?'	Swamp of Fred Biggs (Beckett 1957:71) and his 'boundary' (1957:75)

10	*Kukuma-puwan*	stumps-with	/'tɪbərə/ /ˌtɪbəˈru/	Tibora	'Tibora. Named from high trees, climbed to make observations' (Cameron 1899). Also pre-1901 Cookenmabourne (Map 4). Archie King gives *Thipara* for 2nd pronunciation, 'a station 40m from Tiarra'	
11	*Kunam-pityaal*	turd-with big	/'traɪdə/	Trida	'Koonamgal. Dirty place (a lot of dung about)' (Cameron 1899)	Main camping place of Trida mob
12	*Kunu-puru*	sick of-balls	/ˌkunəˈbʌrə/	Koonaburra		
13	*Kuulan-kirran*	?-big one		Irish Lords	Ngiyampaa mostly say 'Irish Lord'. '... named by its first owner said to be the son of an Irish Lord' (Glover 1989:16)	'Sung out at the *purrpa*' by Uncle Pluto Williams, said Sarah Johnson, to show his association with the place. Song about going to *Kuulankirran* to be counted
14	*Kuuna*		/'mʌrəmˌbɒŋ/	Murrumbong	'Goonah. Grey water snake'. 'Murrumbong. Good place or good game' (Cameron 1899). Cameron appears to have transferred the name Murrumbong on moving to a part of Conoble from Moolbong Stn, south of Willandra Creek, where there are tanks named Murrumbong. Ngiyampaa people say the name is the Wiradjuri word for 'good' and pronounce it *marrumpaa*, without the Wiradjuri final *ng*	
15	*Manytyaakata*		/mən'dʒɛəkətə/	Munjagadah	Cameron: 'Munjackata A kind of pipe clay' (white paint is *manha*). Liza Kennedy remembers a	

					new selector changed the homestead name to Tasman. Then the tank name was changed to match (though not on the map)	
16	*Maringapanini*		/ˌmʌrɪgəˈbʌnjə/	Murrigo-bunyah		
17	*Marra-puwan*	?-with		Scrubby Tank		
18	*Milarru*			Nine Mile Marlowe	Archie King comments 'a big lake the other side of Marlowe, fills up when the rivers are flooded. They [Ngiyampaa people] call Marlowe *Milarru* too?' (half to one mile away)	Old Parkes' Swamp (Beckett 1957:75)
19	*Minhthiirrinya*		/mɪnˈtɪnərɪ/	Mintinery		
20	*Mukaatha*		/ˈwɒŋgrəm/	Wongram		
21	*Mula*	pus	/ˈmulə/	Moolah		
22	*Mulamanhtha*			Paddington	Cameron: 'Milamanda lazy fellow'. Eliza Kennedy comments that *mulamanhthiyi* (*tyu*) means '(I) couldn't get there (to where I set out for)'. And *mulamanhthaay* (*mayi*) means '(person) who can't get there'. Such a person might be lazy	
23	*Munanytyili*, also *Munanytyil wakaytyili*			Blue Tank		
24	*Muumpal-paa*	cassia-world	/ˌdʌbəljuˈtri/	W-Tree	'Box tree [there] with W cut in it'	
25	*Ngarrkaa-puru*	?-balls	/nəˈkabou/	Neckarboo		
26	*Nhinti*	clitoris	/ˈnɪntɪ/ /ˈboukə/ /əvˈoukə/	Ninty Avoca	Archie King: 'Bob Morant called the country round Ninty Tank Avoca'	
27	*Paampulu*		/ˈbɛəmbələ/	Bambilla		

28	*Pakuwanata*		/ˈjælək/	Yallock	Cameron: 'Bugwanada Stunted pine with spreading branches'. Eliza Kennedy comments 'not many pine trees around there, mostly dense mallee'	Mamie King thinks that *Kulangkirri* was ceremonially 'sung out' as the name for (a place on?) Yallock
29	*Palpata*		/ˌbuləbəˈda/	Bulbadah	Cameron: 'Boolbadah Where a camp was being made, but was for some reason abandoned'	
30	*Pamirri*		/ˈbumərɪ/	Boomery Marfield		Main camping place of the Marfield mob
31	*Parangapa*		/bəˈræŋɪˌba/ /ˈbrɪŋɪˌba/	Berangabah		
32	*Pathurru-kiparra*			Mountain Tank		
33	*Pilaarr-puwan*	belah trees-with	/bəˈlɛərəbən/	Belarabon	Balarambone on an early map	
34	*Pithakiirra*		/ˌbɪθəˈgɪərə/	Bithegera		
35	*Puraantaam*			Browns Dam	This is the name Brown's Dam made to 'fit [Ngiyampaa] mouths', except for irregular final *m*	
36	*Purrakayara-puwan*	?-with	/ˈheɪmɪjən/ /ˈeɪmɪjənz/	Amiens	Eliza Kennedy recalls that 'Catholic people changed the black name to Amiens', presumably after World War One	A swamp of 'Old Parkes' (Beckett 1957)
37	*Purriinyini*		/ˈbrɪnɪnɪ/	Borraninna		
38	*Thaltaakurri*		/tɪlˈtagərɪ/ /tɪlˈtagərə/	Tiltagara (Lake etc)	Marfield mob camped there. Many spoke Paakanytyi 'as well as their own': cf. Paakanytyi *thiltakara* 'permanent lake' (Hercus 1993:120)	
39	*Thanpin-pita*	?-balls		Lambing Camp		
40	*Thinku*	needle-wood	/ˈtɪŋkoʊ/	Tinco		
41	*Thiyaali*		/tɪˈjɛərə/	Tiarra		
42	*Walan-kirran*	hard-big one	/ˌwɑlənˈgɪərə/	Wallangarra Wallangery		

43	*Warriyarr-puwan*	wild currant bushes-with	/wʌl'wɛərə/	Wallwera Walewira		Song where the singer heads there to earn 'shilling-a-day' (probably rabbiting)
44	*Wii-nini*	fire(wood)-?	/'wɪnɪnɪ/	Winnini	c.f. **Wiiny-tyalapaa* T4,4	
45	*Wuupakirri*		/'oʊbə/	Obah	Liza Kennedy comments '*wuupa* might have been a mouse of one kind or another'	
46	*Yaarrali*		/'jæθɒŋ/	Yathong	Donaldson (1980:317) gives Yathong, corrected in 1984 to 'by Sandy Creek' whether as location of *Yaarrali* or story is not clear	Site of Redbreast & W*aaway* story (and/or by *Karaawi*) (Donaldson 1980:317)
47	*Yakarraray*		/'kiwɒŋ/	Yugaruree (swamp) Keewong (station, lake with dam etc.)	'Yikarreri Stagnant green water'. 'Keewong Where the moon camped' (Cameron 1899), 'moon' in Wiradjuri	Main camping place of the Keewong mob
48	*Yurruutu*		/'roʊtoʊ/	Roto	Liza Kennedy comments 'I don't know if the whites got it off the blacks or the blacks got it off the whites'. The Ngiyampaa regularly made English words beginning with r 'fit their mouths' by adding *yu-* e.g. *yurraapat* 'rabbit'. Obviously Aboriginally derived names to the east begin with /ju/, e.g. Uranaway, Euabalong, from *yu*, sometimes *ngu*	
49	*Yuula*		/'julə/	Ulah		

Table 2: Ngiyampaa names with known equivalents which do not appear on 1975 1:100,000 maps

	Ngiyampaa name	Literal meaning	Pronunciation of equivalent name	Spelling of equivalent name	Location	Significance of place
1	*Kunu-kimpa*	sick of-?	/ˌmʊŋɪ'dʒaɪrə/	None found	Near Trida	
2	*Paynparra*			Corner Pen Tank		
3	*Paympulaat*			Pine Flat	In Trida territory. For Pine Flat Tank, see Map 2	Mamie King's birthplace 1918 or 1919
4	*Pulparri*		/'bulbərɪ/			Nickname of Harry Kelly (brother of Jack) who was born there
5	*Puluparampa*		/pʊ'lipə,rʌmpə/ /pʊ'lupə,rʌmpə/		Tank on Marfield? Perhaps near Trangie Tank?	
6	*Punhthita*		/bʌn'dɪtɪ/		Tank on Marfield 'with ruiny hut'	

Table 3: Ngiyampaa names without known equivalents

	Ngiyampaa name	Literal meaning	Location	Comment and information	Significance of place
1	*Kalpakalpaka-puwan*	taggy (trees)-with	Near Murrumbong	Liza Kennedy says *kalpa* is a 'jutting tag on a branchy tree'	
2	*Kukum-puwan*	erection-with	Near Keewong		
3	*Kulam-puwan*	tree fork-with	Keewong area	'Koolagin forked sticks stuck in the ground' (Cameron 1899)	
4	*Kulu-warra*	seed pod-?	A good crabhole or gilgay by the back gate on W-Tree, between W-Tree and Carowra	Liza Kennedy recalls 'We used to camp there often'	Song starting *'Kuluwarraku thu manapiyaka'*, 'I'll go hunting to Kuluwarra'
5	*Maki-minyi*	raw-?	Near Murrumbong		Swamp of 'Old Parkes' (Beckett 1957)
6	*Malka-ngurrunha* *Malka-ngurrunhu*	mulga-crabhole	A 'good holding' crabhole not far from Koonaburra on the way to Carowra Tank	Both *ngurrunha* and *ngurrunhu* are usual for 'crabhole'	
7	*Mukarr-kaa*	porcupine grass-?	On Keewong	Porcupine grass and all burrs and prickly plants are *mukarr*	

8 *Mukarr-puwan*	porcupine grass-with	Tank near Keewong Crossroads (station)	2 Ngiyampaa places with the same name? Unlikely, because Archie King gives *Muunkatha* for Crossroads Stn	
9 *Muthaangkirri*	a beetle?	Out from Murrumbong		Swamp of 'Old Parkes' (Beckett 1957)
10 *Ngarrampa-puwan*	a creeper?-with	Out from Murrumbong		Swamp of 'Old Parkes' (Beckett 1957)
11 *Palapalaka-puwan*	?-with	Near Crossroads, near Carowra		
12 *Pimpimpi*		A government tank and gypsum deposit '7m from Trida'		
13 *Pithaaki*		Near Trida		
14 *Puka-tharaa*	dead, rotten-?	Murrumbong area		
15 *Pumpu-puru*	?-balls			
16 *Punpirr-kirran*	crooked-big one	An unlocated place called after a 'big old bended tree'		
17 *Punpirr-puwan*	crooked-with	Tank between Keewong and Yallock; closer to Yallock	According to Liza Kennedy the name refers to an old crooked tree	Fred Biggs sang a song about tracking the son of a boundary rider, Jim Robertson, who got lost out from W-Tree and 'trotted straight through *Punpirrpuwan*'
18 *Puru pinytyi-la-nhi*	balls clashed (with each other)	Tank on Keewong or Kaleno		
19 *Thalaynytyaa*		Dam near Keewong and swamp with camping areas		
20 *Thantaay puma-naarra*	frogs getting killed	Tank and well half a mile off the railway track, out from Trida on the Ivanhoe side		Song about the train coming to *Thantaay pumanaarra* in 1919 (Donaldson 1980:335). People camped there during the flu pandemic
21 *Thirrama-puwan*	hills-with	Near Corinya		
22 *Waa-puru*	so there's-(the) balls!	Near Koonaburra		
23 *Warru-puwan*	hornets-with	Irish Lord(s) area		
24 *Wili-wurrunmiya*	lips-?	Out from Trida	*wurrunmiya* has the form of a present tense verb	
25 *Winiyan-kupa*	?-cooba tree	Just out from Irish Lord(s)		Swamp of 'Old Parkes' (Beckett 1957)
26 *Wirrpinyi*		Tank on Keewong		Liza Kennedy's birthplace. 'Uncle Red Tank always put his hand on my head and called m nothing but that'

Table 4: Ngiyampaa names reconstructed by Ngiyampaa people with the help of Cameron (1899)

	Ngiyampaa name	Literal meaning	Pronunciation of equivalent name	Spelling of equivalent name	Comment and information
1	*Murrupa-ra*	thunder-s	/məˈrupə/	Marooba	'Marooba. Like thunder'. (Cameron 1899). Liza Kennedy comments *'panturr murrupara* means 'the thunder is making its noise'
2	*Punan* or *Punan-puwan*	dust dust-with	/bəˈnunə/	Bonuna	'Boonoona. White ground' (Cameron 1899). Liza Kennedy comments 'White dust coming up off a clay sort of ground'
3	*Puumpil*	emu lure	/ˈkumbɪ/	Coombie	'Coombie. A sort of drum used to decoy emus' (Cameron 1899)
4	*Wiiny-tyalapaa*	firewood-without	/ˈwɪtʃɪlɪˌba/ /wɪˈtʃɪlɪˌba/	[part of?] 'Conoble' (Beckett 1957)	'Wee-in-jellebah. A camp where wood was scarce' (Cameron 1899). Liza Kennedy gives anglicised pronunciations without recognising their Ngiyampaa source

3 NGIYAMPAA AND THE NEWCOMERS' NAMING PRACTICES

3.1 Which names get anglicised equivalents?

Major Thomas Mitchell, New South Wales' first Surveyor-General, responsible for the policy of perpetuating convenient Aboriginal names, had this to say in his instructions of 1851 about how to use them:

> Map names, though derived originally from Aborigines, are for the use of Englishmen and once adopted it matters little what they mean. Our use of them, when they fit our mouths, is to distinguish geographical features. (in Heathcote 1965:34)

How does the newcomers' use of them to 'distinguish geographical features' sit with Ngiyampaa naming practices?

It will be clear from a glance at the 'Literal meaning' columns of the tables that analysable Ngiyampaa placenames do not as a rule make reference to a geographical feature, X Lake, X Hill and so on. There is one exception, *Malka ngurrunha* 'Mulga crabhole' (T3, 6). There are also occasionally names that describe a place as having certain geographical characteristics, e.g. *Thirramapuwan*, 'There is/are hill(s) here/there' (T3, 21), in the same way as others have certain plants (*Warriyarrpuwan*, T1, 43) or other characteristics. The principle is to characterise the places in terms of something there, or some event associated with the place (*Thantaay pumanaarra*, T3, 20), rather than to give whatever feature is there a name, (like 'Here be dragons' rather than like 'Dragon Lake'). Hence the general absence of feature specifications for anglicised names in the tables' pronunciation, spelling and location columns. The Ngiyampaa people giving what they recognise as pronunciations 'in the whites' of their own names use them in their own way, regardless of whether the anglicisers have attached environmental or built-feature specifications to them or not. (Reminiscing about their days camped at 'old Carowra', they would of course include a built-feature name when talking specifically about, say, the sinking of Carowra Tank, or the likelihood of its running dry, the Aborigines Protection Board's reason for moving 'the Carowra (Tank) mob' to Menindee in 1933.)

The majority of recorded names whose geographical reference is known denote swamps or other water sources that were central to the Ngiyampaa way of life (see, for instance, the 'Significance of place' columns at Tables 1 and 3). These were precisely the places the newcomers were interested in mapping and using to establish themselves and their pastoral regimes. For contemporary corroboration 'in the Lachlan district', see Donaldson (1985b:77). A 'Map of New South Wales showing stock routes, tanks, wells, and trucking stations by D. Macdonald, CE, MGSA' published in 1888 marks only three names in the *ngurrampaa*: 'Paddington' (station), 'Balarambone Mt.' and 'Corowra swamp', later the largest government tank in the *ngurrampaa* (*Kirraawara* T1, 7). As the newcomers built homesteads, made tanks and drilled bores, these additional features tended to be named after the Ngiyampaa names of the swamps they were dependent on or closest to. The non-physical areal colonial subdivisions of the land — paddocks, stations as shown at Map 3, runs, leases, and parishes — might likewise be named after salient swamps and water sources within them. Figure 1 shows all the names listed on one map, Neckarboo, together with the physical features they mark, and all the parish names. The map sheets themselves, and the larger areal entities, such as counties, shires and electorates, might also end up with homestead or property names based on water source names: thus Calare electorate and an eponymous pastoral station both get their names from *Kaliyarr*, the Wiradjuri and Ngiyampaa name for the Lachlan River.

Regardless of whether the name anglicised was a swamp name, or whether the name was replaced with a different equivalent, a Ngiyampaa name could end up being represented by equivalents at several spots on a map, each being the location of a different cartographic feature: the two dots in the bottom left corner of the 'Neckarboo' part of Map 2 indicate different physical features, a homestead and a tank, both named by an equivalent for *Mantyaakata* (T1, 15), one based on the Ngiyampaa name, one quite different.

Gedda Aklif (1999:127) warns at the start of her Bardi placenames inventory: 'there were no resources available to ascertain the precise location and boundaries of these places', and writes the names on her Bardi only maps without any dots. Nor have there been any such resources here. The presence of dots in association with Ngiyampaa placenames on Map 2 in this study should not be taken as any sort of precise location of the scope of reference of the Ngiyampaa names in traditional Ngiyampaa usage any more than the use of Ngiyampaa derived names for, say, parishes should encourage people to imagine that the scope of the original Ngiyampaa names' reference conformed with parish or any other introduced boundaries. We are dealing with the mapping of equivalent names, not of identically conceived places.

3.2 Which names get replaced by what different ones, when?

Roughly three out of every five equivalents 'in the whites' for Ngiyampaa names are based on the Ngiyampaa names made to 'fit [English-speaking] mouths'. But the creators of the remainder had other priorities than using pre-existing local names. Sometimes they chose names on grounds whose history will not be explored here, as with Wing Ding (T1, 7), defined by the *Macquarie Dictionary* as '*colloq.* a wild party' or Irish Lords (T1, 13). Often they identified a feature such as a tank in terms of a conventional range of widely used kinds of characteristics — McDonald's, Blue, Mountain, Scrubby, Nine Mile, Lambing and so on. But as often, and more importantly for this study, they chose pre-existing placenames originating elsewhere.

Map 1 shows the four most significant camping places in the lives of the Ngiyampaa people who participated in this placenames study. Map 2 shows how in 1975 some local features, but none of the homesteads at these places, are named by anglicised versions of their Ngiyampaa names, *Pamirri* (T1, 30), *Yakarraray* (T1, 47), *Kunampityaal* (T1, 11) and *Kirraawara* (T1, 7).

Some of the introduced names are British placenames, typically already in use elsewhere in Australia (e.g. Paddington T1, 22). One is French, Amiens, associated with the 1914–18 World War (T1, 36). Some are names of places elsewhere in Australia 'derived originally from Aborigines' there (e.g. Yallock, from Victoria, T1, 28).

Place Names

Belarabon (Hills)
Belarabon Tank (& Wind Pump)
'Berangabah' Homestead
Berangabah Tank
Blind Creek (Intermittent)
Bullbi Tank
Campbells Tank
Collins Swamp (Perennial)
Keewong Well
'Kiama'
Kolstad Bore (Wind Pump)
'Koonaburra'
Kulki Bore (Wind Pump)
'Kulwin'
Mailbox Tank
Mirrabooka Bore (Wind Pump)
Munjagadah Tank
'Mount Doris'
'Neckarboo'
Red Swamp (Intermittent)
Sammys Bore (Wind Pump)
Sandy Creek (Intermittent)
'Tasman'
Tergarenara Tank
Timonga Lake
Tomki Tank
'W-Tree'
W-Tree Bore (Wind Pump)
Wallywigram Tank
'Yallock'

Parish Names

BARCHAM
BERANGABAH
BURRINGANNI
BURRINYANNI
COORUBA
FINLEY
GOONABURRA
KOORINYA
MARRIBUTA
PINGUNNIA
TEMOUNGA
WALLA WIGRA
YALLOCK

Notes:

1 All underlined names are spellings of names said by Ngiyampaa people to be equivalents 'in the whites' of names in Ngiyampaa, either anglicised versions of the Ngiyampaa name or changed names.

2 Note the number of names of likely Aboriginal origin, especially parish names, which Ngiyampaa people did not recognise as versions of Ngiyampaa names or as names of familiar places, despite many exploratory pronunciations, and discussions of the area.

Figure 1: All names on 1975 1:100 000 map 'Neckarboo, 7833'

When the names have been 'derived ... from Aborigines' from neighbouring groups and transported only a short way, there is room for considerable confusion of various kinds unless the history of both derivation and bestowal is available, as it is in the case of Murrumbong (*Kuuna*, T1, 14). There are two other names 'in the whites' that end in 'ng', representing a sound permissible word-finally in neighbouring Wiradjuri but not in the Wangaaypuwan Ngiyampaa of my teachers: Yathong and Keewong. Here too we should beware, in the absence of the history of bestowal, of simply assuming that these places both had different Wiradjuri names as well as Wangaaypuwan Ngiyampaa ones. Yathong (a State Forest not marked at Map 2, and homesteads) for *Yaarrali* (T1, 46) is also the name of a homestead in Queensland, but I have no proof of any connection in either direction. Even quite elaborate linguistic evidence may be inconclusive without the bestowal story, as with Keewong for *Yakarraray* (T1, 47). Is Keewong, said to mean 'Where the moon camped', a name based on a spelling of the Wiradjuri version, recorded from a Wiradjuri speaker, of a Ngiyampaa word for the (full?) moon, *kiwaay*? Or is 'Keewong' based on truncating the locative form of this word, *kiwaayngka*, meaning 'at' or 'by' the 'moon'? R.H. Mathews (1904:358–361) recounts a 'Wongaibon' story about 'Giwa' the moon, 'a corpulent old man', prevented by two young men from climbing an (unlocated) rock they had magically raised 20 feet. Brad Steadman (pers. comm. 2000) points out that the (1904) Ngeumba [*sic*] vocabulary in which Mathews gives 'wongaia' (*wangaay*) for 'I don't think so' (i.e. 'no') and 'giwir' for 'moon' was compiled at least in part through correspondence with the policeman at Byerock north of Cobar, D.H. Wilson. Wilson consulted half a dozen named Aboriginal people locally, of whom Billy Coleman was the authority from whom the others insisted he should learn. Pilaarrkiyalu speakers of Wangaaypuwan Ngiyampaa, in my experience, as opposed to the people from further north, from whom Mathews most likely learnt the story, use *kapataa* for 'moon' and *kiwaay* to refer to *puntay kiwaay* 'kneecaps' and to grindstones. For my teachers' comments on parts of this vocabulary as possibly from the 'Brewarrina mob', see Donaldson (1980:8). Brewarrina is even further north, but was also a town with a 'mission' to which many people from elsewhere went or were sent.

3.3 Which names get ignored by newcomers?

While roughly three in five equivalents 'in the whites' for the almost 60 Ngiyampaa names listed at Tables 1 and 2 are based on the Ngiyampaa names, nearly 30 more Ngiyampaa placenames, listed at Table 3, have no such equivalents that my teachers could remember. If there were any based on the Ngiyampaa names, interaction involving their use was insufficient for Ngiyampaa people either to learn or to remember them. I have not systematically searched old sources for likely candidates. Nor have I searched for anybody more knowledgeable about English speakers' placenames locally, in case there might be someone with a focus on equivalents 'in the blacks' as strong as the Ngiyampaa speakers' focus on equivalents 'in the whites'. What Table 3 does show is that these 30 Ngiyampaa placenames at least had an independent life of their own for Ngiyampaa people of my teachers' generation. The existence of equivalents 'in the whites' was not an essential part of how they learned or remembered them.

3.4 Which names get forgotten by Ngiyampaa people?

Figure 1 shows names of likely Aboriginal origin which Ngiyampaa people were unable to recognise. There is no way that their derivation can be pursued further without extra information. Of more interest is Table 4 where Cameron's (1899) etymologies open up the possibility for Ngiyampaa people to propose Ngiyampaa forms that may have been the later forgotten basis for anglicised placenames. Until Liza Kennedy had the opportunity to back-translate Cameron's etymology, no-one imagined that *Puumpil* might underlie Coombie (T4, 3). But in fact only two quite common mishearings need to be assumed in order to establish a link between *Puumpil* and *Kuumpi* (as Coombie would be spelt Ngiyampaa fashion) — a final *l* not heard, and a peripheral stop consonant made at the front of the mouth, *p*, mistaken for one made at the back, *k*.

The 'Comment and information' and 'Significance of place' columns in the tables give glimpses of a limited range of mainly ancestral Ngiyampaa linguistic and cultural values and show how ceremony and song-making conventions provided traditional mnemonics for remembering swamp and other placenames. The first contact period was one of life-threatening as well as cultural practice threatening upheaval, as people had to fit in with or dodge the re-organisation of their country and the disorganisation of their lives by outsiders. Their routes, easy camping and hunting places and means of survival all changed, and with them the numbers and kinds of people who survived where. The Marfield, Keewong and Trida mobs are known by the introduced names of the stations, never by the Ngiyampaa names of the places, which suggests they are localised social groupings formed after the pastoralists' arrival. The Aborigines Protection Board intervened, taking censuses and distributing rations from certain places (Map 3), and later institutionalising the three mobs together with a manager, first at Carowra (a reserve from 1909), then outside their *ngurrampaa* at Menindee in 1933 (Maps 1 and 2). We can only guess at the total impact all this may have had on previous Ngiyampaa knowledge of Ngiyampaa places and Ngiyampaa use of Ngiyampaa placenames — and on how earlier generations understood their *ngurrampaa*. But a contemporary song about going to *Kuulankirran* (T1, 13) to be counted in a census kept that placename alive, and gives a sense of the times. John King ('Tap') described it as *paapimpaangku kuthi*, a 'song of his (classificatory?) grandfather's', when he sang it for Luise Hercus in the 1960s (Archive tape 1595, side 2 item 8, Australian Institute of Aboriginal and Torres Strait Islander Studies). Archie King helped me establish the following basic text for this performance. For the poetics and performance conventions associated with such texts, see Donaldson et al. (1998).

Kuulankirran-ku	*yaay!*	*Ngathi-tyan=likii*		*nganaay.*
K.-DAT	thus	in.that.(place)-LING.EVID=1OBL+DU		that.way

So! They say (they want) us (to go) up there, to *Kuulankirran*.

Minya-ngalmayng-kaa	*mayi*	*ka-ra?*
what-QUANTITY-IGNOR	person	be-PRES

How many people are there?

Yingkal-may-kalay		*marrayipiya*	*mayi!*
same-QUANTITY(as fingers held up)-only		old	person

Just a handful of old people!

Purraay-tya-la-kalay   *pungku*   *wii-nya.*
child-LING.EVID-THEN-ONLY   many   sit-PRES
Supposed to be just a lot of kids now living (there).

Kurrum-pi-kirri-piyal[1]   *nganaay*   *maying-ku.*
go in-BEHIND-PURP-CONT   that way   person-DAT
We gotta keep going, go in after, together with the people.

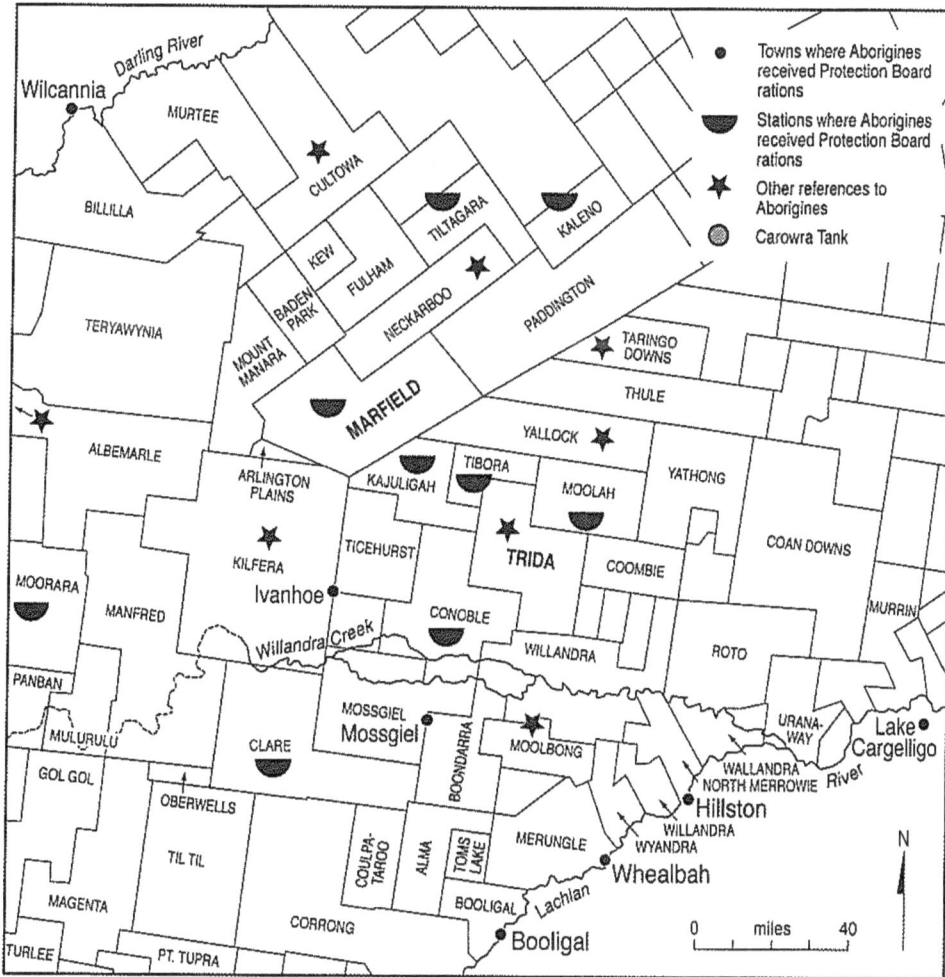

Map 3: Aboriginal people in the north-eastern Willandra Creek area, 1883–1915, as mentioned in Aborigines Protection Board records. Reproduced from part of Figure 8.1, Hope et al. (1986:109), designed by Jeannette Hope and used with her permission. *Cartographer: Kay Dancey*

---

[1]   The more usual Ngiyampaa form for *-piyal*, used a lot by Wiradjuri speakers, is *-ngila*.

3.5 Are there new Ngiyampaa names and places?

Some of the new English placenames were taken back into Ngiyampaa for instance *Puraantam* (T1, 35), and *Paympulaat* (T2, 3), taught me by Mamie King, the youngest of my main teachers, who was born at *Paympulaat* in 1918 or 1919.

It is not always immediately clear which way the traffic went. *Yurruutu* is predictable as a Ngiyampaa version of Roto (T1, 48). But could English speakers have failed to hear the initial *Yu-* when anglicising *Yurruutu*? How did the name Roto (with one gazetted occurrence in Queensland) arise? Also, compare *Karraapu* (T1, 2) and *Wiinytyalapaa* (T4, 4). One Ngiyampaa and one imported name for the same place? Could Canoble be a pre-existing placename transported from somewhere else, provoking the possibly originally joking etymology *Karraa-pu* 'Wait on!' given by Manny Johnson? Or were there two close Ngiyampaa places with two Ngiyampaa names selectively remembered by different Ngiyampaa people? This is simply more likely, on the negative evidence that there is no Conoble or Canoble gazetted elsewhere in Australia in the first Australia-wide gazetteer, based on the 1975 CMA 1:100 000 map series used in this study.

Puraantam and *Paympulaat* retain the English final consonants *m* and *t* of Browns Dam and Pine Flat, not normally permissible word-finally in Ngiyampaa. Other consonants in anomalous positions, however, are not the result of recently taking English names into Ngiyampaa. The otherwise unknown sequence *nng* in *Kirranhannga* (T1, 8) can be taken as a sign of the probable antiquity of the name (Donaldson 1980:48).

There is a range of other, non-phonotactic, evidence for the antiquity of many of the names. This includes the lack of evident literal meaning for many of them, even when they contain intelligible parts (e.g. T1, 17), or grammatically recognisable forms (T3, 24). There are inherited stories from, literally, 'absolutely long ago' (*marrathal-pu*) linked to some places. *Waaway* was killed at *Yaarrali* (T1, 46) (Donaldson 1980:317). Eagle took water to the sky from *Pamirri* (T1, 30) (Berndt 1947:81–82). Other such stories may explain why some of the literal meanings are as they are. A widely known story survives with versions in which Porcupine drops a talking turd (Donaldson 1980:319–322). Did this happen at *Kunampityaal* (T1, 11)? (Cameron, by contrast, with his 1899 interpretation: 'Koonamgal. 'Dirty place (a lot of dung about)'[*sic*], seems to have thought the place so named or newly renamed because of striking changes brought about by introduced animals.) There are also names with narrative meanings without remembered fuller stories to explain them. Who, for instance, are the 'Frogs getting killed' at *Thantaay pumanaarra* (T3, 20)?

Also, are all the names alluding to the plant world of equal age? *Punpirrpuwan* refers to a particular crooked tree (T3, 17), *Muumpalpaa* to a prevalent species (T1, 24). Patterns of vegetation alluded to in some of the names would have remained basically stable, depending as they did chiefly on the underlying geology until some ecological changes began as a result of radical changes in land use, such as those leading to the listing of the *Acacia loderi* [nelia] Shrublands, with 'major stands' near Ivanhoe, as an endangered ecological community under the New South Wales *Threatened Species Conservation Act 1995* (New South Wales Government Gazette No. 131, 6 October 2000: 10957–8). The group names by which the Wangaaypuwan Ngiyampaa designate themselves, as 'nelia tree', 'belar tree' and, further north, as 'stone/rock' people (Donaldson 1984), are consonant with gross differences in the underlying geology of their groups' traditional areas in so far as these areas are known. There is no reason why this principle of basic stability of significant vegetation depending on underlying soil structure should not apply on smaller scales too (Victor Gostin pers. comm. 2000). The issue could be pursued with the help of the land system map series whose sheets correspond to the 1:250 000 CMA topographic maps. These maps integrate the representation of geology, soils and vegetation. A useful source of very detailed vegetation information from early pastoral days in the *ngurrampaa* is the pastoral leasehold area maps of the Western Lands Commission, one of which is reproduced as Map 4. The types of tree and plant they indicate belong to the same range that occurs most frequently in the Ngiyampaa placenames.

While most of the placenames may be of a great age, many of the principles of placename formation are quite transparent, at least on a grammatical level (see §2.2). The analysable suffixes involved are highly productive elsewhere in the language, readily recognisable and available for use in the creation of new placenames (Donaldson 1980, especially Chapter 4). Did Ngiyampaa people keep creating new names, for instance in response to changes wrought by the newcomers? Or simply make a few English ones 'fit [Ngiyampaa] mouths'?

Table 4 shows Liza Kennedy not creating, but reconstructing, a placename in a time-honoured way, proposing **Punanpuwan* as a possible source for (T4, 2). The comitative suffix *-puwan* 'with' is also translatable as 'having', '(It) has', 'There is/are X (there)' etc. It is the most widely used suffix not only in placename formation but in name formation generally. The language name Wangaaypuwan means literally 'no-with', that is '(the language) with *wangaay* as its word for 'no''. More relevantly, *-puwan* was still being used in the formation of personal nicknames during the period that I was learning placenames (King 1990). We have already seen how a birthplace placename could become a personal name (*Wirrpinyi*, T3, 26). When people adopted English-style forenames and surnames, they often continued to use or be referred to by placenames as surnames. Liza Kennedy, born at *Wirrpinyi*, like many of her relations of the Keewong mob, took the maiden surname Keewong. She also appears with her sisters in a 1909 list of Ngiyampaa children in need of schooling (Kabaila 1996:98) with the surname Yallick [*sic*] (T1, 28). And some people's nicknames continued to reflect birthplaces (T2, 4) and names of places with which people had other associations. Were people creating new placenames too?

People were on occasion creating etymologies with recent reference. In Ngiyampaa, *yathara* means '(It) is sticking'. Ida Singh said she had been told a place (outside her own *ngurrampaa*), which I had found spelt 'Yathara' on a local map, had been named when a whitefella asked a blackfella to open a gate and the blackfella answered: '*Yathara!*' Was the place with a defective gate already a place before it got a gate? In other words, would it help to rephrase the question, 'Were people creating new placenames?' and ask: 'Were Ngiyampaa people creating or recognising new places?'

Ad hoc descriptions of spots using placename formation stratagems, like adding *-puwan* to a characteristic of the place, were plentiful when I went hunting with my teachers. On one occasion Sarah Johnson built a tiny *nganu* (which she translated 'miamia') to show me how. Ten years later a group of us chanced on it again, still intact. Thereafter, discussing where to go hunting, we might orient ourselves in relation to *nganum-puwan* 'miamia-with'. I imagine it was ever so, that there were fleeting ad hoc names for places of temporary importance to small numbers of people. Some early surveyors, according to Heathcote (1965:34), appear to have formed the impression, no doubt for a variety of other reasons as well, that Aboriginal placenames were often unreliable 'ephemeral titles ... changing from year to year'. Mapmakers have to have formal processes for deciding what is a place for a given map, what constitutes a name worthy to begin with a capital letter. The main method available for compiling the tables presented here, comparison of Ngiyampaa recollections and recent practices with names on newcomers' maps, gives us only partial insights into the full social meaning of places and place naming for my teachers' forebears in an all-Aboriginal world.

TIBORA HOLDING

Map 4: Leasehold area plan, Tibora Holding, showing pre-1901 placenames and vegetation.
Reproduced from the original map in State Records New South Wales: Western Lands Commission: Pastoral
Lease Files CGS 14570. (SRNSW ref: 10/43877). Land District: Hillston North, Leasehold Area: Tibora, Lease
Number 150, with the permission of State Records New South Wales.

4 RELATIONS BETWEEN NGIYAMPAA NAMES AND THOSE BASED ON THEM

4.1 Can we predict from the form of Ngiyampaa names how they might be anglicised?

The answer is yes, but only when the Ngiyampaa name already 'fits [English] mouths' so closely that the impact of 'making them fit' is only to introduce a hint of an English speaker's 'accent'. It is hard to see how *Mula* (T1, 21) could have been heard or pronounced by English speakers in any other way than it has been. Each of the sounds involved corresponds to a single phoneme in English (both *u* in this stressed position and *uu* get interpreted as the 'oo' in 'moon' rather than 'book'). The phonotactics are fine in English, so is the stress placement on the first syllable. This means that there are unambiguous ways of spelling it according to English spelling conventions, so that it will be read by those who have never heard it said in exactly the same way the writer intended. It could have been spelt Mooler instead of Moolah, but that might have encouraged interpretation in terms of a verb to 'mool' or an adjective 'mool' rather than as a conventional name with a final vowel like that in Dinah or Kogarah. The spelling 'moolah' also represents an existing colloquial English word for 'money' with its own entry in the *Macquarie Dictionary*, a mnemonic for the pronunciation of the name regardless of whether users are also aware of the word's (Ab)original meaning.

By the same token, spelling the name 'Mula', using the phonological spelling system devised for writing Ngiyampaa, could have consistently resulted in the same pronunciation when read aloud by English speakers, unless they chose to pronounce the 'u' as in 'mute' rather than like the 'oo' of 'moot'.

The moment any of these formal coincidences between Ngiyampaa and English does not apply, any one of a range of things may happen at each stage of converting the name — hearing, saying, writing, reading — with each choice leading to a different range of possibilities at the next stage, perhaps responded to differently by different participants in the processes involved. To make matters more complex the 'stages' are of course not likely to occur in a strict linear progression but to loop back, with each repetition of a stage able to influence each of the others, though 'once adopted' as map names, the pronunciations in Aboriginal languages of Aboriginal placenames have usually mattered as little as 'what they mean' did according to Mitchell. The fate of *Kityuulitya* (T1, 9) includes confusion resulting from competing pronunciations, with a reading pronunciation of the map spelling conflicting with, and gradually gaining ground on, the pronunciation in local oral tradition, which is much closer to the Ngiyampaa one.

A game to enable people to explore the effects of these trickier anglicisation processes is described in Donaldson (1995). A version of it was played at the Adelaide placenames meeting at which versions of some of the papers in this book were first presented. The idea was to mimic extreme situations requiring impromptu decision making on the part of lone individuals in the writing of Ngiyampaa names using English spelling resources and in the pronunciation by other lone individuals of the names thus written. I asked people to imagine themselves as builders of the first homestead at a Ngiyampaa place, paintbrush in hand to write its name on the gate, asking a passing (non-literate) Ngiyampaa person to say the local placename for them and painting it on. I acted the passer-by, dictating a name from Table 1, which they wrote on slips. Later I asked them to imagine themselves metropolitan radio announcers without pronunciation research help, reading news of a rural tragedy occurring on a pastoral station whose name is appearing before them for the first time, in writing only. They each picked a slip (not of their own writing) and pronounced what they saw written there. There were very few identical spellings. And there were even fewer identical readings, whether of the same or different spellings of the one name.

I chose *Karrarra* (T1, 3) and *Kirraawara* (T1, 7) as names to dictate because their anglicised pronunciations today are so close that they are frequently confused. The anglicising switch of stress from the first to the second syllable of *Karrarra* and the maintenance of the Ngiyampaa stress on the long second syllable of *Kirraawara* has meant that the first vowel in both names has been pronounced by English speakers as unstressed English /ə/, then written 'a' (Karrara, Carowra), which in turn continues to get read as /ə/. However, the only thoroughly consistent result of the game was that all the painters of names on gates, without exception, having just listened attentively to the Ngiyampaa pronunciation, chose the spellings 'a' or 'u' and 'i' respectively to write the vowels in the first syllables of *Karrarra* and *Kirraawara*.

Sometimes anglicisers will stabilise the pronunciation of their spellings by using the spelling of a similar-sounding English word, thereby sometimes masking the word's Aboriginal origin, provided the word chosen does not attract too much attention by its bizarreness in the contexts in which it is used. Ngiyampaa *warriyarr* appears as 'warrior bush' as well as 'currant bush' in Cunningham et al. (1981:337), giving rise to little change in pronunciation except anglicisation of the first *rr* to r, and loss of the second. The comitative suffix *-puwan* 'with' appears as an English placename element '-bourne' as in Melbourne, intended to be pronounced the Queen's English way, presumably, in Cookenmabourne (Map 4) for *Kukumapuwan* (T1, 10) and as '-bon' or '-bone' commonly elsewhere (amongst other spellings).

But sometimes the effect of 'Ngiyampaa whispers' and its orthographic counterpart appears to have so overwhelmed the angliciser that the result looks and sounds like an act of flailing panic. The anglicised version of the placename *Warriyarrpuwan* (T1, 43) is nothing like Warriorbourne. It is written, unpredictably, 'Wallwera' and, in a probably later form, 'Walewira'.

4.2 Can we reconstruct from the form of anglicised versions what the original Ngiyampaa name was?

Faced with the spelling 'Wallwera' and no other information except how it is pronounced, what are a linguist's chances of reconstructing a source form **Warriyarrpuwan*? Much like a Ngiyampaa speaker's when asked what the Ngiyampaa name for the place with the map name Bonuna is before hearing that A.L.P. Cameron had published a meaning for it in 1899. Liza Kennedy, given the meaning, suggested **Punan* or **Punanpuwan* (T4, 2). Given a meaning involving 'wild currant bush' or 'warrior bush', a speaker or a linguist with a dictionary and a familiarity with placename formation strategies in Ngiyampaa could offer a similar pair of possibilities **Warriyarr* or **Warriyarrpuwan*. If we look at Wallwera as an attempt to represent *Warriyarr*, it suddenly becomes quite easy to deduce what has probably happened. Its spelling involves exactly the same difficulties in representing *rr* sounds as the 'warrior' spelling does. It looks as if the first to write the name have recognised that there is something un-English about the first lateral sound *rr*, so that they have started writing 'll', and have added a 'w' to account for its strangeness. Likewise, they have added a neutral final vowel 'a', to make sure that the strange final *rr* (usually voiceless and sometimes described as 'whistled' in this position) is also pronounced — but as /rə/, rather than being dropped altogether as in the Australian-English pronunciation of 'warrior'.

Comparison of the two spellings 'warrior' and 'Wallwera' alone, without access to the source form *warriyarr*, but with the knowledge of their pronunciations and that they both mean the same thing, could lead to a suspected original form **warriyarr*. For detailed examples of how to reconstruct Aboriginal words through comparing (as many as possible) anglicised spellings, see Austin and Crowley (1995).

In this sort of work with placenames, it is important to spot and use wherever possible the anglicised spellings which look like English names. In the Araluen Valley between Canberra and the coast is an ex-Post Office. When it was opened in 1894 the name Sherarba was proposed, 'as that was the native name of the place' (Cremer nd:1). Behind it rises Shellharbour mountain. The daftness of a landlocked inland mountain being called after a harbour alerts one to the likelihood of its being an alternative attempt at making this native name in Mitchell's words 'fit our mouths'. Indeed Cremer reports that the Post Office was named Araluen Post Office because the name Sherarba was considered to be 'too similar to Shellharbour', and so, presumably, likely to cause confusion with the post office at Shellharbour on the Wollongong coast. Assuming the same inventory of phonologically significant sounds, if not the same phonotactics, for the language formerly spoken in the valley as for Ngiyampaa, and employing the spelling system used for Ngiyampaa, it is possible to suggest an original name with seven segments, with the following possibilities for each:

$$* \left\{ \begin{matrix} ty \\ th \end{matrix} \right\} \; i \; \left\{ \begin{matrix} r \\ rr \\ l \end{matrix} \right\} \left\{ \begin{matrix} a \\ aa \end{matrix} \right\} \left\{ \begin{matrix} nothing \\ rr \end{matrix} \right\} p \; a$$

But one cannot plump with certainty for, say, **Thirarrpa* as opposed to other possible combinations in the absence of more direct evidence from speakers. Even a few more spellings might help to constrain the generalisations in the formula. One thing, however, is certain: the original name never meant anything to do with shells or harbours.

This has been a scanty exploration at §4 of the sound relationships, as mediated in talk and writing, between Ngiyampaa placenames and their anglicised equivalents. The material presented at the tables would allow pursuit of the topic at much greater depth. It should also be helpful to others considering reconstruction where there is less information available, or of fewer kinds.

CONCLUSION

Reacting to the way in which George Augustus Robinson recorded the personal names of the Tasmanian survivors whom he gathered up, then gave them new ones, Inga Clendinnen writes: 'Real names differentiate and connect; they have social meaning' (Clendinnen 2000:216).

We have seen how Ngiyampaa placenames, whatever the stories encrypted in their literal meanings, can provide names for people whose social meaning is to connect them as a group, via their personal histories, to their *ngurrampaa*. We have seen how the introduced placenames which have replaced many of the Ngiyampaa ones on maps often connect these places with other places which have been meaningful in their bestowers' lives elsewhere.

Major Mitchell may have been right that the meanings of anglicised Aboriginal placenames are irrelevant to their usefulness for cartographic purposes. But the business of which names get anglicised, how, and which get replaced, how, also involves the generation of all sorts of social meanings, all sorts of differentiations and connections between people and people and between people and places. Stories arise, often without reference to the history of the namer's bestowal, if known. They can be 'real' enough to their proponents to be capable of overriding strictly phonological processes in determining pronunciations and spellings. Kim Mahood tells five vividly competing stories about the naming of Mongrel Downs (later Tanami Downs), a station in the Tanami Desert of the Northern Territory (Mahood 2000:47). Her own story is that it was in fact named by Joe Mahood (her father) and his friend Bill Wilson 'as a kind of up-yours gesture' in response to local views that they 'must be mad to try to create a cattle station on "that mongrel bit of country out there"'. She also writes that:

> An artist friend in Alice who works for the sacred sites department tells me that the local Aborigines believe Mongrel is a distortion of Monkarrurpa, … the traditional name for Lake Ruth. In a Brisbane gallery I picked up a catalogue of work by Balgo and Tanami artists, and featured in it was a work by a woman artist from 'Mongrelupa'. (Mahood 2000:47)[2]

Here I have tried to show, with data from the 1880s to the 1980s, a century of placenames competing and collaborating to create many worlds, both different and connected, out of this western New South Wales Wangaaypuwan Ngiyampaa drylanders' 'camp-world' — as many worlds as there have been people to use these names to structure and communicate aspects of their knowledge about and relations to the land.

Figure 2: View of the *ngurrampaa* from Mount Manara, with the hills at *Karriinya* on the horizon
Photo: Tamsin Donaldson, 1997

---

[2]   The former Mongrel Downs pastoral lease is now the Mangkurukurrpa Aboriginal Land Trust. According to the Aboriginal Land Commission (1992:18), '*Mungkururrpa* [is a] Dreaming [site] close to the … homestead'.

REFERENCES

Aboriginal Land Commission, 1992, *Tanami Downs Land Claim*. Report by the Aboriginal Land Commissioner to the Minister for Aboriginal and Torres Strait Islander Affairs and to the Administrator of the Northern Territory. Report No. 42. Canberra.

Aklif, Gedda, 1999, *Ardiyooloon bardi ngaanka One Arm Point Bardi Dictionary*. Halls Creek: Kimberley Language Resource Centre.

Austin, Peter and Terry Crowley, 1995, Interpreting old spelling. In Nicholas Thieberger, ed., *Paper and Talk: a manual for reconstituting materials in Australian indigenous languages from historical sources*, 53–102. Canberra: Aboriginal Studies Press.

Beckett, Jeremy, 1957, unpublished field notebook 6 'TRAD. K.' 87pp. Manuscript.

— 1959, Further notes on the social organisation of the Wongaibon of western New South Wales. *Oceania* 29(3):200–207.

Berndt, Ronald M., 1947, Wuradjeri [*sic*] magic and 'clever men'. *Oceania* 18(1):60–86.

Cameron, A.L.P., 1899, Aboriginal names of places. *Science of Man* 2(10):195.

Clendinnen, Inga, 2000, Reading Mr Robinson. In Inga Clendinnen, *Tiger's Eye: a memoir*, 191–218. Melbourne: The Text Publishing Company.

Cremer, Vann L., n.d. Araluen post office: Brief historical notes. 3pp. Supplied in 1986 by the NSW Historical Officer, Historical Section, Australia Post. Sydney: Australia Post GPO.

Crotton, G.W. and C.B., n.d., *Crotton's Australia: Entered according to Act of Congress in the year 1855*. Map. 172, William Street, New York.

Cunningham, Geoff, Bill Mulham, Peter Milthorpe and John Leigh, 1981, *Plants of Western New South Wales*. Sydney: New South Wales Government Printing Office.

Donaldson, Tamsin, 1980, *Ngiyambaa: the language of the Wangaaybuwan*. Cambridge: Cambridge University Press.

— 1984, What's in a name? An etymological view of land, language and social identification from western New South Wales. *Aboriginal History* 8:21–44.

— 1985a, From speaking Ngiyampaa to speaking English. *Aboriginal History* 9(2):126–147.

— 1985b, Hearing the first Australians. In Ian Donaldson and Tamsin Donaldson, eds, *Seeing the First Australians*, 76–91. Sydney: George Allen and Unwin Australia.

— 1995, What word is that? A hearing writing reading game. In Nicholas Thieberger, ed., *Paper and Talk: a manual for reconstituting materials in Australian indigenous languages from historical sources*, 43–52. Canberra: Aboriginal Studies Press.

— 1997, *Ngiyampaa wordworld 1: thipingku yuwi, maka ngiya, names of birds & other words*. Canberra: the author with assistance from the Australian Institute of Aboriginal and Torres Strait Islander Studies.

Donaldson, Tamsin with Margaret Gummow and Stephen A. Wild, 1998, Traditional Australian music: south-eastern Australia. In Adrienne L. Kaeppler and Jacob W. Love, eds, *Australia and the Pacific Islands*, Volume 9 of *The Garland Encyclopaedia of World Music*, 439–443. New York and London: Garland Publishing Inc.

Glover, H.M. (Noni), 1989, *A Town Called Ivanhoe: a history*. Hay: The Riverine Grazier.

Heathcote, R.L., 1965, *Back of Bourke: a study of land appraisal and settlement in semi-arid Australia*. Melbourne: Melbourne University Press; London and New York: Cambridge University Press.

Hercus, Luise A., 1993, *Paakantyi Dictionary*. Canberra: the author with assistance from the Australian Institute of Aboriginal and Torres Strait Islander Studies.

Hope, Jeannette, Tamsin Donaldson and Luise Hercus, 1986, A history of the Aboriginal people of the Willandra Lakes Region. Unpublished report, NSW Department of Planning (now NSW Department of Urban Affairs and Planning).

Johnson, Doreen, Eva Longmore, Eliza Kennedy, Horace King, Lena Parkes and Mamie King, 1982, *Ngiyampaa Alphabet Book*. Dubbo: Western Region Country Area Program Western Readers. Revised ed; 1985. Dubbo: Development and Advisory Publications, Aboriginal Heritage.

Kabaila, Peter Rimas, 1996, *Wiradjuri Places: the Lachlan River basin*. Canberra: the author.

Kennedy, Eliza and Tamsin Donaldson, 1982, Coming up out of the *nhaalya*: reminiscences of the life of Eliza Kennedy. *Aboriginal History* 6:4–24.

King, Mamie, 1990, Text 1, page 24. In Tamsin Donaldson 'Patakirraparaaypuwan in western New South Wales', 21–27. In Peter Austin, R.M.W. Dixon, Tom Dutton and Isobel White, eds, *Language and History: essays in honour of Luise Hercus*, 21–27. Canberra: Pacific Linguistics.

The Macquarie Dictionary, 1981. Arthur Delbridge, editor-in-chief. St Leonards, NSW: Macquarie Library.

Mahood, Kim, 2000, *Craft for a Dry Lake*. Sydney: Anchor.

Malouf, David, 1985, The only speaker of his tongue. In *Antipodes: stories by David Malouf*, 68–72. London: Chatto & Windus/The Hogarth Press.

Mathews, R.H., 1904, Ethnological notes on the Aboriginal tribes of New South Wales and Victoria. *Journal and Proceedings of the Royal Society of New South Wales* 38:203–381.

ASSIGNING AND REINSTATING PLACENAMES

17 CREATING ABORIGINAL PLACENAMES: APPLIED PHILOLOGY IN ARMIDALE CITY

Nicholas Reid

1 ARMIDALE CITY COUNCIL'S POLICY FOR LOCAL PLACENAMING

In a move progressive for local councils at the time, Armidale City Council (henceforth ACC) initiated a draft policy in 1994 for placenaming in the City. An interim policy commenced in January 1994, and the policy proper was adopted in June 1996. The push to develop a policy for placenaming was born of dissatisfaction with the ad hoc and inconsistent bases by which names were being applied.[1]

The stated aims of the ACC's 1996 policy are to:

* provide a consistent approach to placenaming;

* increase the use of placenames with local historical, botanical and zoological associations;

* provide the local community and intending developers with clear information on Council's requirements for placenaming;

* simplify administration of road naming by introducing a procedure for such matters to be dealt with by Council officers under delegation.

As this last point makes clear, the policy is intended to be used mostly in respect of road naming, but the guidelines within it can 'be applied to any geographic placenaming in Armidale', including buildings, hills, parks, etc. The policy does not involve plans to rename any existing places, but rather is intended to form the basis for coordinated future naming practices.

An important part of the ACC's policy is the development of a Register of Placenames. The register is a list of words and names of local regional significance. All future placename choices in Armidale must be drawn from this list. The Register is organised into the following eight categories:

---

[1] Often the first names of the spouses and children of developers.

L. Hercus, F. Hodges and J. Simpson, eds, *The Land is a Map: placenames of Indigenous origin in Australia*, 241–254. Canberra: Pandanus Books in association with Pacific Linguistics, 2002.
© Nicholas Reid

Aboriginal	Medical Associations[2]
Architects, Surveyors	Vietnam Veterans[3]
Educational, Cultural	WW1 Ex-Servicemen
Miscellaneous Historical	WW2 Ex-Servicemen

The procedures involved in developing the Policy and the Register of Placenames were these:

- Linguist to investigate the known sources of Anewan, and provide a report to ACC including a list of 'words suitable for placenaming purposes'.

- The draft report is distributed to a range of groups for comment, including AIATSIS, Armidale Aboriginal Cultural Centre and Keeping Place, Northern Tablelands Regional Aboriginal Lands Council, Geographic Names Board of NSW. The report is also given to the Aboriginal Community Committee (henceforth AbCC) for comment and further distribution to local Aboriginal organisations.

- The Register scrutinised by ACC and list revised in view of feedback received from community and relevant groups. Register further revised at an ACC placenames workshop.

- Register of placenames is posted in Armidale for six weeks for community feedback, before being adopted as policy in June 1996.

- Future applications for street names to follow procedures set out in policy. Once proposed names are approved by AbCC, ACC notifies Geographical Names Board of NSW, Australia Post, Telstra, Dept of Conservation and Land Management, Land Titles Office, Dumaresq Shire Council and local emergency services. Placenames other than street names to be referred to the Geographical Names Board of NSW for approval and gazettal.

The Policy endorses the following criteria to be used in assessing proposed placenames (this is a reduced version of 18 criteria, many of the others are not relevant to the Anewan project and are thus not given here). Names should be:

- appropriate to the physical, historical or cultural character of the locality;

- not duplicates, or too similar to names used in surrounding shires, or elsewhere within the shire;

- thematic names with an historical, botanical or zoological background are generally preferred;

- names derived from the language of the Indigenous population of the district are also welcomed, and should be approved by the Aboriginal Community Committee prior to use;

- euphonious names (i.e. pleasing to the ear, easy to pronounce) shall be used.

[ACC Draft Policy document, 1994]

---

[2] This refers not to any organisations, but to the names of local individuals (doctors, surgeons, etc.) who have contributed to the region in the area of medical service.

[3] Veterans of the Korean War seem not to have been considered, highlighting a media debate about their invisibility at the time of writing.

2 THE ANEWAN PROJECT

Armidale City Council asked me to undertake research into the local language and provide a list of 'relevant placenames'. The language traditionally associated with this area is fairly clearly Anewan (Nganyaywana). Although by the time any records were made the Anewan people had already suffered massive social disruption, most early sources fairly uniformly describe Anewan as covering an area of which Armidale sits in the centre. In 1903 Mathews notes that 'The remnants of the Anewan tribe are scattered over the southern half of what is known as the "tableland" of New England, including Macdonald River, Walcha, Uralla, Bendemeer, Armidale, Hillgrove, and other places' (Mathews 1903:251). These days, descendants of speakers of Gamilaraay, Gumbayngir, Dhungutti and Bundjalung also have long-standing established histories with the region. I expected to find interest from the AbCC in extending the range of languages used to assemble the list, but there was general agreement that Anewan was the appropriate language.

2.1 Anewan language sources

There are no longer any Anewan descendants with active speaking knowledge of this language, though there is knowledge of certain words. The available written sources of information are mostly the writings of Europeans, and generally the work of people who were enthusiastic amateurs at best. The bulk of material comes from contributions to journals like *Proceedings of the American Philosophical Society*, *Proceedings of the Royal Australian Historical Society*, *Mankind*, etc. by R.H. Mathews and J. MacPherson in the first two decades of this century. A full listing of the sources is provided at the end of this paper.

The single significant exception is Crowley's (1976) article 'Phonological Change in New England' which focuses on phonological changes that for a while disguised the relatedness of Anewan to the rest of Australian languages. While Crowley's work is a rigorous and careful piece of scholarship, it is of course limited by the sources he is dependent on. He predominantly uses Mathews, also relies on MacPherson, and provides a phonemicised list of some of Mathews' vocabulary.

The existing data reveal no placenames as such (i.e. words that Anewan speakers used to name places in the specific locality where the City of Armidale now stands). The nearest thing is a very short list of 'Walcha District' names in the unidentified 1904 source, but these were excluded for lack of immediate local significance. In the absence of 'real placenames', the Aboriginal Community Committee decided that the next best thing would be to document known Anewan words and make them available as potential placenames.

The existing records allowed us to produce a phonemicised list of 271 words. From this list we needed to choose words to go on the register, and consider how best to represent them. Sections 3 and 4 of this paper deal with each of these issues in turn. However, before addressing these, we need to complete our background information by considering the sound system of Anewan.

2.2 Sounds of Anewan

The inventory of Anewan phonemes contains the following 14 consonants and 3 vowels.

		consonants				vowels	
stops		*b*	*d*	*j*	*g*	*i*	*u*
nasals	*m*	*n*	*ny*	*ng*			
laterals	*l*	*ly*					
trill		*rr*				*a*	
continuant		*r*					
semivowel	*w*		*y*				

Anewan has aroused some interest among linguists because it has undergone some interesting historical sound changes. Notably, it has lost the initial syllable, or initial consonant of many words. The language today is known as Anewan, but is reconstructable as Nganyaywana. For a while it was considered that Anewan was not demonstrably related to other Australian languages, but this notion has been effectively disproved. Demonstrating the sound changes that have taken place in Anewan, and thus proving that it is in fact a typical Australian language, is the main thrust of Crowley (1976). One interesting by-product of these phonological changes is an unusually high number of homophones in the short list of available Anewan words, for example:

janda	'goanna'	*janda*	'bull ant'
rrula	'male'	*rrula*	'stone'
rrala	'hair'	*rrala*	'gumtree species'

Another by-product of initial loss is the fairly high frequency in Anewan of word-initial nasal/stop clusters, for example:

| *mbunja* | 'bandicoot' | *nduda* | 'spear shield' |
| *nyjunda* | 'down there' | *nggurra* | 'leaf' |

and also other initial cluster types unusual by Australian standards, for example:

| *lwanyja* | 'geebung tree' |

3 CHOOSING WORDS TO GO ON THE REGISTER

In drawing up my report to ACC I provided a list of every known Anewan word, together with its meaning, a Phonemic Representation, and a Recommended Spelling. As a linguist and highly conscious of the preciousness of each and every word of a language about which so little is known, I found myself reluctant to eliminate any words at all. This reluctance was further heightened by a certain uneasiness about the highly subjective basis of choosing 'appropriate' and 'euphonious' words. So I had a two-way bet. I included every word, but chose to additionally provide a pronunciation guide for only a subset of 57 words that seemed to meet the expressed expectations. This in effect provided a reduced subset of words, but it came with an explanation that the process was subjective, and that any other words meeting the collective fancy of the ACC could readily be given pronunciation guides. (See the extract of the full list attached as Appendix 1.)

I now want to address the kinds of criteria I employed in selecting this subset of 57 words. The original brief from the ACC focused on sounds only, requesting that words be 'pleasing to the ear, easy to pronounce'. On discussion with various councillors and members of the Aboriginal Community Committee, I took this highly subjective direction to be sensitive to:

sound: broadly fit English phonotactics

visual look: words that are not too long or complicated

3.1 Choosing words by sound

The permissible patterns of sounds that can occur next to each other in Anewan are quite different from those permissible in English. Phonotactic considerations led us to set aside words with initial velar nasals (*ngana* 'fast') and initial nasal/stop clusters (*nggada* 'fog', *mbwi* 'swim').

3.2 Choosing words by size

In the initial brief it was stressed by ACC that words should not be too long or complicated-looking. After receipt of the report, the AAC workshop rejected the two words *imboodoonga* and *roowalgoonda* for violating these criteria. I am intrigued by this rejection, for it operates against a dominant notion operating in the minds of almost everyone I spoke to about this project: that Aboriginal words should be long. Many people (both Aboriginal and White) expect good Aboriginal placenames to have a four-syllable CVCVCVCV pattern where C1 and C3 are peripheral consonants, and C2 and C4 are voiceless stops, laterals or rhotics, tend towards reduplication, and should ideally sound like 'Parramatta' or 'Muttaburra'. Appreciation of such placenames is an important theme in Anglo–Australian culture, and I recognise this notion in myself as much as in others.

In Armidale there are many street names based on monosyllabic words (Hay St, John St, Crest Rd, etc.). Those subjective demands of being 'euphonious' and 'melodious' seem not to apply to these English street names. No-one associated with this project was at all interested in monosyllabic Aboriginal placenames. Setting aside the word class issues (most Anewan monosyllabic words in the list are verbs, this/that, yes/no), possibilities like Na St, Bwi Rd, were rejected all round. Euphony and melody, it seems, demand syllabic weight. Consider these word syllable count statistics across the three stages of trimming the list.

172 nouns, full list	57 nouns, NR report	33 nouns, final register list
1 syllable: 2	0	0
2 syllable: 93	10	6
3 syllable: 76	38	24
4 syllable: 11	9	3

Although 51 per cent of nouns in the data are disyllabic, in the final list of words chosen for the register just six of 33 are disyllabic.

3.3 Choosing words by meaning and wordclass

The original brief from ACC established *meaning* as a criterion to be employed in the choice of words for the register, in only these general terms:

> Thematic names with an historical, botanical or zoological background are generally preferred.

We thus first extracted all non-referential words — mostly pronouns (we all, you two), adjectives (slow, big), demonstratives (this, that) and particles like 'yes' and 'no'. We also removed most verbs (*mbwi* 'swim', *dwa* 'cry') on the grounds that even though they typically had lexical content, they failed the referentiality criterion.

All people involved on the ACC and AbCC were unified in their belief that only nouns make good potential placenames. This favouring of nouns seemed justifiable; the bestowal of a name is the evocation of some referent. And it is thus the very non-referentiality of words from other wordclasses, that makes them unsatisfactory placenames. Do you find English examples such as Slow St, What St, Then St, etc. to be satisfactory street names? Nevertheless I had some qualms about throwing out every non-noun, and slipped a few 'euphonious sounding' ones in to see how they fared under ACC scrutiny (two pronouns, three adjectives). Not well; they were promptly removed by the AbCC, the final list includes no pronouns, adjectives or verbs. Every word considered suitable for the register is a noun referring to an animal (just over half of the 33 words), plant, artefact or landscape feature, plus one word *indaralla* 'totem of medicine man', which even to people with little knowledge of Aboriginal cultures (or maybe especially to them) sounds satisfactorily ripe with Aboriginal cultural meaning.

Note that this focus on nouns as being the best tokens of placenames is quite at odds with traditional patterns of placenaming. Hercus and Simpson (this volume) comment on the widespread use of verbal expressions, often full clauses, in naming places in terms of activities that were carried out at them.

Word meaning was a criterion for the rejection of a few words considered 'undesirable'. While R.H. Mathews' careful recording of the verb 'masturbate' in nearly every word list he provides is no doubt a valuable inclusion, I suspected it would not pass muster with the City Councillors. I also unilaterally weeded out *dunya* 'penis' and *bula* 'anus' on the assumption that these would not be seized upon as appropriate placename material.

3.4 Further pruning the list

The list of 57 words that I provided to Council in my report underwent three further prunings. The Aboriginal Community Committee removed the following words on the grounds that they were inappropriate. No details were provided, but note that these words (including the three adjectives and two pronouns that I slipped in) all have poor referentiality.

Gwanga	children
Aroonba	good
Irabilla	mother
Argana	section name

Noombadja	slow
Lidjirana	small
Nunnyaburra	we (PL INC)
Nanambinga	we (PL EX)

The Council officer responsible for street naming removed the following words because of similarity to other words, as specified below. This action was explained in terms of practicalities, similarities in names potentially causing problems for emergency services.

Igina	too similar to *Igana*
Imboonda	too similar to *Imbandja*
Imboondja	too similar to *Imbandja*
Oowurra	too similar to *Oorala*
Lyburra	too similar to *Lumburra*
Roogala	too similar to *Rujala*
Woongala	too similar to *Rujala*
Naiya	too similar (in pronunciation) to *Jundja*[4]

The Council officer also removed one word from the list due to concerns about its similarity to an English word:

Loona	'too easily bastardised to Loony'

And two four-syllable words were removed because of their perceived length:

Imboodoonga	'too long'
Roowalgoonda	'too long'

The following words were removed at a Council workshop as they were considered 'inappropriate given the English translation' (ACC).

Ilgaiwa	'summer'
Iluna	'darkness'
Jarrwanba	'winter'
Nura	'sun'
Yoongarra	'rain'

The exclusion of these words surprised me, for although the cited reasons for exclusion seemed to be semantic ones, I could see no consistency in excluding *yoongarra* 'rain' but leaving *arribana* 'hail' and *igana* 'snow' on the list, nor in excluding *nura* 'sun' but leaving *jundja* 'moon'. However, by this stage of the process the input from the linguist was completed.

---

[4] If the perceived similarity surprises you, then that makes two of us.

4 DECIDING HOW TO WRITE WORDS

4.1 Phonemic and non-phonemic spelling

In the full list words are provided in three forms: a 'Phonemicised' form, in a 'Recommended Spelling' form, and a 'Pronunciation Guide'. The intention of the last was to provide Council with a resource that will be of assistance in handling inquiries from the public about the pronunciation of these words.

Phonemicised spelling refers to a writing system that uses an invariant symbol for each contrastive unit of sound in the language. The phonemicised forms are based on Crowley's (1976) reconstruction of the Nganyaywana (Anewan) phonemic inventory.

For placenaming purposes we decided that a truly phonemic writing system for Anewan words would be inappropriate, given the intended purpose and audience. The ACC made it clear that placenames must be easily interpretable and their pronunciation self-evident to native speakers of English, stressing the practical importance of avoiding ambiguity in, for example, directing an ambulance to a certain address, etc. The target audience in this case is the residents of Armidale who are predominantly speakers of varieties of Australian English, including Aboriginal English.

The writing system used in the 'Recommended Spelling' list differs from the phonemic representation. The rationale behind this writing system is, on the one hand, to preserve the phonemic system where possible. On the other hand it has been influenced by the fact that the 'users' of these words are likely to be speakers of standard English-only, or in some cases of varieties of Aboriginal English. It is desirable therefore that these users be able to look at these words in written form and be able to guess at an obvious and fairly uniform pronunciation for them. To this end, the 'obvious pronunciation' principle tends to win out over the 'maintain phonemic accuracy' principle.

The Recommended Spelling draws on the principles of non-phonemic re-spelling, where the ordinary spelling rules of English are used to best represent pronunciation, rather than systematically using one symbol for each phoneme. In this I drew on work by Helen Fraser who has used non-phonemic English spelling rules to develop pronunciation guides for dictionaries. Her idea is that people frequently malign English spelling for its numerous inconsistencies, but ignore the underlying regularity. They forget that the myriad exceptions to every rule, are just that: exceptions to rules. Using 26 symbols for 44 phonemes may sound disastrous, but using 26 symbols and their spatial arrangements allows for the regular contrast of further sets of sounds, albeit in a non-phonemic way. Consider the regular representation of the [eɪ] – [æ] contrast using the 'silent-e spellings' (hate/hat, pate/pat, fate/fat, mate/mat, etc.), and the regular representation of short vowels through the doubling of following consonant, etc. The influence of these ideas can be seen most clearly in the Pronunciation Guide, but also in the Recommended spellings.

4.2 Making Anewan words more like English

There are some features of Anewan words that make them difficult for English speakers to pronounce. We had already weeded out words that obviously violated the phonotactic rules of English. For example, we excluded words that began with the velar nasal 'ng' sound, and with clusters like that in *mbwi* 'swim'. The phonologies of Anewan and English also differ in other ways. In some respects Anewan makes distinctions that English does not make. For example, in Anewan there are two 'r' sounds, written with a single 'r', and with a double 'rr'. Conversely in other respects English makes distinctions that Anewan does not make. For example, English distinguishes between voiceless stops (like *p*, *t* and *k*) and voiced stops (like *b*, *d* and *g*), while Anewan does not. After testing out various spelling systems, we proposed a number of simplifications that made Anewan words more pronounceable to English speakers. These pronunciations were not entirely innovations. Most of them are changes that have come about in the way that English-speaking people of Anewan descent now pronounce these Anewan words. Let us now consider some of the practices we have adopted.

4.2.1 *Stop voicing*

In Anewan there is no systematic contrast between 'voiced' and 'voiceless' pairs of stops (like *p* and *b*, *t* and *d*, *k* and *g*, etc.), and so only the voiced symbols are used for a phonemic writing system. Sources like Mathews suggest that word-initially stops tend to be voiceless, and in the middle of words they tended to be voiced (post-nasally and inter-vocalically). For the purposes of this project, where users of these words will be speakers of English and/or Aboriginal English, we could have chosen to use the voiceless stop symbols in word-initial position in order to achieve phonetically closer pronunciations. However, for the sake of simplicity, and because the majority of stops in the data are either inter-vocalic or post-nasal, we chose to use just voiced stop symbols for the Recommended Spelling.

4.2.2 *Types of* 'r'

Anewan had a distinction between two rhotics, a continuant represented phonemically as /r/ and a trill represented phonemically as /rr/. In the Anewan words now known to Anewan descendants, who are speakers of standard and Aboriginal English, this distinction between rhotics has been neutralised, with both sounds now realised as the continuant [ɹ]. As this neutralisation has operated under influence from English, we decided that it suits our purposes. The writing system we have chosen uses both 'r' and 'rr', but we intend no pronunciation difference by these. In word-initial position, the trill (e.g. Rrundja 'emu') was considered 'too foreign-looking', so all word-initial trills have been rewritten using a single 'r' (thus Rundja 'emu'). We anticipate that syllable-initial 'r' or 'rr' will always be pronounced as a continuant, and that syllable-final 'r' or 'rr' will be not pronounced at all. Nevertheless, we have used both 'rr' and 'r' in the word list, and feel that this serves another function — to provide an Aboriginal flavour to the visual word. Many well-known anglicised Aboriginal words, such as 'kookaburra', 'Parramatta', etc. feature double *r*'s, and there was general agreement among the AbCC members that this was a salient indication of the Aboriginal origin of Anewan words. So in general we have used 'rr' intervocalically (e.g. *burra* 'kangaroo rat'), except where it may conflict with vowel length considerations (thus, *lara* 'rock').

4.2.3 ng *and* ngg

The sounds /ng/ and /ngg/ present difficulties because of the common failure of English spelling to distinguish between them (e.g., 'singer' versus 'finger', etc.). Systematic ways of representing them tended to be undone by English speakers' inclination to read 'anga' as [ʌŋgʌ]. We have chosen to spell /ng/ with 'ng', and /ngg/ with 'ngg', and hope that reference to a pronunciation guide will provide the correct pronunciation to any interested person. However, we also accept that variant pronunciations may be inevitable.

4.2.4 a *and* u

The vowels /a/ and /u/ present some real difficulties. Spelling 'parakeet'/yimbangga/ as *yimbangga* tends to trigger the [ae] vowel. Trying to force short [ʌ] vowel by writing 'u' and doubling the following consonant works well in straight intervocalic environments (e.g. /arrgana/ as *argunna* 'boomerang'). However, consonant doubling works less well for /ngg/ sequences: writing /yimbangga/ as *yimbungga* worked best — but occasionally yielded [jɪmbuŋgʌ].

4.2.5 *Lost initial syllables*

Some Nganyaywana words begin with clusters of homorganic nasal/stop, and early recorders like Mathews mis-heard these as having an initial vowel. For example, he records 'imboanda' for 'kangaroo', which Crowley reconstructs as /mbanyja/. Reluctant to throw out too many words, we took the decision to 'save' these words by re-introducing an initial vowel, giving *imbandja*. In the case of /mbangga/ 'parakeet', we re-introduced a glide as well, giving *yimbungga*, using knowledge from a neighbouring language to inform our choice of sounds. Two factors made this 'fiddle' easier. If hearing these words with an initial vowel was a 'natural mistake' for the likes of Mathews, then it seemed fair to allow contemporary English speakers the same pronunciation. Secondly, to some members of the AbCC, having learned of the historical fact of initial-syllable loss in Anewan, this seemed to be a historically informed re-introduction, as in a sense 'a vowel really used to be there'.

4.2.6 *Stress*

In addition to non-phonemic re-spelling, we have provided an indication of word stress by bolding certain syllables in the Pronunciation Guide (e.g. 'brolga' is roo-wal-**goon**-da, rather than say **roo**-wal-goon-da, etc.). There is no information on word stress patterns in any of the early sources, so we have no certain knowledge of which syllables of Anewan words were stressed. We could perhaps make intelligent guesses based on patterns found in other NSW languages, but even if we did know this, the English-speaking residents of Armidale would pronounce these words with Australian English stress patterns. We see no point in any attempt to be prescriptive here, so in providing information about syllable stress, we are merely codifying the pronunciation that local Aboriginal people currently give these words, and which we believe Anglo speakers of Australian English will also naturally give these words.

5 ATTITUDES TO PHONOLOGICAL ADAPTATIONS

In coming up with a short list of Anewan words to be used for placenaming purposes, we began by examining all the sources and producing a phonemic listing of all known Anewan words. However, we have gone well beyond serving up Anewan words in their pristine form. As §4 makes clear, we have engaged in a series of practices intended to render Anewan words in a form that both makes them pronounceable to speakers of standard Australian English, and reflects the way in which English-speaking Anewan descendants now say these words. These practices have resulted in quite significant changes, such as the neutralisation of certain phonemic contrasts, the imposition of English stress patterns, and the partially inconsistent representation of some sounds. Needless to say, many of these practices are at odds with contemporary descriptive linguistic practice in Australia. Much of the early description of Aboriginal languages by amateur linguists yielded data hampered by these very features. In a day and age when well-trained linguists bring higher levels of skill to the task, how can we excuse such sloppy work?

In dealing with this question, we need to consider whether this work is descriptive, i.e. is it involved in describing Anewan words, or is it about commodification — creating from Anewan materials resources that did not exist in Anewan itself. We also need to consider what are feasible expectations when it comes to pronunciation. Can English-speaking Australians, of both Anglo- and Aboriginal descent, be expected to pronounce these words in the same way that a speaker of Anewan would have? Clearly not! Anyone whose first language is English, and who does not know Anewan, will pronounce these words in a non-Anewan way, and a host of articulatory factors contributing to accent will determine this — including word stress, and allophony, and some of the other factors discussed in §4. Clearly, the expectation that English speakers will pronounce the words of any other language just as native speakers of those languages would is not feasible.

Indeed there are some very public failures in this regard. Two highly successful authors, Ngaio Marsh and Ngaire Thorpe, have written in English and carefully and conscientiously maintained their Maori first names. However, English speakers have proved very poor at saying these names with initial velar nasals. While writing this paper, I rang around several bookshops out of curiosity, and in all cases these authors were described to me as [nəgeɹi] or [nəgeə] (for Ngaire) and as [nəgaɪoʊ] (for Ngaio).

It is probably true that loanwords tend to exhibit greater variation in pronunciation than do other words. For example, some English speakers say [ɹɛstəɹã] for 'restaurant', while others say [ɹɛstəɹɒnt]. There are various motives underlying this variation within a speech community with regard to the pronunciation of loanwords, ranging from a display of knowledge of the source language, to more socio-political motives such as the signalling of solidarity, etc. There is evidence that widespread intentional accommodation in matters of pronunciation is possible across a speech community where such motivations exist. For example, while speakers of Australian English say [mauɹi] for 'Maori', speakers of New Zealand English very distinctively say [mauɾi] with a flap instead of a continuant. Similarly, Julia Roberts (pers. comm.) notes the rapid increase in the last two years among young white South Africans of pronouncing language names like Xhosa with an initial click, rather than an initial velar stop. The increased Western interest in ethnicity over the last decade has also possibly led to an increased interest in the 'correct pronunciation' of words known to come from another language. Whether or not such targeted pronunciations are actually like native speaker ones is probably not the point. Even those people who deliberately adapt their phonology to best replicate the source of the loanword, for whatever reasons, typically do not achieve this without some knowledge of the source language. However, perhaps it is enough to be seen to be trying. This kind of position would appear to lie behind Australian Broadcasting Corporation policy on the pronunciation of words and names from other languages. Often these are not quite as they should be in the source language, but they do represent a distinctive departure from the typical Australian English pronunciation.

In this area where socio-political motivations intersect with phonological competencies, little seems to be known about how far such 'accommodation' might go. It will be very interesting in coming years to follow the success of attempts to establish within contemporary Australian English loanwords from Aboriginal languages which significantly violate the phonotactic and phonological rules of English. For example, the Kaurna word *Ngamatyi* which is a name for the Adelaide place known in English as 'Victoria Square' (Amery & Williams this volume) can be 'reinstated' and even have its correct pronunciation modelled by a Kaurna speaker. However, ultimately its fate and its pronunciation as a loanword into English are not in the hands of prescriptivists who wish to impose a 'correct' pronunciation. Ultimately the wider speech community of Adelaide residents will settle on a pronunciation of such words, a pronunciation that is largely shaped by their mother tongue phonology.

Many of the contributions to this volume detail attempts to carefully record and preserve Aboriginal names for places, and are characterised by a concern for faithfully describing the meanings and pronunciations of identified toponyms. This project is rather different, and some of the steps we have taken may well strike you as being a little cavalier, involving as it has: guessing at word classes; neutralising some phonemic distinctions; imposing the word stress patterns of another language; and selecting a group of words whose syllabic length and phonotactic structure are not representative of Anewan as it once was. And to pre-empt an English pronunciation of these Anewan words, and to deliberately employ a spelling convention that best represents that pronunciation, has been an act that has drawn criticisms of tokenism[5] from some quarters. But this work is not descriptive, it is not about describing Anewan. Rather it has involved serving up words of Aboriginal origin to an English-speaking audience, to create a commodity from Anewan resources that did not exist in Anewan itself. It is language engineering not language description.

---

[5] Indeed at the Canberra workshop where this paper was first given, one commentator used this exact expression.

Table 1: Final 33 words placed on the Placenaming Policy Register

Meaning	Recommended spelling	Pronunciation guide	Phonemicised spelling
'apple tree'	Doonba	**doon**-ba	dunba
'bird (gen)'	Bilunnya	bi-**lunn**-ya	bilanya
'blacksnake'	Alinnya	a-**linn**-ya	alinya
'black cockatoo'	Wilara	wi-**lah**-ra	wilara
'boomerang'	Argunna	ar-**gunn**-a	arrgana
'camp'	Oorala	oo-**rah**-la	urala
'crab'	Jumbunna	jum-**bunna**	jambana
'curlew'	Rilwinoo	ril-**win**-oo	rrilwinu
'eaglehawk'	Lumburra	lum-**burra**	lambara
'flying fox'	Ramana	re-**mahn**-na	rramana
'forest oak'	Riwilla	re-**willa**	rriwila
'goanna (black)'	Rujala	roo-**jah**-la	rrujala
'grass tree'	Doonboora	doon-**boo**-ra	dunburra
'hail'	Arribunna	ah-ri-**bun**-na	arribana
'ironbark'	Girunba	gi-**run**-ba	girranba
'kangaroo'	Imbandja	im-**ban**-dja	mbanyja
'kangaroo rat'	Burra	**bah**-ra	barra
'koala'	Lawunnya	la-**wunn**-ya	lawanya
'moon'	Jundja	**jun**-dja	janyja
'mountain ash'	Oowinba	oo-**win**-ba	uwinba
'parakeet'	Yimbungga	yim-**bung**-ga	mbangga
'pelican'	Wooyara	woo-ya-ra	wuyara
'plover'	Jaringga	jah-**ring**-ga	jarringga
'rock (large, flat)'	Lara	**lah**-ra	lara
'snow'	Igana	ee-**gah**-na	igana
'swan'	Juwoola	dju-**woo**-la	juwula
'totem of medicine man'	Indaralla	in-da-**rahl**-la	ndarrala
'tree'	Dala	**dah**-la	dala
'turtle (stinking)'	Wirra	**wi**-ra	wirra
'uphill'	Indabaiyee	in-da-**bai**-yee	ndabayi
'wallaroo'	Lamala	la-**ma**-la	lamala
'water'	Yoogoonda	yoo-**goon**-da	ugunda
'white cockatoo'	Yirrbadja	yirr-**bah**-dja	irrbaja

REFERENCES

Buchanan, F.J.,1901, Aboriginal words and meanings from Ee-na-won. *Science of Man* 4(4):64–65.

Crowley, T., 1976, Phonological change in New England. In R.M.W. Dixon, ed., *Grammatical Categories in Australian Languages*, 19–50. Canberra: AIAS.

Fraser, H., 1996, Guy-dance with pro-nun-see-ay-shon: a consideration of how the everyday spelling system can be used more effectively than phonetic symbols in English pronunciation guides. *English Today* 12(3):28–37.

Hoddinott, W., Fieldnotes on Anewan. AIATSIS MS 2126/1. Item 8.

Laves, G., n.d. [probably 1929], A single text in 'Anewan' by Jack Malone at Bowraville. Laves Papers AIATSIS MS 2189 Box 2.

MacPherson, J., n.d., Comparative vocabulary of 'Enneewin', Ngarbal, Yugambal, Bigambal, Guyambal, Ngarrabul, and Bundel(la)'. MS.

— 1904, Ngarrabul and other Aboriginal tribes. *Proceedings of the Linnean Society of New South Wales,* 677–684.

— 1930, Some Aboriginal place names in northern New South Wales. *Royal Australian Historical Society* 16:120–129.

— 1931, Some Aboriginal animal names. *The Australian Zoological Society* 6(4):368.

— 1934, Some words from the New England vocabularies. *Mankind* 1:235–236.

Mathews, R.H., Notebook: 2:35–42.

— Notebook: 4:11–14.

— Notebook: 7:82–88.

— 1903, Languages of the New England Aborigines, NSW. *Proceedings of the American Philosophical Society* 42:249–263.

Unidentified 1904, Word list headed 'Walcha District' in *Science (of Man)* 7(4):59–60.

18 RECLAIMING THROUGH RENAMING: THE REINSTATEMENT OF KAURNA TOPONYMS IN ADELAIDE AND THE ADELAIDE PLAINS

Rob Amery and Georgina Yambo Williams

Some placenames on the Adelaide Plains, such as Yankalilla, Myponga, Aldinga, Willunga and Waitpinga, have always been in use.[1] These localities have always been known by their Indigenous names and only by their Indigenous names. Yankalilla, for instance, was in use by sealers based at Kangaroo Island before colonisation and also recorded by George Augustus Robinson on 2 June 1837 in an interview with Kalloongoo, a Kaurna woman who had been kidnapped from the district some years previous (see Amery 1996). The name appeared in Colonel Light's journals and was in frequent use before the establishment of a settlement in the district. However, while these names are used in the same vicinity of the original place, it is not clear whether any of these names now refer to the same place that they did 200 years ago. For instance, Myponga now refers to a town, a district, the Hundred[2] of Myponga and a reservoir. The district of Myponga is probably not co-extensive with Maitpangga as known to Kaurna people in the early nineteenth century. Teichelmann and Schürmann (1840) indicate that it then referred to a plain. The remaining entities (the town, reservoir and Hundred) of course did not even exist in those days.

Even so, few South Australians know the meanings of these names and the meanings promoted by local councils and books on placenames are often highly suspect. In some cases (e.g. Yankalilla or Onkaparinga), the name is the key to understanding the landscape from a Kaurna perspective. So there is a need, not only to reinstate the names themselves, but to rehabilitate the understandings behind existing names.

George Gawler, South Australia's second governor (1838–1841), actively promoted the application of Aboriginal placenames where these were known:

---

[1] Paper presented at the 'Placenames of Indigenous Origin in Australia: An Interdisciplinary Workshop', 31 October 1999, Department of Linguistics, Australian National University, Canberra.

[2] 'Hundred' means a major division of land, the purpose of which is to provide unique identities for properties in conjunction with the section numbers. The actual size varies from one to another (Bill Watt pers. comm.).

L. Hercus, F. Hodges and J. Simpson, eds, *The Land is a Map: placenames of Indigenous origin in Australia*, 255–276. Canberra: Pandanus Books in association with Pacific Linguistics, 2002.

> In regard to the minor features of the country to which the natives may have given names, the Governor would take the present opportunity of requesting the assistance of the colonists in discovering and carefully and precisely retaining these in all possible cases as most consistent with property and beauty of appellation.
>
> All information on this subject should be communicated in precise terms to the Surveyor-General who will cause memoranda to be made of it and native names, when clearly proved to be correct, to be inserted in the public maps. (Government Gazette, 31 October 1839, cited in Cockburn 1990:xviii)

Not only did Gawler exhort the colonists to seek information relating to Indigenous placenames, but he also appears to be responsible for the reinstatement of at least one Kaurna name. T&S (1840) give Pattawilya as the Kaurna name for Glenelg. Patawalonga, the name of the waterway that drains into the sea at Glenelg, is derived from Pattawilya + -*ngga*. However, the plan prepared by Light, Finnis & Co. in 1836 and approved by Governor Gawler shows the Patawalonga as the River Thames (Cockburn 1990:87). It appears that the English name was never officially adopted and the Indigenous name prevailed. Gawler was also responsible for the reinstatement of Onkaparinga in 1838. This watercourse appears as Field's River on Colonel Light's charts (Manning 1986:157).

There are possibly other examples earlier this century where Kaurna names may have been reinstated. The suburb of Taperoo in the Port Adelaide area was laid out in 1925 by Wilkinson Watkinson. Before that, the area was known as Silicate. According to Manning (1986:202), Taperoo 'is aboriginal for calm', though Kaurna *tapurro* 'possum skin drum' seems a more likely etymology. It is unclear whether Taperoo is a reinstated original Kaurna name, or even if it is Kaurna at all. Pooraka is another puzzling case. 'In 1916 the Yatala District Council discarded the name "Dry Creek" as applied to the old post office, in favour of Pooraka, a native word meaning "dry creek"' (Cockburn 1908:56; 1990:63) while Praite and Tolley (1970:148) say it means 'dry waterhole', also found in Endacott (1955:48), a source which draws primarily on Victorian materials. The origins of Pooraka are obscure. It bears no resemblance to documented Kaurna words for 'dry' or 'creek'. Yet it was named at a time before the appearance of books promoting the use of Aboriginal words from anywhere and everywhere. Perhaps it was in fact the original name for the Dry Creek watercourse itself. However, Pooraka does appear in Ingamells (1955) and Reed (1967) with the same spelling, where it is identified as a New South Wales word meaning 'turpentine tree'.

There are certainly more recent examples where non-Aboriginal people or government agencies have applied or reinstated Kaurna names without reference to the Kaurna community. For instance, the Osmond Terrace Drug and Alcohol Rehabilitation Centre was renamed Warinilla in 1985 after the original homestead there (Robin Brandler, pers. comm., 7 November 1997). The name Warinilla was certainly in use in the mid-1920s or early 1930s when the property was sold by the Holden family and may well date back to the early years of the colony.

Warekila Lodge was named in about 1992–1993 after the original name for Happy Valley, where it is located. The name Warekila was obtained from Cooper (1962), in which it is said to mean 'place of changing winds' (Gwen, pers. comm., Warekila Lodge; Paul Springthorpe, November 1997).

1 BELAIR NATIONAL PARK

There is just one locality in Kaurna country today where Kaurna names almost dominate the map. Even though all the main roads and several other features are named with English names and several words from other Aboriginal languages have been imported, the main natural features — the creeks and ridges within Belair National Park — bear Kaurna names as follows:

Willa-willa Ridge	(cf. T&S Willawilla 'Brownhill Creek')
Warri-Parri Ridge	(cf. T&S Warriparri 'Sturt River')
Yulti-Wirra Ridge	(cf. T&S *yulti* 'stringybark'; *wirra* 'forest; bush')
Perroomba Creek	(cf. T&S *purrumba* 'flower, blossom'; Wyatt (1879) *perroomba*)
Kurru Creek	(cf. T&S *kurru* 'grass tree')
Tilti Creek	(cf. T&S *tilti* 'native cherry')
Minno Creek	(cf. T&S *minno* 'wattle, wattle gum')
Tarnma Creek	(cf. T&S *tarnma* 'honeysuckle')
Tarpurro Creek	(cf. T&S *tapurro* 'possum skin, drum')
Karka Creek	(cf. T&S *karka* 'sunset, dusk'; *karko* 'sheoak'; Wyatt (1879) *karkoo* 'sheoak')
Workanda Creek	(cf. T&S *workanda* 'cataract, cascade')

On the face of it, it would seem that the names may be original retentions. The country down from Willa-willa Ridge is in fact the head reaches of Brownhill Creek, known to the Kaurna as Willawilla (T&S) and Warri-Parri Ridge similarly drains into Warriparri, the Sturt River. There are indeed numerous *tilti* 'native cherry' near Tilti Creek and there is a waterfall on Workanda Creek etc. However, in this case we can establish without doubt that the names were obtained from Kaurna wordlists. Cordes (1983:2) quotes a 1909 publication *The National Parks of South Australia* in which the Commissioners of the Park observed:

Nomenclature —

A number of graceful native names have been given to the gullies and ridges of the Park. They are appropriate words, chosen from the vocabularies of aboriginal languages, but are not directly inherited from the wandering blackfellow, who has left no linguistic traces in this spot. They have superseded some of the names given by early settlers in the 'Tiers'. In the days of Government Farm Workanda was known as 'Waterfall Gully', Tarnma as 'Honeysuckle Gully', ... and the lower part of Minno Creek bore the homely title of the 'House Paddock Creek'.

Even though the quote indicates that the words were taken from 'the vocabularies of aboriginal languages', all the 'Native Names' listed in this publication are in fact clearly identifiable Kaurna words. Several of the spellings are, however, at variance with T&S and other known Kaurna sources — Karka is given as 'sheoak', probably a simple typographical error, and Tarpurro is glossed as 'opossum' by the Park Commissioners, not 'possum skin drum' as in T&S (1840) and Cawthorne (1844). Is there another Kaurna wordlist in existence that we have failed to locate?

The map of Belair National Park included in Cotton (1953) lists these same Kaurna words in the names of ridges and creeks, though Perroomba appears as Peroomba and Tarpurro appears as Tarpurra. In addition, a number of arbours and other features have been added to the map, some of them adopting these same Kaurna names as follows:

Peroomba Arbour

Workanda Arbour

Minno Arbour

Tarnma Arbour

Karka Arbour

Karka Pavilion

Karka Oval

Tilti Arbour

Titti Dam <this is probably a misprint for Tilti. It is located at the head of Tilti Creek>

Some time after this, walking trails were added to the park. Three were named Workanda Track, Tilti Track and Yulti Wirra Track drawing on Kaurna names listed above. Additional names drew on both English and Indigenous names. The following appear to have been drawn from Aboriginal languages :

Moorowie Track	(cf. T&S *murre-* 'go, walk, travel' + *kauwe* 'water', i.e. 'running water' or *murro* 'dust' + *kauwe* 'water', a meaning more consistent with Cooper (1962:23) 'discoloured waterhole' and Endacott (1955:38) 'cloudy water')
Nookoo[3] Track	(cf. T&S *nguko* 'a species of owl'; *muka* 'egg') (cf. Cooper (1962:26) *Nookoo* 'valley')
Carawatha	(cf. T&S *karra* 'red gum tree'; *karrawadlo* 'shrub generally') (cf. Tyrrell (1933) *Carawatha* (NSW) 'pine trees'; Cooper (1962:10) and Endacott (1955:17) *Carawatha* 'place of pine trees'; Sugden (n.d.) *Karawatha* 'place of pines')
Wypanda Track[4]	(cf. Cooper (1962:39) *Wypanda* 'echo'. Note the proximity of Echo Tunnel under the railway line which the track passes.)
Wilyawa Track	(cf. T&S *wilya* 'foliage') (cf. Cooper (1962:37) *wilyawa* 'scrub country')
Kaloola Track	(cf. T&S *kudlilla* 'rainy season, winter') (cf. Cooper (1962:15) *kaloola* 'to climb')

---

[3] Also spelt Mookoo on some maps.

[4] The track, running alongside the railway line in the north of the park, is marked in the Street Directory, but not on the map issued at Belair National Park itself.

Berri Werri Track	(cf. Tyrrell (1933) *Berri-Werri* (NSW) 'a crossing place') (cf. Cooper (1962:8) and Endacott (1955:11) *Berri-werri* 'crossing place')
Thelmytre Track	
Curta Track	(cf. Wyatt *kerta* 'a forest')
Yungurra Track	(cf. T&S *yangarra* 'wife') (cf. Cooper (1962:41) *yungura* 'crested pigeon')

While many of these words could have been taken from Cooper (1962), not all of them are listed there. A source has not been identified for Thelmytre. Perhaps it is not an Indigenous word at all.

It would appear that these names have been added to the map since 1953, but before 1968 (letter from Dene Cordes, 10 January 1996). However, it seems to be impossible to find out who placed them on the map, or indeed to be certain of the origins of some of these words. My own researches and enquiries with Belair National Park, the Department of Environment and Natural Resources and the Friends of Belair National Park have proved fruitless. Cordes suggests that it may have been Professor J. B. Cleland.

While none of the Indigenous names in use at Belair is an original retention, clearly the methods used by the Park Commissioners at the turn of the century to apply Kaurna names were well thought-out and well motivated. This contrasts with the importation of names from out of the area in the 1950s or 1960s, a practice that is quite inappropriate and superficial in the eyes of Kaurna people.

However, these activities are not the main focus of this paper. Rather, we intend to investigate the use of Kaurna names by Kaurna people themselves, or by government agencies and other bodies who are in consultation with and working at the direction of the Kaurna community. We are interested in investigating naming activity that serves to increase awareness of the Adelaide Plains as Kaurna country with an Indigenous history, so often overshadowed and obliterated by colonisation, urbanisation and development.

2 KAURNA NAMES PROMOTED BY THE INDIGENOUS COMMUNITY

In the case of Kaurna, little knowledge of the language is retained within the community. Knowledge of Indigenous placenames of the Adelaide Plains not already on the map is derived solely from archival material, though it is possible that some additional information may be held in the oral traditions of certain Nunga families.

Two decades ago Kaurna people started to turn to their own languages as a source of names for a variety of purposes. The establishment of Warriappendi Alternative School in 1980 was the first instance of the bestowal of a Kaurna name that was instigated and controlled by Kaurna people. Since then, Kaurna people, and others, have been turning to the historical materials increasingly as a source of names. This activity has extended to the reinstatement and institution of toponyms in Kaurna country, including metropolitan Adelaide.

Warriparinga, on the Sturt River, Marion, is an important case. This site is highly significant for the Kaurna people (see, for instance, Williams & Chapman 1995:89) because of its centrality in the Tjilbruke Dreaming story which tells of the creation of springs and other sites to the south of Adelaide and of the iron pyrites deposits at Brukunga. Warriparinga marks the site where Kulultuwi, Tjilbruke's nephew, was killed. As such, it is recognised as the start of the Tjilbruke Dreaming track. It was from here that Tjilbruke collected his nephew's body, wrapped it up and carried it down the coast. The Kaurna name was recorded by several observers as follows:

Warriparri	the Sturt River (T&S 1840:73)
War-rey par-rey	the creek that runs from the hills into Holdfast Bay (Williams 1840)
wari pari	Sturt River meaning 'wind river' or 'river of the west wind' (Black 1920:82)
War:pari	Sturt Creek (Tindale 1987:8)
Walpari	on Sturt Creek at Marion (Berndt & Berndt 1993:233)
Warriparri	Sturt River meaning 'creek fringed with trees' (Cockburn 1990:209)
Warri-Pari	Sturt River meaning 'The Throat River' (Webb 1936–37:308)
Warreparinga	Sturt Creek district 'windy river place' (Cooper 1952:28)
Warriparri	Sturt Creek 'windy river' (Cooper 1952:28)

Black's (1920) recording probably indicates that it was still known by Ivaritji, recognised as the so-called 'last speaker' of Kaurna, whilst Berndt's and Tindale's recordings indicate that the name was known to their Ngarrindjeri informants in the 1930s and 1940s.

Until recently the Warriparinga site has been officially known as Laffers Triangle, whereas the river itself has been, and still is, known as Sturt River. However, the name Warri Parri Drive has been applied to a street in the nearby suburb of Flagstaff Hill. This street is adjacent to the Sturt Gorge Recreation Park which straddles the Sturt River behind Flinders University. Incidentally, the nearby suburb of Warradale was named in 1917 by George Hamilton, whose family were early residents in the district. He knew that Warraparinga was the rightful name for the Sturt River in that vicinity, but considered it 'too long to become popular and substituted "dale" for "paringa"' (letter written by A. Hiscock, 4 July 1940).

In 1992, a proposal[5] was put forward by Paul and Naomi Dixon for the establishment of the 'Warriparinga Interpretive Centre', though the site itself was still being referred to as Laffers Triangle (see Boynes 1992). A local environment group initially known as The Friends of Laffers Triangle became known as Friends of Warriparinga in 1992 or 1993. The name was soon applied to the entire reconciliation project planned for the site, no longer just the planned interpretive centre. At some time after 1992, the name was applied to the entire site. An article reporting on an interview with Paul and Naomi Dixon in the Salisbury, Elizabeth, Gawler *Messenger* (19 October 1994:5) refers to the 'Warriparinga area'. The site is now officially known as the Warriparinga Reserve and the adjacent wetland established in 1998 was named Warriparinga Wetland.

The site in 1999 was virtually only referred to as Warriparinga by the Kaurna community, environment groups, the Marion City Council and the general public alike. Today, many are

---

[5] This proposal was taken from Georgina Williams' application to Susan Lenehan for the protection of Tjirbruki sites and the sustaining of the family clan group and the Kaurna peoples.

not even aware of the existence of any other name. The name has been fully accepted as *the* name for the site almost totally precluding the use of other names.

In April 1997 a new wetlands reserve not far from Warriparinga was named Tartonendi[6] by the Marion City Council and a plaque erected reading: 'Tartonendi. This Reserve is named Tartonendi which is a Kaurna word meaning "transforming the land into wetlands". The Kaurna people are the original inhabitants of the Adelaide Plains'.

3 ADELAIDE CITY CASE STUDY

The Adelaide City Council area, consisting of the city square, North Adelaide and the surrounding parklands, is a good case study in which to investigate issues associated with Indigenous placenames and their reinstatement.

Within this area there are just three names on the map of Indigenous origins in official use — Moonta Street, Morialta Street and Medindie Road, all of them minor thoroughfares. Moonta derives from Yorke Peninsula, having Narungga origins. Morialta is taken from Morialta Falls in the eastern foothills (Morialta possibly derives from *mari* 'east' + *yertalla* 'cascade') while Medindie Road is named after the adjoining suburb to which it leads. Its meaning is unknown, though it may derive from *mettindi* 'to steal'. Tindale (1974:213) refers to the 'Medaindi (horde living near Glenelg)' with a variant spelling Medaindie, though he does not suggest an etymology. So none of the names in use (ignoring the names applied to buildings and organisations, most of which have been adopted recently)[7] is a name employed by Kaurna people to refer to a natural feature in the area.

The only possibility of an original retention still in use is that of Pinky Flat near the Adelaide Oval which may derive from Kaurna *pingko* 'bilby' (Praite & Tolley 1970:139ff.), though it and some other sources say it derives from English. According to Manning (1986:169) 'during the depression years of the 1930s, it was used as a camp by unemployed, and cheap wine called "Pinky" was consumed there'. Pinky or pinkie appears in the *Australian National Dictionary* meaning both 'bilby' and 'cheap or home-made (fortified) wine' (Hughes 1989:411), giving further credence to both etymologies.

However, names for a number of geographical features in the area were recorded by several observers. The Torrens River was recorded as Karrauwirraparri (T&S 1840:75), the 'redgum forest river'. Wyatt (1879) records Korra weera, yerta and perre 'Adelaide, and the Torrens': so it would appear that the land surrounding the Torrens was known as Karrawirrayerta.

---

[6] The word *tarto* 'low, swampy country' appears in T&S. The word *tartonendi* was formed by myself through the addition of the inchoative suffix. That is, *tartonendi* literally means 'becoming low, swampy country' (i.e. being transformed into wetlands).

[7] A number of Aboriginal organisations bearing Kaurna names and names drawn from other Indigenous languages including Ngarrindjeri and Pitjantjatjara are located in the Adelaide City Council area. These include Tandanya, Kumangka, Patpa Warra Yunti and Nunkuwarrin Yunti. Educational institutions, notably the University of Adelaide and the University of South Australia, have adopted a number of Kaurna names for various buildings and centres, including Wilto Yerlo, Mattanya Housing, Wirranga Health Service and the Yungondi Building. Several businesses in the area also bear Indigenous names. These include Ngapartji multimedia consortium and Wirranendi urban ecology centre.

Other names recorded include:

Kaurna name	meaning	English name/location
Tandanya	'red kangaroo rock'	Adelaide south of the Torrens
Piltawodli	'possum home'	'Native Location'
Tambawodli	'plain house'	'Emigration Square'
Tinninyawodli	'rib house'	'the Ironstores'
Kainkawirra	'redgum forest'	North Adelaide
Kainkawirra	'redgum forest'	lake in Botanical Gardens
Ngamaji	unknown	GPO
Walinga	'house place'	city of Adelaide

At least one of these names clearly refers to a post-colonial imposition on the landscape. Tinninyawodli (lit. 'rib house') was the name given to the 'Ironstores' or Colonial Store. It appears that the word for 'rib' was extended to 'iron',[8] probably because iron was imported into the colony in lengths resembling a rib. Other names, such as Walinga 'house place' and Tambawodli 'plain house' referring to Emigration Square where there were rows of temporary huts and tents to house new arrivals in the colony, are probably similar.

4 THE ADELAIDE CITY COUNCIL PLACENAMING PROPOSAL

In December 1996 I (Amery) was approached by the Adelaide City Council to research original names and the history of the Kaurna people within the Adelaide city and North Adelaide areas under the jurisdiction of the Council. As part of the reconciliation process, a naming proposal was developed which suggested Kaurna names for the 29 parks within the parklands and names for the seven city squares. In 1997 a set of naming principles was developed (see Appendix 1) which promoted the use of original names. Where these names were not recorded, I proposed names that related to a particular plant species found in the area or known to have existed there. I also proposed several names that related to current use of the park. For instance, I suggested naming Park 02 Padipadinyilla 'swimming place' because of the presence of the Aquatic Centre there. Furthermore I proposed that the squares and four high profile parks be named after prominent Kaurna individuals, such as Mullawirraburka 'King John', while the golf course greens at the Piltawodli 'Native Location' site on Park 01 could be named after Kaurna children and adults who were known to have lived there. (The full proposal is included in Appendix 2 of this paper.)

While few of these names are original retentions, the names applied do give a Kaurna perspective and serve to remind us of an Indigenous heritage through the names of prominent leaders known to have frequented the area, and the names of Indigenous plants that formerly dominated the landscape. It is important, though, that the signs be accompanied by information, so that residents and visitors have the opportunity to appreciate deeper understandings of Kaurna culture and history.

---

[8] Wyatt (1879:20) gives *tinninye* 'iron' and *tinninye werle* 'an iron store', though he does not list a word for 'rib'. T&S give *tinninya* 'rib' and *Tinninyawodli* 'the Ironstores'.

The Adelaide City Council considered five Kaurna names at its meeting of 13 March 2000 (communication from General Manager City Strategy to the Strategy and Policy Committee on 13 March 2000 re Kaurna Naming [C] (1999/00996)). The Council made a formal request to the Geographical Names Board to dual name the Torrens River as Karrawirraparri (lit. 'redgum forest river'). Karrawirra Parri was officially adopted in November 2001. Four additional names of parks (previously unnamed) were also adopted by the council as follows:

Piltawodli	Park 01 — North Adelaide golf course	(lit. 'possum house')
Karrawirra	Park 12 — near University footbridge	(lit. 'redgum forest')
Wirranendi	Park 23 — includes West Terrace cemetery	(lit. 'being transformed into forest')
Tambawodli	Park 24 — Glover Ave and West Terrace	(lit. 'plain house')

The issue was reported in *The Advertiser* (15 March 2000:11) in an article entitled 'Reconciliation's sign of the times' which quoted Kaurna Elder Lewis O'Brien. Signage for these four parks was erected and unveiled on 14 November 2001 with the Council announcing the adoption of a further 19 Kaurna names for the remaining unnamed parks. Dual naming of Victoria Square as Tarndanyangga was adopted by the Council on 27 May 2002. However dual naming of the remaining squares and parks that already bear English names has yet to be adopted.

5 PILTAWODLI

Piltawodli, the site of the 'Native Location' on the banks of the Torrens River, is perhaps the most important of the five names adopted by the Adelaide City Council in 2000. Even before its official adoption on 13 March 2000, this name was becoming better known within the community. This was especially so after the Journey of Healing reconciliation event, attended by several thousand people on 26 May 1999. Shortly afterwards, an article entitled 'Piltawodli: remembering the "possum house"' appeared in the *Adelaidean* (14 June 1999:4).

The Piltawodli site, currently a golf course with few visible signs of the former existence of the 'Native Location', was given some recognition in the literature by Foster (1990) and since that in Amery (1998; 2000), Hemming and Harris (1998) and Harris (1999). I first took a group of students who were studying Kaurna language at Para West Adult Campus on an excursion to Piltawodli in 1997. Since then Lewis O'Brien, Cherie Watkins, Karl Telfer and I have accompanied students from the University of Adelaide, Tauondi College, Kaurna Plains School and Inbarendi College, a large group of teachers, as well as small groups of international visitors. In this way it has been gaining increasing informal recognition.

Piltawodli is a particularly important site, for it was here that almost everything we know of the Kaurna language was recorded. At this site we also see the beginnings of policies in South Australia which separated Aboriginal children from their families — policies that gave rise to the 'stolen generations'. Because of its importance, a plaque now marks the site (see Appendix 3 for the text of the plaque). A mock-up prepared by design staff within the Adelaide City Council was put on display at the site on 26 May 1999. The Council set aside $5,000 for its establishment and an equivalent amount has been donated by the Lutheran Church. The plaque was unveiled by Kaurna Elder Doris Graham and incoming Lord Mayor Alfred Huang on 26 May 2000 to mark the anniversary of National Sorry Day during the Journey of Healing reconciliation event. Those who attended included Kaurna people, descendents of the German missionaries, representatives of the South Australian Government, the Lutheran Church and the Adelaide City Council, a large number of school children and members of the public.

The Piltawodli site has become an important focus for the Kaurna language program at Kaurna Plains School and the name is becoming known within the Kaurna community.

6 REINSTATING KAURNA NAMES

As the Adelaide City Council Placenaming Proposal demonstrates, only a certain amount can be done in terms of reinstating the rightful original names. Owing to the paucity of records, the potential for reinstating authentic Indigenous toponyms on the Adelaide Plains is limited. However, there are a number of important Kaurna names known from the historical record, where dual naming would be a possibility. These include:

Yertabulti	'land of sleep or death'	Port Adelaide
Kaleeya ~ Kaleteeya	?	Gawler
Pattawilya	'gum tree foliage'	Glenelg
Karraundongga	'redgum chest place'	Hindmarsh
Warkowodliwodli	'? houses'	Klemzig
Putpayerta	'fertile ground'	Lyndoch Valley
Yerltoworti	'? tail'	Hindmarsh Valley
Parriworta	'river behind'	Hutt River
Witongga	'reed place'	The Reedbeds (Fulham)

There are basically two approaches we can use in the reinstatement of Kaurna names. We can seek to make them official by working through the relevant authority having jurisdiction over naming in that area for the purposes proposed. For instance, we can lobby for their placement on the map by the Geographical Names Board, local council etc. Alternatively, we can simply begin to use them. Of course both processes can work together, but the latter is much more meaningful. As Bill Watt pointed out on ABC Radio (25 October 1999), the technical process of changing a name is easy. More demanding is the need to conduct the appropriate consultation and develop a sense of ownership of the changes within the community. In 1918, many German placenames in South Australia were removed and replaced with English and Indigenous names. However, in most cases the community did not embrace the changes and following the cessation of hostilities many of them reverted to their German names.

It is evident that many people in the community would like to see the restoration of Indigenous names, but there are strong indications that they are still a minority. Over the last few years with the restructuring and amalgamation of local councils, names have been needed for the new entities. Several times I have been asked for advice by councils themselves, councillors and residents who are wanting to promote Kaurna names, and especially original names from the respective areas. Alternatives have been put forward, but in every case non-Indigenous names have won the day. Five years ago, the Australian Labor Party suggested that the name for Adelaide should be changed to Kaurna in the year 2000 (*The Advertiser,* 24 November 1994:1–2), but only a small minority interviewed by *The Advertiser* supported the proposal. Similarly, when Mark Brindal MP proposed changing the name of Victoria Square to Tandanya, most callers to the ABC talkback program (ABC Radio, 25 October 1999) did not support the idea. So a lot of groundwork needs to be done before Kaurna names can be reinstated in a way that is acceptable to and owned by the community at large.

Figure 1: Cartoon from *The Advertiser*, 16 March 2000, courtesy of Michael Atchison

We would argue that the best way to ensure the reinstatement of Kaurna names, where these are known, is for us to simply begin to use them in our teaching programs, our writings and in our everyday conversations, and to familiarise the public with these sites and their history. I (Amery) recall back in 1974 during an acute accommodation crisis at the Australian National University, some of us were allowed to occupy rooms at Toad Hall before its completion and before it had been named officially. It was then referred to by the administration as 'the Fourth Hall of Residence'. We made a rough official-looking sign out of polystyrene and had mail sent to us at Toad Hall. In the end, the administration had no option but to follow suit.

If Kaurna names are to be accepted officially, then we need to do a lot of work to promote them within the community. To totally replace high-profile existing names, such as Victoria Square, is exceedingly difficult. Of course it can be done as was the case with St Petersburg becoming Leningrad and Batavia becoming Jakarta, but these were renamed following revolution and expulsion of a colonial power. Far less threatening than total replacement is the application of a Kaurna name alongside an existing name. Fortunately in South Australia such legislation exists, known as the Dual Naming Policy, which has been applied in the Flinders

Ranges, and only recently within Kaurna country.[9] Under the Dual Naming Policy, the Indigenous name may eventually gain more popularity than the existing English name, as is the case with Uluru which was initially introduced alongside of Ayers Rock in the face of considerable resistance.

In 1991 the Victorian Labor government adopted Geriwerd, the Jardwadjali word for 'mountain' for the Grampians National Park. The name was dropped by the newly elected Kennett Liberal government in 1992 following the tabling of a petition of some 57,000 signatures. Despite this, according to *The Age* (15 March 2000:6), 'the park's ranger-in-charge, Mr Graham Parkes, said yesterday that the park's management, locals and tour operators continued to use Geriwerd'.

Of course, for locations such as Tambawodli or Piltawodli, which currently do not have an established English name, the process is much easier, but even here resistance can be expected.

7 AUTHENTICITY AND INTEGRITY OF NAMES

In reinstating or placing Indigenous names on the map, we need to be sure that they are correct and appropriate. We have already seen in the Adelaide City Council area and Belair National Park that a number of names have been drawn from out of the area and from other Indigenous languages. Norman Tindale, of the South Australian Museum, was a prolific writer and compiled placename card files and annotated maps from many parts of Australia, including the Adelaide Plains, with Indigenous names. Many of his published papers on ethnography and Dreaming stories are filled with placenames. So Tindale's work is a logical source of Indigenous placenames. However, much care needs to be taken with Tindale's materials. Firstly, there is an orthographic problem. Tindale used a modified IPA orthography which uses 'ŋ' where others use 'ng'; and 'j' where other writers and the general public would write 'y'. Tindale's orthography is not designed for or suited to signage and necessarily causes mispronunciation of the words. In a display at Mount Lofty Summit established in 1997, a panel telling of the Urebilla story talks of '*Jureidla*, the two ears of the great ancestral giant Urebilla (pronounced Yura-billa)'. Will visitors link Jureidla with Uraidla? Certainly the visitor information desk was not aware of the connection when I (Amery) rang on 4 November 1997. Perhaps more serious in relation to names on the Adelaide Plains, is the fact that most of his sources were Ngarrindjeri people from the Lower Murray and Coorong. Consequently, many of the names he imposed on the Adelaide Plains are Ngarrindjeri names or Ngarrindjeri adaptations of Kaurna names. For instance, he records the name for Glenelg as Pattawilyangk with the Ngarrindjeri locative suffix -*ngk* as opposed to Pattawilyangga with the Kaurna suffix -*ngga*. Many other names begin with 'r' or 'l' (see Amery 1998:199–202, 207–208; Tindale 1987). Yet Kaurna words, as we know them from all of the Kaurna sources, never commence with 'r' or 'l' and they clearly have Ngarrindjeri origins. These words should be avoided or attempts should be made to ensure that these words are Kaurnaised — Kaurna suffixes should be used to replace the Ngarrindjeri suffixes. The Kaurna sources record a number of names, such as Ngalta 'the Murray River' and Parnka 'Lake Alexandrina'

---

[9] The adoption of Karrawirra Parri for the Torrens River by the Adelaide City Council was ratified by the Geographical Names Board in November 2001. This is the first application of the Dual Naming Policy involving a Kaurna name.

(T&S 1840:75) for locations deep inside Ngarrindjeri territory, and no-one would suggest that these Kaurna names should be accepted as the official names for those locations.

CONCLUSION (WILLIAMS)

These final remarks are my views on placenames and their rehabilitation. As a Kaurna person, I have grown up with oral traditions and understandings about places and their significance that do not necessarily appear in the historical record. As a child, I was in the presence of old people and my grandfather who shared the stories of the spirit beings. As a younger woman I travelled over the country with my dad to places on the Adelaide Plains and surrounding hills. He explained these things to me as he remembered stories which had been passed on to him.

The placenames that have survived in an Anglicised form are part of the story, law and lore of the land formations and places in which they are situated. These placenames are the 'skeletal remains' of the historical surviving reality of Kaurna First Nation peoples, once a peaceful and intact body of lore/law of the land.

The word Yankalilla, derived from *yernkandi* 'to hang down, on; to join; impart; infect, as with a disease; to depend' (T&S 1840:61) + *-lya* + *-illa* 'place' (i.e. 'place of the fallen bits' Manning 1986:237; see Amery, this volume, for further details), is part of the law ceremony mortuary ritual and is of immense importance in reclaiming, through language recovery, the understanding of Kaurna Aboriginal people's rightful place and function in the framework of one's identity with the law of the land, and the spiritual and custodial responsibilities in everyday life. Tjirbruki, carrying the remains of his dead nephew's deteriorating and flaking body, falls to the ground at Yankalilla.

The word Onkaparinga derives from *ngangkiparringga* (lit. 'woman river place' or 'women's river') which leads into deeper insider understandings that we will not go into here. Ngurlongga or Horseshoe Bend on the Onkaparinga River is probably derived from *nurlo* 'curvature, corner' + *-ngga* 'place'. From these pieces and fragments of the remains of a hunter–gatherer society can be seen the horrific effects of the dispossession process.

I believe that the placing of Kaurna names to places today creates identity of a superficial nature unless they are relating to some source of relationship to the land and through this to the spirit of the land. The language of the Kaurna, and all other Aboriginal language names, should stay true to the original nature of the land and spirit relationship or we are contributing to even further dispossession by putting anything anywhere because some people might think this is a 'nice' gesture of remembrance to a now extinct people to meet a fashionable and acceptable fetish of the day.

When we name a building, we draw on what the building might be used for. The same would apply to an organisation — the name would have meaning and relate to the purpose, function or place. For example, Yaitya Makkitura 'Indigenous mirror' for the new Indigenous film, screenwriters and multimedia incorporated body in South Australia, founded by David Wilson, a Kaurna descendant/survivor. This has been how we have approached the way to best use the Kaurna language for the purposes of being a part of our life and identity in the contemporary Nunga world today.

By using our language reclamation in this way, we identify with the purposes served by the language in the reconstruction of identity from the roots remaining. Naming activity that is not rooted in the land and in the people of the land runs the risk of being a shallow and meaningless activity that misappropriates our language and culture.

APPENDIX 1

STRATEGIES FOR NAMING PARKS

Rob Amery

1 ORIGINAL NAME

The first choice for a park name is obviously to use the original name (e.g. Ngamatyi for 'Victoria Square') where that name is known. In the case of Ngamatyi, the Kaurna name could be used alongside the English name according to the Dual Naming Policy. Unfortunately, few original names survive within the Adelaide City Council area. They include Tarndanya, Karrawirraparri, Piltawodli, Tambawodli, Tinninyawodli, Ngamatyi, Kainka Wirra and possibly Pinky Flat.

2 NAME RELATING TO A PARTICULAR TYPE OF FLORA FOUND IN THE AREA OR KNOWN TO HAVE EXISTED THERE

Advice has been sought from Waldo Bushman regarding the indigenous flora in different localities within the Adelaide City Council area. In a few cases there are remnant trees and plants surviving, but in many cases the name is based on a guess taking topography and soil types into account.

A number of food plants are believed to have been widely distributed across the Adelaide Plains. It is not possible to identify the precise species of many of these recorded plant names, as the descriptions provided are inadequate. Some of these plant names have been used to name various parks, though we cannot be certain that the said plant was prevalent in the particular area covered by the park.

All of the plant terms used (except for *kurra* 'blue gum' and *tandotitte* 'native lilac') were recorded by Teichelmann and Schürmann (1840) and some of these terms are confirmed by other sources. Teichelmann and Schürmann lived at Piltawodli opposite the Adelaide Gaol. There is a high probability that the terms they recorded are local terms.

3 NAME RELATING TO CURRENT USE OF THE PARK

Several suggested names relate to current use or activities habitually carried out within that area of the parkland. Such names include Bakkabakkandi 'to trot', Nanto Womma 'horse plains', Padipadinyilla 'swimming place' and Wikaparndo 'netball park'.

4 NAMED AFTER KAURNA INDIVIDUALS

Names of Kaurna individuals have been suggested for four high-profile parks, the wives of the three *burka* 'elders; leaders' at the time of colonisation and Pangki Pangki, a trusted Kaurna guide and tracker who accompanied Moorhouse and Tolmer up the Murray River to Lake Bonney and the Rufus River. These names could be used in addition to existing names.

It is also suggested that the names of the three *burka*, Mullawirraburka 'King John', Kadlitpinna 'Captain Jack' and Ityamaiitpinna 'King Rodney', and the names of two prominent Kaurna women, Ivaritji and Kudnarto, be applied to the squares under the Dual Naming Policy.

In addition, the greens in the golf course in Park 01 could be named after children who were known to have attended the 'Native School' at Piltawodli located within the current precincts of the course and with the names of Kaurna adults known to have lived at the 'Native Location'. Refer to the paper on Kaurna individuals.

A plaque could also be erected on the site of Tinninyawodli 'The Ironstores' shown on Kingston's (1842) map, as a memorial to the two men, Bakkabarti Yarraitya and Parudiya Wangutya, who were hung there on 31 May 1839.

APPENDIX 2

1997 Proposal for park names in Adelaide City Council area[10]

No.	Kaurna name	English name	Notes
01	Piltawodli 'possum place'	Prev. 'Native Location' Montefiore Park	City of Adelaide Golf Links. Par 3 of Golf Links is the actual Piltawodli site. *pilta* 'possum' + *wodli* 'house; camp; place' Tinninyawodli 'The ironstores' (*tinninya* 'rib' + *wodli* 'house'), site of 1839 hangings, is also located within the South Course.
02	*Padipadinyilla 'swimming place'		Location of Adelaide Aquatic Centre. *padipadinya* 'swimming' < *padendi* 'to swim' + *-illa* 'LOC'
03	Kandarilla 'kandara root place'		*kandara* T&S 'native vegetable resembling radish' + *-illa* 'LOC'
04	Kangattilla 'kangatta berry place'		Croquet Lawns; adjacent reservoir *kangatta* T&S 'a kind of berry eaten by the natives' + *-illa* 'LOC'
05	Ngampa Yerta 'ngampa root ground'		*ngampa* T&S 'a kind of native vegetable' Wyatt *umba* 'edible root' [microseris] *yerta* 'ground'
06	Nanto Womma 'horse plain'		Horse agistment, playground, tennis *nanto* 'horse' + *womma* 'plain'
07	Kuntingga 'kunti root place'		*kunti* T&S 'a root of red colour and bitter taste, which the natives roast and eat' + *-ngga* 'LOC'

---

[10] 23 of these names have now been officially adopted by the Adelaide City Council. In the process some minor changes were made to this proposal. For example, the names for Parks 24 and 25 were swapped around and Park 15 was changed to Ityamaiitpinna.

08	Barnguttilla 'barngutta root place'		*barngutta* T&S 'native root; potato' This tuber most closely resembled the European potato.
09	Tidlangga 'tidla root place'		Prince Alfred College Sportsground *tidla* T&S 'a bulbous root eaten by the natives'
10	Warnpangga 'bulrush root place'		University of Adelaide Sportsground; archery club; Soldiers Memorial Gardens; ACC Nursery; northern bank of Torrens *warnpa* 'bullrush root', a staple food source prolific along the Torrens and other waterways; + -*ngga* 'LOC'
11	Kainka Wirra 'Eucalypt forest'	Botanic Park	includes zoo; botanical gardens. Original name from Ivaritji for waterhole, now the main lake in the botanical gardens. River red gums would have been the dominant eucalypt species there. Perhaps *kainka* is a synonym for *karra* 'river red gum' + *wirra* 'forest'
12	Karra Wirra 'River red gum forest'		precise location unclear. ? includes Grundy Gardens and Angas Gardens; University footbridge. Karrawirraparri 'red-gum forest river' was the original name for the Torrens which flowed through the *karra wirra* 'red gum forest'
13	Mogata wife of 'King John' or Mullawirraburka	Rundle Park	Light Horse Memorial *mogata* (meaning unknown)
14	Tangkaira wife of 'King Rodney' or Ityamaiitpinna	Rymill Park	Adelaide Bowling Club *tangkaira* T&S 'a species of fungus' N.B. Tangkaira was a signatory of the 1841 letter. She was also Ivaaritji's mother.
15	Wauwe wife of Kadlitpinna or 'Captain Jack'		CBC Oval; Glover Playground; centre of Grand Prix circuit *wauwe* 'female kangaroo' Her name is spelt Wahwey in the original source.
16	Bakkabakkandi 'to trot; a term applied to horses'	Victoria Park Racecourse	
17	Tuttangga 'grass place'		croquet; tennis; oval; south-east corner some original native grasses preserved here. *tutta* 'grass; hay' + -*ngga* 'LOC'
18	Witangga 'Peppermint gum place'		Osmond Gardens; Himeji Gardens *wita* 'peppermint gum' + -*ngga* 'LOC'

19	Pityarilla 'marshmallow root place'		Glover Playground; Bowling Club; Equestrian Area *pityarra* Teich 'edible root of the ngunna marshmallow'; Wyatt *peecharra* 'mallow' (a shrub) + *-illa* 'LOC'
20	Kurrangga 'Blue gum place'		picnic areas; playgrounds *kurra* 'blue ground' + *-ngga* 'LOC' [?? original source. Coora in Tauondi book]
21	Walyo Yerta 'walyo root ground'		Veale Gardens; Rose Garden; Conservatory; restaurant; tramline *walyo* 'edible white root resembling a radish' + *yerta* 'ground'
21 W	Minno Wirra 'golden wattle grove'		Wattle Grove; Princess Elizabeth Playground; Lundie Gardens; oval *minno* 'golden wattle; wattle gum' + *wirra* 'forest' Minno was a staple food for the Kaurna.
22	*Wikaparndo Wirra 'netball park'		SA United Church Netball Assoc. courts *wika* 'net' (wallaby or fish net) + *parndo* 'possum skin ball' (used as a football) + *wirra* 'forest; park'
23	*Wirranendi 'to become wirra'		West Terrace Cemetery; oval playground; Kingston Gardens; several surviving indigenous plant species including native apricot trees in West Terrace Cemetery. *wirra* 'forest' + *-nendi* 'to be transformed into'
24	Narnungga 'native pine place'		Adelaide High School; Ellis Park; oval *narnu* 'native pine' + *-ngga* 'LOC'
25	Tambawodli 'plain house'	prev. Emigration Square	sportsground; ANI oval *tamba* 'plain' + *wodli* 'house; camp; place'
26	Wilyaru Yerta 'Initiation Ground'	Adelaide Oval	cricket; tennis; Creswell Gardens; Pennington Gardens. The Adelaide Oval itself could be named Tarndanya Womma 'Adelaide Plain' *wilyaru* 'the final stage of initiation which includes cicatrisation' + *yerta* 'ground'
27	Tainmundilla 'mistletoe place'	Bonython Park	Adelaide Gaol; Police Barracks; SES HQ *tainmunda* 'mistletoe' + *-illa* 'LOC' Mistletoe was prevalent wherever redgums were located, especially along the Torrens.
28	Pangki Pangki name of Kaurna tracker and guide	Palmer Place	*pangki pangki* (meaning unknown) Pangki Pangki accompanied Moorhouse to Lake Bonney and the Rufus River in 1841.

29	Tandotittingga 'native lilac place'	Brougham Place	The native lilac flowers on the shortest day of the year — a sign of hope. Note proximity to Adelaide Children's Hospital. *tandotitte* 'native lilac' + *-ngga* 'LOC'
	†Tarnda Kanya 'red kangaroo rock'	Elder Park	*tarnda* 'red kangaroo' (principal totem of the Adelaide clan) + *kanya* 'rock' being the likely source of Tarndanya, the Kaurna name for Adelaide.

Squares

Suggested Kaurna name	Current name	Notes
Ngamatyi (meaning unknown)	Victoria Square[#]	Ivaritji was the source of the name Ngamatyi, spelt Ngamaji in the original sources. She said it was the site where the GPO now stands, also Victoria Square.
Mullawirraburka 'King John'	Hindmarsh Square	*mulla* 'dry' + *wirra* 'forest' + *burka* 'elder' More is known about Mullawirraburka than any other Kaurna person last century. see Gara (forthcoming). Mullawirraburka had four wives and his authority was recognised to some degree by the colonists.
Kadlitpinna 'Captain Jack'	Hurtle Square	*kadli* 'dingo; dog'; + *-itpinna* 'father of' Kadlitpinna was also well known to the colonists. His portrait was painted by Angas.
Ityamaiitpinna 'King Rodney'	Light Square	*ityamaii* 'name of student at Piltawodli' + *-itpinna* 'father of' Ityamaiitpinna was one of the first Kaurna to meet the colonists. He was Ivaritji's father.
Ivaritji Amelia Savage	Whitmore Square	*ivaritji* 'misty rain' Ivaritji was most likely the last speaker of the Kaurna language. She died in 1929.
Kudnarto Mary Ann Adams	Wellington Square	*Kudnarto* 'third born if a female'. Kudnarto married Tom Adams in 1848 and died in 1855. She is the ancestor of many Kaurna people living today.

*indicates that the term is a neologism (new term) constructed by myself.

† Tarnda Kanya has not been recorded in this form, but has been arrived at through a process of inference and interpretation of sources. The word *kanya* 'rock' is not actually documented as an independent word in Kaurna sources, though it appears in derivations and certainly exists in related languages such as Adnyamathanha. Numerous quarries were located on the south bank of the Torrens, including where the Railway Station and the Festival Theatre are now located. It is likely that one of these sites was the Tarnda Kanya 'red kangaroo rock'.

Victoria Square has since been dual named Tarndanyangga 'red kangaroo rock place'. Tarndanyangga was promoted during the Journey of Healing in 2001 and officially adopted by the Adelaide City Council on 27 May 2002.

APPENDIX 3

Figure 2: Plaque erected at Piltawodli, May 2000

The text of the Piltawodli plaque is as follows:

Piltawodli (possum home)

This plaque is dedicated to the Kaurna people

Wanti nindo ai kabba kabba? Ningkoandi kuma yerta. ('Where have you pushed me to? You belong to another country.') This Kaurna song was sung by Ngurpo Williamsie in 1844 in protest at the invasion of his country.

In May 1837 the Kaurna people led the 'Protector' of Aborigines to a site across the river to begin the first 'Native Location' in South Australia. In late 1838 it was relocated to this place, known to the Kaurna people as Piltawodli 'possum home'. It included cottages for Kaurna families, cottages for missionaries, gardens, and, by late 1839, a permanent school. In the 1840s Piltawodli was a fenced area extending over 14 acres.

Piltawodli is especially significant for Kaurna people, because it was here that almost everything we know about the Kaurna language was written down by German missionaries and others. This information was provided by the Kaurna people, particularly Mullawirraburka ('King John'), Kadlitpinna ('Captain Jack') and Ityamaiitpinna ('King Rodney'). The missionaries, Schürmann and Teichelmann, published a grammar and vocabulary in 1840 and, along with missionary Klose, taught in the Kaurna language. Each day the children at the school sang or recited hymns, the Ten Commandments, a prayer and some Bible stories. They wrote several letters in Kaurna which still survive. In 1845, the children were relocated to the English-only 'Native School Establishment' on Kintore Avenue and their houses at Piltawodli were destroyed by soldiers.

These children became the first of the 'stolen generations'. On 26 May 2000 many of the descendents of these children returned home to this site.

[Erected by the South Australian 'Bringing Them Home' committee on the second anniversary of National Sorry Day, 26 May 2000.]

Proudly Sponsored by:
Lutheran Church in South Australia
Corporation of the City of Adelaide

REFERENCES

Amery, Rob, 1996, Kaurna in Tasmania: a case of mistaken identity. *Aboriginal History* 20:24–50.

— 1998, Warrabarna Kaurna! Reclaiming Aboriginal languages from written historical sources: Kaurna case study. PhD thesis, University of Adelaide.

— 2000, *Warrabarna Kaurna! Reclaiming an Australian language.* Lisse, The Netherlands: Swets & Zeitlinger.

Berndt, Ronald and Catherine Berndt with John Stanton, 1993, *A World That Was: the Yaraldi of the Murray River and the Lakes, South Australia.* Carlton, Vic.: Melbourne University Press at the Miegunyah Press.

Black, J.M., 1920, Vocabularies of four South Australian languages, Adelaide, Narrunga, Kukata, and Narrinyeri with special reference to their speech sounds. *Transactions of the Royal Society of South Australia* 44:76–93.

Boynes, Michael, 1992, Submission to Science Park Board of Management for Warriparinga Interpretive Centre prepared by the Marion Environment Team, Marion City Council, October 1992.

Cawthorne, W.A., 1844, Rough notes on the manners and customs of the natives [Archives Department]. Published in *Proceedings of the Royal Geographic Society of Australia, SA Branch*, Session 1925–26 (1927).

Cockburn, Rodney, 1908, *Nomenclature of South Australia.* Adelaide: W.K. Thomas & Co.

— 1990, *South Australia. What's in a name?* Adelaide: Axiom Publishing (Revised edition. First published in 1984).

Cooper, H.M., 1952, *Australian Aboriginal Words and Their Meanings.* Second edition, revised and enlarged. (1962, 4th edition). Adelaide: South Australian Museum.

Cordes, Dean, 1983, *The Park at Belair: a social history of the people whose struggles and visions gave South Australia the National Parks and Wildlife Service we cherish today. Vol. 1. The pioneering decades.* Adelaide: the author.

Cotton B.C., ed., 1953, *National Parks and Reserves: an account of the national parks and reserves situated near Adelaide, South Australia.* Adelaide: Commissioner of the National Park.

Endacott, Sydney J., 1955, *Australian Aboriginal Words and Place Names and Their Meanings.* Melbourne: Acacia Press (9th enlarged edition).

Foster, Robert, 1990, The Aborigines location in Adelaide: South Australia's first 'mission' to the Aborigines. In Gara, ed., *Aboriginal Adelaide, Journal of the Anthropological Society of South Australia* 28(1):11–37.

Harris, Rhondda, 1999, Archaeology and post-contact Indigenous Adelaide. Honours thesis, Flinders University, Adelaide.

Hemming, Steve and Rhondda Harris for the Kaurna Aboriginal Community Heritage Committee, 1998, *Tarndanyungga Kaurna yerta: A report on the Indigenous cultural significance of the Adelaide parklands.* Adelaide Parklands Management Strategy prepared for Hassell Pty Ltd and the Adelaide City Council.

Hughes, Joan, ed., 1989, *Australian Words and Their Origins.* Melbourne: Oxford University Press.

Ingamells, Rex, 1955, *Australian Aboriginal Words: Aboriginal—English, English—Aboriginal.* Melbourne: Hallcroft.

Manning, Geoffrey H., 1986, *The Romance of Place Names of South Australia.* Adelaide: the author; Gillingham Printers.

Praite, R. and J.C. Tolley, 1970, *Place Names of South Australia.* Adelaide: Rigby.

Reed, A.W., 1967, *Aboriginal Place Names.* Sydney: A.H. & A.W. Reed.

Sugden, Joah H., n.d., *Aboriginal Words and Their Meanings.* Sydney: Dymock's Book Arcade.

Teichelmann, C.G. and C.W. Schürmann, 1840, *Outlines of a Grammar, Vocabulary, and Phraseology, of the Aboriginal Language of South Australia, spoken by the natives in and for some distance around Adelaide.* Adelaide: published by the authors at the native location. (Facsimile edition, 1962, State Library of South Australia. Facsimile edition, 1982, Tjintu Books, Adelaide. A copy annotated by Teichelmann was sent to Grey in 1858 and is held in the Sir George Grey Collection, South African Public Library, Cape Town.)

Tindale, Norman B., [Assorted papers] held in the Tindale Collection, Anthropology Section, South Australian Museum, Adelaide.

— Kaurna place names card file held in the Anthropology Section, South Australian Museum, Adelaide.

— 1974, *Aboriginal Tribes of Australia: their terrain, environmental controls, distribution, limits and proper names.* Berkeley: University of California Press.

— 1987, The wanderings of Tjirbruki: a tale of the Kaurna people of Adelaide. *Records of the South Australian Museum* 20:5–13.

Tyrrell, James R., 1933, *Australian Aboriginal Place-names and their meanings.* Sydney: Simmons.

Webb, Noel Augustin, 1936–37, Place names of the Adelaide Tribe. In *Municipal Year Book, City of Adelaide.* Printed at *The Advertiser*, Adelaide, 302–310.

Williams, Georgina and Don Chapman, 1995, The Warriparinga interpretive centre: Aboriginal and non-Aboriginal people working together. In *Creating Heritage Partnerships: a selection of papers from the Creating Heritage Partnerships Conference, 21–23 August 1995,* 89–98. Canberra: National Museum of Australia.

Williams, William, 1840, The language of the natives of South Australia. *The South Australian Colonist* 1(19):295–296.

Wyatt, William, 1879, Some account of the manners and superstitions of the Adelaide and Encounter Bay tribes. In J.D. Woods, ed., *The Native Tribes of South Australia,* 157–181. Adelaide: Government Printer (Original manuscript with corrections in BSL Special Collection).

APPENDIX

GUIDELINES FOR THE RECORDING AND USE OF ABORIGINAL AND TORRES STRAIT ISLANDER PLACE NAMES

Committee for Geographical Names in Australasia

---

1 MAIN OBJECTIVE

To ensure that Aboriginal and Torres Strait Islander place names are recognised by all Australia as being part of Australian heritage and need to be preserved.

2 SECONDARY OBJECTIVES

2.1 The names of places as given by Aboriginal and Torres Strait Islander people be recognised initially by place names authorities and ultimately by all Australians.

2.2 Aboriginal and Torres Strait Islander place names be preferred as the name to be used for any feature that does not have a name recognised by the relevant place name authority.

2.3 Aboriginal and Torres Strait Islander communities be consulted on all dealings concerning Aboriginal or Torres Strait Islander place names in their areas of current occupation and traditional association, in line with self-determination policies. (This includes any proposals to assign new names, alter spellings, institute a dual naming system etc.)

2.4 The Australian Institute for Aboriginal and Torres Strait Islander Studies be used as a resource to assist in the development of a writing system for any specific language which does not have an existing system. (The Australian Institute of Aboriginal and Torres Strait Islander Studies can assist with advice regarding linguists/anthropologists who have worked with the language group, previous surveys etc.)

2.5 Nomenclature authorities to undertake when possible to educate the general community in the use and pronunciation of Aboriginal and Torres Strait Islander place names. (This can be started by the use of authorised names on maps, wide distribution of policies, taking opportunities to speak to appropriate and interested groups, various media releases etc., all of which can be very beneficial without the need to be involved in costly programs.)

2.6 Nomenclature authorities be committed to the continuing development of appropriate procedures to facilitate the recording and use of Aboriginal and Torres Strait Islander place names and State and Federal governments recognise the need to provide funding.

2.7 Nomenclature authorities to seek the involvement of other interested/concerned groups (e.g. land councils, local government authorities, language centres).

3 GUIDELINES

3.1 Recognition

3.1.1 Any use of names of Aboriginal or Torres Strait Islander origin should be made following consultation and with appropriate recognition.

3.1.2 A recognition of the self-determination concept and its importance in contributing to place names issues.

3.1.3 A recognition to be given to the use of traditional names for places and localities bearing an officially recorded name from another source.

3.1.4 A recognition that more than one Aboriginal or Torres Strait Islander place name may exist for any particular feature, both within a specific language group and from two or more language groups.

3.1.5 A recognition that Aboriginal and Torres Strait Islander place names were in use prior to European occupation.

3.1.6 A recognition that the oral recording of place names in Aboriginal and Torres Strait Islander culture has equal standing with written recording.

3.1.7 A recognition that some Aboriginal and Torres Strait Islander place names may be subject to restrictions that must be respected. (This may apply to some names in common usage which are of a very sensitive nature — either sacred or offensive. Names in this category will be revealed following the establishment of good relations between the communities and the nomenclature authorities, and should be negotiated on an individual basis.)

3.1.8 A recognition of Aboriginal and Torres Strait Islander cultural expectation. (This particularly applies to methods of contact, community structures, respect for community wishes etc.)

3.2 Preferences

3.2.1 That Aboriginal and Torres Strait Islander place names be preferred for those features that do not have a name recognised by the nomenclature authorities or the local community.

3.2.2   That a preference be stated for the spelling and accenting of place names to agree with the rules of the written form of the language (if one exists) from which the place names originate.

3.2.3   That there should be no interference with established Aboriginal or Torres Strait Islander place names without the consent of the relevant community. (This applies to names in any location.)

3.2.4   That during the development and after the adoption of the guidelines, there be an assurance of the involvement of participants representing a diversity of interests, including government, non-government and voluntary organisations.

3.2.5   Adjustments may be made to the spelling of place names in consultation with the local Aboriginal or Torres Strait Islander community and their linguist if the current form is under threat of mispronunciation by the wider community or has been previously incorrectly represented.

3.2.6   Appropriate follow-up be made with Aboriginal and Torres Strait Islander communities to show the results of any specific field work or project.

3.2.7   Linguists should be consulted to maintain standards of excellence in written form.

3.2.8   Local government bodies, National Parks and Wildlife agencies, heritage bodies etc. should be consulted as required.

3.2.9   Authorisation is to be obtained from the relevant community for the use of any Aboriginal or Torres Strait Islander name or word taken from any source in official naming. (This refers to the use of names or words for new naming proposals, e.g. suburbs, conservation parks etc. The proposed use of the name or word may not be appropriate.)

3.2.10   Questions of copyright/ownership of information collected during any fieldwork or investigation must be resolved prior to the survey being conducted or prior to the names being used in a public domain.

3.2.11   The wishes of the Aboriginal or Torres Strait Islander community must be respected in relation to names and related information associated with areas of land currently occupied or areas of traditional association.

3.2.12   Consultation must try to meet the expectations of all involved parties, however, failing complete agreement, a consensus of opinion is to be aimed for. Various methods of consultation must be tried for.

3.3 Writing system

3.3.1   Where a writing system already exists and is in use by the community, that system should be used (e.g. Pitjantjatjara).

3.3.2   Where no writing system exists, the Australian Institute of Aboriginal and Torres Strait Islander Studies should be contacted as a reference source for the development of a writing system.

3.3.3 Ease of pronunciation be a criteria for the writing of Aboriginal or Torres Strait Islander place names.

3.3.4 English generic terms may be used if considered necessary to specify the type of feature involved.

3.3.5 Those researching languages that are no longer spoken will need to seek the assistance of a linguist to enable accurate renditions of the names to be determined.

3.3.6 The language source of each place name is to be noted if it is known or can be determined.

3.4 Education

3.4.1 A commitment by nomenclature authorities to undertake where possible an educative role in popularising correct spelling and pronunciation of Aboriginal and Torres Strait Islander place names.

3.4.2 Nomenclature authorities and the Committee of Geographical Names in Australasia to undertake an educative role to develop positive international perspectives of the use of indigenous names in Australia.

3.4.3 Assist in the education of the wider community about Aboriginal and Torres Strait Islander culture and the importance of place names to that culture.

3.4.4 Impart a realisation that Aboriginal and Torres Strait Islander place names represent a gift from another culture, the sharing of which imposes ethical obligations on the users. (This covers such areas as respect for restrictions, acknowledgment of sources, authorisation for use etc.)

3.4.5 To create an awareness among Aboriginal and Torres Strait Islander people of the importance of being able to preserve culture through place names and minimise the encroachment of new European names on the landscape, particularly for features of high cultural significance.

3.4.6 Foster a knowledge among Aboriginal and Torres Strait Islander people that their wishes will be respected.

3.4.7 Educate nomenclature authority support staff in appropriate consultative mechanisms.

3.4.8 Create an increased awareness among nomenclature authority staff of Aboriginal and Torres Strait Islander language and culture.

3.5 Procedural

3.5.1 A dual naming system or use of alternative names may be used as a management and educative tool for the naming of physical and environmental features of significance to the local Aboriginal or Torres Strait Islander community when an official name already exists and when a name change is not possible or acceptable.

3.5.2   Aboriginal or Torres Strait Islander names or terms from one area not to be applied to other areas for official naming purposes.

3.5.3   Local historical and cultural information relating to the meaning and origin of the place names should be collected whenever possible.

3.5.4   Previous relevant surveys by anthropologists, linguists, land councils, Aboriginal and Torres Strait Islander traditional owners and others be used as a resource prior to any fieldwork.

3.5.5   Names and spellings may be changed to avoid duplication of names, present a better vehicle for correct pronunciation and provide for better cultural retention.

3.5.6   Roman characters should be used in preference to other syllabic forms.

3.5.7   Where alternative spellings of a specific Aboriginal or Torres Strait Islander name exist, only one official spelling should be used following consultation with the relevant community.

3.5.8   Aboriginal and Torres Strait Islander place names are to be actively sought with the assistance of the State/Territory and Federal Governments.

3.5.9   State/Territory authorities agree to cooperate in undertaking joint field projects where common State/Territory boundaries have no meaning to local Aboriginal and Torres Strait Islander culture and language.

For further information about the National Policy Guidelines for the Recording and Use of Aboriginal and Torres Strait Islander Place Names, please contact:

Executive Officer
Committee for Geographical Names in Australasia
PO BOX 2
BELCONNEN ACT 2614
AUSTRALIA

INDEX OF PLACES AND PLACENAMES

Warri-Pari
   See also Warriparri, 260
Warriparinga
   See also Laffers Triangle, Marion, Sturt
      River, Warreparinga, Warriparri, 259,
      260
Warriparinga Reserve, 260
Warriparinga Wetland, 260
Warriparri
   See also Walpari, Wari pari, Warri-Pari,
      wari pari, Warpari, War-rey par-rey,
      Warripari, Warriparinga, Sturt Creek,
      Sturt River, 174, 260
Warri-Parri Ridge, 257
Warrirdila wayi-wuyi, 116
Warriyarrpuwan, 215, 224
   See also Wallwera, 234
Warrumbal River, 183
Wartalyunha
   See also Wertaloona, 151
Wartapa
   See also Wortupa Spring, 150, 151
Wartapanha
   See also Weetoolta Gorge, 150, 151
Warturli Widanha, 147
   See also Mount Lyndhurst, 148
Warturl-ipi Yurru
   See also Warturlipinha, Angepena,
      Mount Constitution, 149
Warturlipinha
   See also Warturl-ipi Yurru, Angepana,
      Mount Constitution, 149
Warupunju, 20
Warurrgu yib-ganyi, 116
Watchem, 161
Wategat, Wateegat, 70
Waterfall, 78
Waterfall Gully
   See also Workanda, 257
Waterhole, 190
Waterthurinna, 205
Wati -warakanha, 20
Wattie Creek, 52
Waukawoodna Gap
   See also Wakarla-udnanha Inbiri, 145
Wave Hill, 52, 53, 60
   See also Jamanku, Jinparrak, Number
      One, Kalkarriny, 51

   See also Lipananyku, 53
   See also Lipananyku, Karungkarni, 51
Wayalayala
   See also Wayarlayarla, 147
Wayarlayarla
   See also Wayalayala, 147
Wayingk Island
   See also Wayingk Thiikanen, 82
Wayingk Thiikanen
   See also Wayingk Island, 82
Weenem Aweyn
   See also Uthuk Aweyn, Big Lake, 79
Weenem Eelen
   See also Uthuk Eelen, Small Lake, 79
Weetaliba Station, 185
Weetaliba Waterholes, 185
Weetalibah Creek, 185
Weetootla Gorge
   See also Wartapanha, 150, 151
Weetowie Waters
   See also Urta Awi, 145
Werlatye Therre, 36
Werre Therre, 36
Wertaloona, 143
   See also Wartalyunha, 151
West McDonalds Tank, 216
Weyamunu, 119
Whirily
   See also O'Reilly, Wirrimbirchip
      pastoral run, 157
White Hill
   See also Parakujjurr, 14
White Jagur Swamp, 158
Wickham River, 54
Widapa Awi
   See also Widapawi Creek, 150
Widap-awi
   See also Bendieuta Creek, 150
Widapawi Creek
   See also Widapa Awi, 150
Wiinytyalapaa
   See also Conoble, 230
Wiitin, 12, 19
Wildu
   See also Atuwarapanha, Mount Serle,
      146
Wildya Vari
   See also Balcanoona Creek, 145, 146

INDEX OF LANGUAGES AND LANGUAGE GROUPS